Converter-Based Dynamics and Co Modern Power Systems

Converter-Based Dynamics and Control of Modern Power Systems

Antonello Monti
Institute for Automation of Complex Power Systems, RWTH Aachen University, Aachen, Germany

Federico Milano
School of Electrical and Electronic Engineering, University College Dublin, Dublin, Ireland

Ettore Bompard
Department of Energy "Galileo Ferraris", Politecnico di Torino, Torino, Italy

Xavier Guillaud
Laboratory of Electrical Engineering and Power Electronics, École Centrale de Lille, Lille, France

ELSEVIER

ACADEMIC PRESS
An imprint of Elsevier

Academic Press is an imprint of Elsevier
125 London Wall, London EC2Y 5AS, United Kingdom
525 B Street, Suite 1650, San Diego, CA 92101, United States
50 Hampshire Street, 5th Floor, Cambridge, MA 02139, United States
The Boulevard, Langford Lane, Kidlington, Oxford OX5 1GB, United Kingdom

Library of Congress Cataloging-in-Publication Data
A catalog record for this book is available from the Library of Congress

British Library Cataloguing-in-Publication Data
A catalogue record for this book is available from the British Library

ISBN 978-0-12-818491-2

For information on all Academic Press publications
visit our website at https://www.elsevier.com/books-and-journals

Publisher: Joe Hayton
Acquisitions Editor: Lisa Reading
Editorial Project Manager: Chiara Giglio
Production Project Manager: Prem Kumar Kaliamoorthi
Cover Designer: Christian Bilbow

Typeset by SPi Global, India

Contents

Contributors

Numbers in parentheses indicate the pages on which the authors' contributions begin.

Mohammed Ahsan Adib Murad (149,273), School of Electrical and Electronic Engineering, University College Dublin, Dublin, Ireland

Claudia Battistelli (91), Institute for Automation of Complex Power Systems, RWTH Aachen University, Aachen, Germany

Ettore Bompard (1,31,67), Department of Energy "Galileo Ferraris", Politecnico di Torino, Torino, Italy

Antoine Bruyère (7), Laboratory of Electrical Engineering and Power Electronics, École Centrale de Lille, Lille, France

Junru Chen (149), School of Electrical Engineering, Xinjiang University, Urumqi, China

Massimiliano Chiandone (273), Department of Engineering and Architecture, University of Trieste, Trieste, Italy

Valentin Costan (7), Lab Paris-Saclay—Energy System Economics, Functioning and Studies (EFESE), EDF, Palaiseau, France

Xavier Guillaud (1,7,193), Laboratory of Electrical Engineering and Power Electronics, École Centrale de Lille, Lille, France

Sriram K. Gurumurthy (305), Institute for Automation of Complex Power Systems, RWTH Aachen University, Aachen, Germany

Muyang Liu (149), School of Electrical Engineering, Xinjiang University, Urumqi, China

Andrea Mazza (31,67), Department of Energy "Galileo Ferraris", Politecnico di Torino, Torino, Italy

Federico Milano (1,125,149,273), School of Electrical and Electronic Engineering, University College Dublin, Dublin, Ireland

Antonello Monti (1,91,235,305), Institute for Automation of Complex Power Systems, RWTH Aachen University, Aachen, Germany

Terence O'Donnell (149), School of Electrical and Electronic Engineering, University College Dublin, Dublin, Ireland

Álvaro Ortega Manjavacas (125,149), Instituto de Investigación Tecnológica, Escuela Técnica Superior de Ingeniería ICAI, Universidad Pontificia Comillas, Madrid, Spain

Guilherme Pereira (7), Lab Paris-Saclay—Energy System Economics, Functioning and Studies (EFESE), EDF, Palaiseau, France

Taoufik Qoria (193), Laboratory of Electrical Engineering and Power Electronics, École Centrale de Lille, Lille, France

David Raisz (235), Institute for Automation of Complex Power Systems, RWTH Aachen University, Aachen, Germany

Giorgio Sulligoi (273), Department of Engineering and Architecture, University of Trieste, Trieste, Italy

Lucian Toma (31,67), Department of Electrical Power Systems, University Politehnica of Bucharest, Bucharest, Romania

Chapter 1

Introduction

Antonello Monti[a], Ettore Bompard[b], Xavier Guillaud[c], and Federico Milano[d]
[a]*Institute for Automation of Complex Power Systems, RWTH Aachen University, Aachen, Germany,* [b]*Department of Energy "Galileo Ferraris", Politecnico di Torino, Torino, Italy,* [c]*Laboratory of Electrical Engineering and Power Electronics, École Centrale de Lille, Lille, France,* [d]*School of Electrical and Electronic Engineering, University College Dublin, Dublin, Ireland*

1.1 Introduction

Power Grids have evolved in the last 100 years without substantially changing the basic principles of operation. Since the end of the famous "War of Currents," the key components have been the same and the principles of automation did not change significantly.

Alternative current (AC) has been the main choice with few exceptions, and correspondingly AC generators, induction motors, and transformers have been the key components of such a system.

The main pillar of the power grid has been for about a century the synchronous machine and, consequently, the dynamics of the grid have been substantially determined by the physical characteristics of this component. At the same time, the prevalent use of thermal and hydro-power plants has pushed more and more towards a concentration in large generation units driven by consideration of efficiency and hence economy of scale.

Thanks to large thermal-driven power plants, it was possible to structure the study of a power system by separating some key categories of dynamics:

- Thermal dynamics: from minutes to hours
- Electromechanical dynamics: from milliseconds to seconds
- Electromagnetic dynamics: in the range of milliseconds
- Switching events and lightning: from microseconds to milliseconds

The availability of clear categories allowed the engineers to tackle the complexity of the power systems by selectively simplifying the models to target the

Converter-Based Dynamics and Control of Modern Power Systems
https://doi.org/10.1016/B978-0-12-818491-2.00001-8

specific timescale of a category of dynamics. Such a process of model reduction has some important consequences:

- A conventional dynamic model for angle and voltage transient stability analysis are reduced to their essentials and, sometimes, an analytical description is possible;
- Even when the analytical description is not viable, the computational burden for numerical analysis is significantly reduced.

This process has been a key element for the success of the development of a power system of the complexity of the continental Europe bringing to synchronism the operation of all the ENTSO-E operators. Growing knowledge in power systems dynamics and automation brought this infrastructure to be an extremely reliable system with an unprecedented level of power quality and security.

The main element of change that appeared in the last 30–40 years is power electronics.

Power electronics started penetrating the power system first of all at the load level. During the 1980s and 1990s most of the industrial processes started inserting variable speed drive as a mechanism to increase the flexibility of operation and also efficiency.

On the load side power electronics expanded then more and more also at the consumer level with the availability of low-cost inverters increasing the efficiency of everyday devices such as washing machine. During the first 10 years of the 21st century, this process has become massive making power electronics an obvious solution for a large majority of loads.

Another parallel development has been the introduction of high voltage direct current (HVDC) connection to reinforce part of the AC grid or to connect systems that are not synchronized. While first commercial implementations happened in the 1950s, the advent of silicon devices has been the major technological change that brought this technology to a new level of maturity in the 1980s. At the end of the 1990s, finally, the first application of voltage source converters (VSC) for HVDC gave a tremendous impulse to the development of this technology opening also the way for possible application of multiterminal solutions.

HVDC represents the first electronic technology able to interact with the electrical grid with an increased level of controllability and flexibility.

The move from current source to voltage source type of inverters allows also the inclusion of HVDC connectors in the provision of reactive power for voltage stability control.

The biggest change though is happening now at the generation level. The process of decarbonization of the energy system, in the framework of the energy transition, is aiming at substituting traditional power plants with renewable driven energy sources (RES). This change brings some significant changes from the power system dynamics perspective:

- RES may be also distributed and then we move away from the concept of having only large centralized power plants;

- RES are interconnected to the grid, in the majority of cases, through power electronics interfaces;
- RES (wind and solar) are driven by changing weather conditions that determine faster changes in the power input and introduce an unprecedented high level of stochasticity in the grid;
- Several novel technologies for energy storage devices are being developed and committed into the grid. Among these, power-electronic-based grid-connected lithium-ion and redox batteries are currently the most promising technologies.

All these three factors point in the same direction: dynamics of the grid is playing a new and more relevant role and the classical separation of categories is losing meaning. Thermal processes are substituted by intermittent and rapid changing patterns of wind and solar generation and "slow" electromechanical dynamics are substituted by "fast" electronic transients.

In a nutshell, future power grids are supposed to become a power-electronic-driven system in which renewable and distributed energy resource interact with smartly controlled (flexible) loads. If we consider that also transformers have the chance to become electronic devices, we have that the future grid will be completely a power electronics power system. While this evolution is also reopening the question DC versus AC, it is also in any case modifying the characteristics of the spectrum of interacting dynamics.

In effect, we can see the situation as a convergence of dynamics. This convergence of dynamics and Eigenvalues is definitely an element of complexity, but power electronics should be also considered as an opportunity. In the new scenario dynamics are the results of controllers that engineers have the freedom to design. This means that it is actually possible to reinvent the basic principle of grid interface with criteria that are totally different from the past.

The scope of this book is to take the reader through this transition. While traditional power systems dynamic is covered, main focus is given to the element of changes and the opportunities opened by the new technologies.

1.2 Book structure

The book can be fundamentally split into two parts. The first part deals mostly with methodologies and techniques that apply also to classical power systems, the second is vice versa concentrated on new or emerging solutions that are directly related to the idea of a power electronics power system.

Chapter 2 introduces the key elements of the classical power system dynamics theory. The key concept discussed is the idea of Swing equation (SE). SE is the cornerstone of power system dynamic theory condensing the interaction of the electromechanical system in a simple dynamic of power imbalance. Understanding the hypothesis of this approach is critical also to appreciate the role and impact of power electronics.

Still in the direction of classical theory, Chapter 3 extends the swing equations to the so-called modal analysis. Modal analysis is a small signal analysis of multibus system, still based on the electromechanical range of Eigenvalues and then still connected to the main hypothesis of the SE.

Chapter 4 moves from dynamics to automation presenting the basic principle of operation for voltage and frequency control. All the basic principles and criteria are reviewed and introduced.

Chapter 5 opens the new scenario of modern power electronics-based power system. It presents a brief introduction to power electronics and focus on the different mode of operation for inverters when used on the generation and load side. This chapter introduces different devices that can play a role in the dynamics of the grid and that are integrating part of the grid infrastructures. These include: FACTS, Smart Transformers, VSC-HVDC.

A key transformation given by the modern dynamics is the fact that also key quantities need new definition and measurement processes. Chapter 6, in particular, describes advanced methods for frequency definition introducing then the key challenges related to grids with low level of inertia.

From the definition to the automation, the following Chapter 7 uses the concept of the previous analysis to introduce modern approaches to frequency control and in particular the possibility to move from a strictly central approach to a distributed intelligence solution.

As mentioned in the previous section, in any case, the big change is introduced by the new dynamics in generation driven by power electronics. Chapter 8 discusses the control structure for different mode of operation of grid-connected inverters (Grid supporting, Grid feeding, etc.). In particular, the role of grid forming converters and of different algorithms and control schema show how to embed a frequency support with power electronics.

Chapter 9 introduces a possible futuristic approach in this direction called linear swing dynamics. The chapter shows how, thanks to the flexibility given by programmable control, the grid interface can dramatically change with respect to the case of a synchronous machine. The idea is to show how in the new scenario it is not necessary to replicate the past, but we can actually think freely and imagine a totally new power system for the future.

Of course, the new characteristics of the grid will affect not only frequency control but also voltage control. The last two chapters presents an overview of advanced method for voltage stability working at two different time scales. Chapter 10 focused on the steady-state characteristics and optimal coordination of control action in distribution networks, while Chapter 11 looks at voltage stability as a power electronic challenge introducing extension to power systems of theorems such as Middle brook criterion.

1.3 How to use the book

The book is intended for engineers that have already a background in power systems at least from the point of view of static analysis of grids.

The book is intended to complement master-level classes on power system dynamics. The content of the book is adequate for a full semester class for students focusing on energy and power grids. A prerequisite of a previous class on power system operation, modeling, and control is recommended.

The book can also represent a useful guide for practitioners to understand how the reality is changing day by day thanks to power electronics. Electrical engineers working for power system operators, for example, can find interesting information about the evolution of the dynamics they experience every day in their job.

Chapter 2

Review of the classical power system dynamics concepts*

Guilherme Pereira[a], Valentin Costan[a], Antoine Bruyère[b], and Xavier Guillaud[b]

[a]Lab Paris-Saclay—Energy System Economics, Functioning and Studies (EFESE), EDF, Palaiseau, France, [b]Laboratory of Electrical Engineering and Power Electronics, École Centrale de Lille, Lille, France

2.1 Introduction

Even with the fast increase of the renewable energy, synchronous machines are still the major way to produce electricity in the grid. They have been used for more than a hundred years for the production of electricity and they are still a very reliable and efficient part of the chain for the conversion of the energy coming from the primary source to the electrical energy delivered to the grid. Developed in the end of the 19th century, the production of electricity was based on the pure electromechanical concept which was emerging in these years. Even if the concept has not evaluated much since the beginning, the size has grown to reach huge power (1750 MW in Taishan power plant commissioned in 2018/2019).

The synchronous machine is composed of a stator made of three-phase windings connected to the AC grid and a rotor connected to a DC source. The rotating dc flux generates induced voltages in the stator windings, which is the base of the electromechanical conversion.

Synchronous machine theory and modeling are vastly addressed in the literature [1–6]. Several types of models have been developed and depending on the required analysis to be carried out, a specific representation may be more suitable than others.

In this chapter, the foundations of the classical EMT representation of the synchronous machine are first recalled and compared with a simplified electromechanical model. The different elements constituting the power conversion system are described. A short overview of the steam turbine is proposed. In

* Figures 2.1–2.12 and Tables 2.1–2.10 were taken from G.S. Pereira, Stability of Power Systems with High Penetration of Sources Interfaced by Power Electronics (Ph.D. thesis).

Converter-Based Dynamics and Control of Modern Power Systems
https://doi.org/10.1016/B978-0-12-818491-2.00002-X

the same time, the way to drive the power with the governor is recalled. Different types of excitation systems have been developed. They are presented and compared in terms of dynamics and performances and linked to the voltage regulation which is inherent to the synchronous machine operation. A Power System Stabilizer may be needed to damp the dynamic behavior of the electromechanical system, its principle is explained in the last part of this chapter.

With each element added to the detailed EMT model, a dynamic analysis is provided and the dominant oscillatory modes are compared with the one obtained from a simplified representation of the synchronous machine. In this model, the synchronous machine is represented by the internal voltage behind the transient reactance and the swing equation is used to describe its electromechanical dynamics.

2.2 Electromagnetic model of synchronous machines

According to [6], a model including two rotor circuits in each axis—one field circuit and one damper in the d-axis and two dampers in the q-axis—is recommended for use in more critical stability studies. The following paragraphs summarize the foundations used to establish this model. More details can be found in [1–3, 7].

Fig. 2.1A illustrates the equivalent electrical circuit of the synchronous machine. As previously mentioned, the rotor circuit is composed of four windings, whereas the stator circuit is composed of one winding for each phase of the system. The terminal voltage of each winding is expressed as the combined effect of an induced voltage and a voltage drop. According to Faraday's law, the induced voltage is due to the variation of the linkage flux across the winding

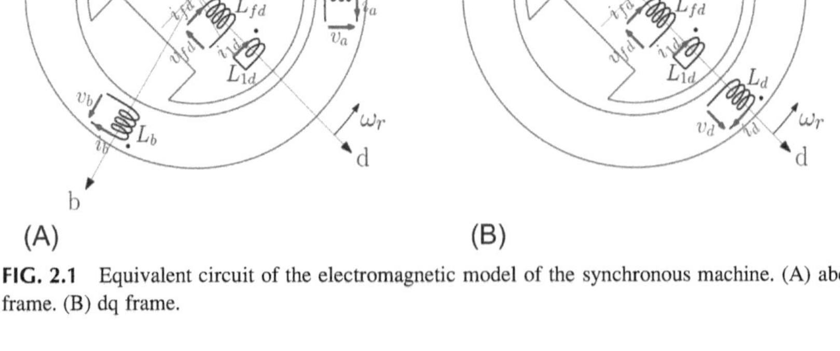

(A) (B)

FIG. 2.1 Equivalent circuit of the electromagnetic model of the synchronous machine. (A) abc frame. (B) dq frame.

over time. From Fig. 2.1A, it can be observed the presence of two reference frames, where one is static and the other rotates over time. The major challenge related to that is the complexity of the equations describing the physical phenomena, which present multiple trigonometric terms.

To circumvent this issue, Blondel has proposed an elegant solution in [8], applying a rotating transformation on the quantities related to the stator. Because this transformation rotates at the angular frequency of the rotor at all times, in the rotor frame, all windings are stationary, which highly simplifies the problem. This transformation is known in the literature as "abc to dq0 transformation" or even, "Park's transformation" after R. H. Park, who has generalized the method in [5]. The winding distribution after the application of Park's transformation is illustrated in Fig. 2.1B. It should be highlighted that, under balanced sinusoidal conditions, the component "0" is null and, since only balanced operations are treated within this work, this component is not represented. Several variants of Park's transformation exist. In the following model, the amplitude invariant transformation is used [1].

Therefore, Eqs. (2.1)–(2.6) express the terminal voltage of each winding of the synchronous machine considering the equivalent circuit illustrated in Fig. 2.1B.

Nota bene: Since the dampers of the synchronous machine are short-circuited, the terminal voltage of these windings are equal to zero. It should be highlighted that all electric equations of the synchronous machine are expressed in *per unit* using the conventions below:

Base power (S_b):	Three-phase rated power
Base voltage (V_b):	Rated line-to-ground voltage (peak value)
Base current (I_b):	Rated line current (peak value)
Base impedance (Z_b):	Base voltage/base current

$$v_d = -R_s i_d + \frac{1}{\omega_b}\frac{d\psi_d}{dt} - \omega_r \psi_q \tag{2.1}$$

$$v_q = -R_s i_q + \frac{1}{\omega_b}\frac{d\psi_q}{dt} - \omega_r \psi_d \tag{2.2}$$

$$v_{fd} = R_{fd} i_{fd} + \frac{1}{\omega_b}\frac{d\psi_{fd}}{dt} \tag{2.3}$$

$$0 = \frac{1}{\omega_b}\frac{d\psi_{1d}}{dt} + R_{1d} i_{1d} \tag{2.4}$$

$$0 = \frac{1}{\omega_b}\frac{d\psi_{1q}}{dt} + R_{1q} i_{1q} \tag{2.5}$$

$$0 = \frac{1}{\omega_b}\frac{d\psi_{2q}}{dt} + R_{2q} i_{2q} \tag{2.6}$$

where v_d, v_q, and v_{fd} are the terminal voltages of the stator and field windings, R_s, R_{fd}, R_{1d}, R_{1q}, and R_{2q} are the resistances of the windings, i_d, i_q, i_{fd}, i_{1d}, i_{1q}, i_{2q}, and ψ_d, ψ_q, ψ_{fd}, ψ_{1d}, ψ_{1q}, and ψ_{2q} are the currents and linkage fluxes across the stator, field and dampers windings, and ω_r and ω_b are the angular frequency and base angular frequency of the rotor.

The linkage fluxes are expressed in Eqs. (2.7)–(2.12). Since all windings along one axis are magnetically coupled, the linkage flux across one winding is composed both of the leakage flux induced by its self-inductance and of the mutual flux linking all inductances along the axis. Because d- and q-axes are perpendicular, there is no linkage between the windings of the axes:

$$\psi_d = -L_{ld} i_d + L_{ad}\left(-i_d + i_{fd} + i_{1d}\right) \tag{2.7}$$

$$\psi_q = -L_{lq} i_q + L_{aq}\left(-i_q + i_{1q} + i_{2q}\right) \tag{2.8}$$

$$\psi_{fd} = L_{lfd} i_{fd} + L_{ad}\left(-i_d + i_{fd} + i_{1d}\right) \tag{2.9}$$

$$\psi_{1d} = L_{l1d} i_{1d} + L_{ad}\left(-i_d + i_{fd} + i_{1d}\right) \tag{2.10}$$

$$\psi_{1q} = L_{l1q} i_{1q} + L_{aq}\left(-i_q + i_{1q} + i_{2q}\right) \tag{2.11}$$

$$\psi_{2q} = L_{l2q} i_{2q} + L_{aq}\left(-i_q + i_{1q} + i_{2d}\right) \tag{2.12}$$

where L_{ld}, L_{lq}, L_{lfd}, L_{l1d}, L_{l1q}, and L_{l2q} are the leakage inductances of the stator, field and the dampers and L_{ad}, L_{aq} are the mutual inductances of d- and q-axes.

Fig. 2.2 illustrates a visual description of the equivalent circuits of the synchronous machine proposed from Eq. (2.1) to Eq. (2.12). From this representation, it is possible to determine the expressions of the so-called standard parameters of the synchronous machines, which "translate" their behaviors during transients, in which, the linkage fluxes are forced into high-reluctance paths by different induced currents, which decay in different periods of time [1, 3]. Depending on the decaying velocity of these components, the parameters associated to the phenomena are called, from the fastest to the slowest, subtransient, transient and synchronous, where the last one represents the steady state of the synchronous machine [1]. These parameters can be determined from experimental data obtained from short-circuit tests on unloaded machines [1, 9].

Regarding the mechanical modeling of the synchronous machine, the turbine-generator shaft is represented by a lumped single mass model, not displaying any torsional effect. Therefore, the variation of the angular frequency of the shaft is caused by the unbalance between the mechanical and electromagnetic torques applied on it. In *per unit*, the motion of the synchronous machine can be expressed as a function of the power unbalance. In the literature, Eq. (2.13) is also known as the swing equation:

$$\begin{cases} \dfrac{d\omega_r}{dt} = \dfrac{1}{2H}\left(P_m - P_e - K_d \Delta\omega_r\right) \\ \dfrac{d\theta}{dt} = \omega_b \omega_r \end{cases} \tag{2.13}$$

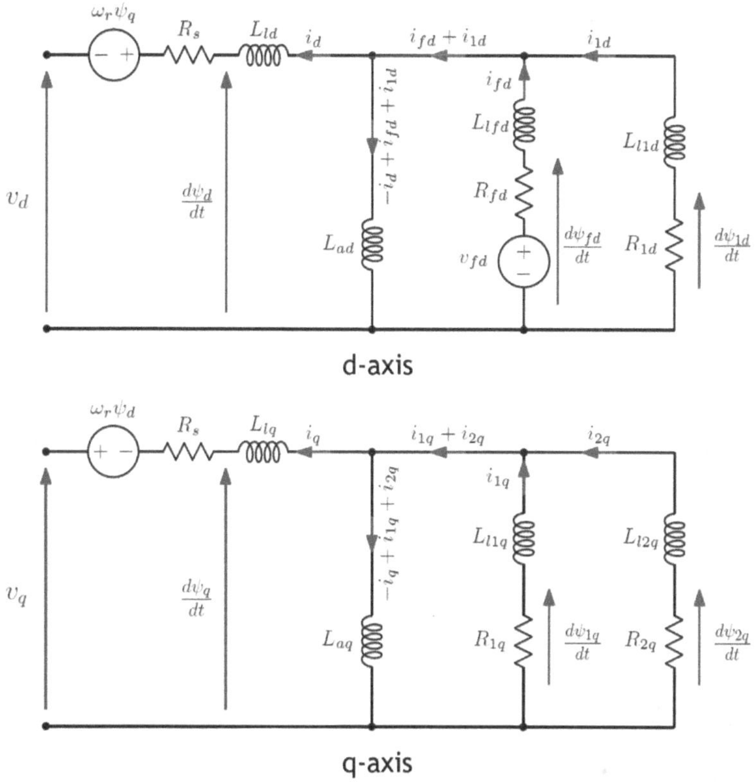

FIG. 2.2 Equivalent circuits of the *dq*-axes of the synchronous machine.

where $\Delta\omega_r$ is the angular frequency deviation of the rotor in relation to the grid frequency (ω_g), H and K_d are the coefficient of inertia and the damping factor of the rotating mass, P_m and P_e are the mechanical and electric powers at the shaft level and θ is the internal angle of the synchronous machine rotor. The coefficient of inertia (H) represents the amount of kinetic energy per rated power (MWs/MVA) stored into the turbine-generator shaft, and it is inherent to the physical construction of the machinery. Typical values of H are between 2.5 and 10 s, depending on the type of generating unit [1]. The damping factor (K_d) represents the damping related to the mechanics of the system (i.e., mechanical losses), and it can often be neglected for practical considerations [2]. To deduce P_e, the electric power at the stator level (P_t) is given in Eq. (2.14).

$$P_t = v_d i_d + v_q i_q \tag{2.14}$$

Replacing Eqs. (2.1), (2.2) in (2.14):

$$P_t = \left(-R_s i_d + \frac{1}{\omega_b}\frac{d\psi_d}{dt} - \omega_r \psi_q\right) i_d + \left(-R_s i_q + \frac{1}{\omega_b}\frac{d\psi_q}{dt} + \omega_r \psi_d\right) i_q \tag{2.15}$$

Rearranging the right side of Eq. (2.15):

$$P_t = \frac{i_d}{\omega_b}\frac{d\psi_d}{dt} + \frac{i_q}{\omega_b}\frac{d\psi_q}{dt} + \omega_r\left(\psi_d i_q - \psi_q i_d\right) - R_s\left(i_d^2 + i_q^2\right) \tag{2.16}$$

The right side terms of Eq. (2.16) correspond, respectively, to the rate of change of the magnetic energy stored in the armature, the electric power at the shaft level (P_e), which is also known as the power transferred across the air gap between the rotor and the stator of the synchronous machine, and the resistive losses at the stator level [1, 5].

2.3 *E'X'H* synchronous machine model

According to Refs.[1, 4], although simplified models of synchronous machines may conceal critical phenomena, they can still be considered acceptable for determined studies. In this section, a simplified model of the synchronous machine is proposed to determine the dominant dynamic of the system composed of a synchronous machine connected to an infinite bus. It is based on the simplest model used to represent synchronous machines during transients usually called "constant voltage behind a single reactance" associated with the swing equation, (2.13). This model is hereinafter referred to as $E'X'H$, referring to its electric and mechanical representations, where $E'X'$ and H are the main parameters of these models, respectively. The hypotheses considered to establish this model are listed below [1–3]:

- Considering that the variations of the angular frequency of the rotor are sufficiently small, ω_r is assumed approximately equal to its nominal value (ω_0), it does not present any impact on the stator voltages.
- The variations of the linkage fluxes of the stator (ψ_d and ψ_q) are also neglected. This neglects the electromagnetic transients of the stator of the synchronous machine, and, according to Ref. [1], the overlook of these phenomena can only be made if the same approximation is applied to the network equations, otherwise, the assumption is inconsistent.
- As previously mentioned, the armature flux is forced into high-reluctance paths outside the field winding by induced currents during transients. Since the subtransient stage decays much faster than the transient stage, the former (and its associated circuit) will be neglected. Furthermore, since the decay of the transient stage is slow—several seconds—the linkage fluxes are considered constant during all the transient period.

From these hypotheses, Eqs. (2.1)–(2.6) can be rewritten as given in Eqs. (2.17)–(2.19):

$$v_d = -R_s i_d - \omega_0 \psi_q \tag{2.17}$$

$$v_q = -R_s i_q + \omega_0 \psi_d \tag{2.18}$$

$$v_{fd} = R_{fd} i_{fd} \tag{2.19}$$

By neglecting the flux in the dampers Eqs. (2.7), (2.9) can be simplified:

$$\psi_d = -L_{ld} i_d + L_{ad}(-i_d + i_{fd}) \tag{2.20}$$

$$\psi_{fd} = L_{lfd} i_{fd} + L_{ad}(-i_d + i_{fd}) \tag{2.21}$$

Defining ψ_{ad} as:

$$\begin{aligned} \psi_{ad} &= L_{ad}(-i_d + i_{fd}) \\ &= \psi_{fd} - L_{lfd} i_{fd} \end{aligned} \tag{2.22}$$

These equations are illustrated by Fig. 2.3. Isolating i_{fd} in the second line of Eq. (2.22) and replacing in the first line, it is possible to rewrite ψ_{ad} as:

$$\psi_{ad} = \frac{L_{ad} L_{lfd}}{L_{ad} + L_{lfd}} \left(-i_d + \frac{\psi_{fd}}{L_{lfd}} \right) \tag{2.23}$$

Also, from Fig. 2.3, the equivalent inductance of the circuit, measured at the stator level, represents the transient inductance of d-axis of the synchronous machine (L_d'):

$$L_d' = L_{ld} + \frac{L_{ad} L_{lfd}}{L_{ad} + L_{lfd}} \tag{2.24}$$

Combining Eqs. (2.23), (2.24) and replacing ψ_{ad} at Eq. (2.20):

$$\psi_d = -L_{ld} i_d + (L_d' - L_{ld}) \left(-i_d + \frac{\psi_{fd}}{L_{lfd}} \right) \tag{2.25}$$

The same can be deduced for the q-axis, *mutatis-mutandis*:

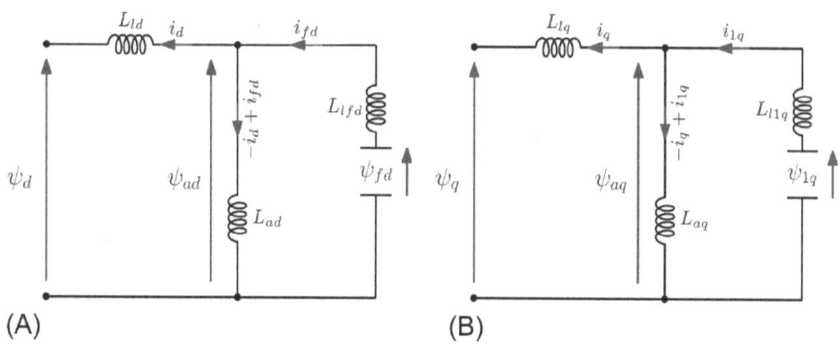

FIG. 2.3 Equivalent circuits of the dq-axes of the synchronous machine illustrating the relationship between fluxes and currents after considering the simplifying hypotheses. (A) d-axis. (B) q-axis.

$$\psi_q = -L_{lq}i_q + \left(L'_q - L_{lq}\right)\left(-i_q + \frac{\psi_{fq}}{L_{l1q}}\right) \tag{2.26}$$

Replacing Eqs. (2.26), (2.25) in (2.17), (2.18), respectively:

$$\begin{cases} v_d = -R_s i_d + \omega_0 L'_q i_q - \omega_0\left(L'_q - L_{lq}\right)\dfrac{\psi_{1q}}{L_{l1q}} \\ v_q = -R_s i_q - \omega_0 L'_d i_d - \omega_0\left(L'_d - L_{ld}\right)\dfrac{\psi_{fd}}{L_{lfd}} \end{cases} \tag{2.27}$$

Defining the transient electromotive force ($e_d{}'$ and $e_q{}'$) as:

$$\begin{cases} e'_d = -\omega_0\left(L'_q - L_{lq}\right)\dfrac{\psi_{1q}}{L_{l1q}} \\ e'_q = +\omega_0\left(L'_d - L_{ld}\right)\dfrac{\psi_{fd}}{L_{lfd}} \end{cases} \tag{2.28}$$

The stator voltage of the synchronous machine ($\vec{v}$) can be rewritten as in Eq. (2.29). For this purpose, the transient saliency of the synchronous machine is neglected, and therefore $L'_d = L'_q$.

$$\left(v_d + jv_q\right) = \left(e'_d + je'_q\right) - \left(R_s + j\omega_0 L'_d\right)\left(i_d + ji_q\right) \tag{2.29}$$

Therefore, from Eq. (2.29), it can be observed that the transient model of the synchronous machine is represented as an ideal voltage source (transient electromotive force, $\vec{e'}$) behind a transient reactance ($X'_d = \omega_0 L'_d$) and, for this reason, this model may present steady state errors at post disturbance conditions. The stator resistance can also be neglected due to its small value compared to X'_d. Fig. 2.4 illustrates the simplified transient model of the synchronous machine connected to an infinite bus bar, where X_t and X_l represent the inductances of a coupling transformer and an inductive transmission line,[a] respectively.

For the $E'X'H$ model, the mechanicals of the synchronous machine are also represented with Eq. (2.13). The electric power at the shaft level of the synchronous machine (P_e) is equal to the active power flow between the synchronous machine and the infinite bus, computed at the synchronization point of the synchronous machine, which is $\vec{e'}$. Therefore:

$$P_e = \frac{E'V_g}{X_{eq}}\sin\left(\theta - \theta_g\right) \tag{2.30}$$

Where E' and θ are the magnitude and phase of the transient electromotive force ($\vec{e'}$), and V_g and θ_g are the magnitude and phase of the infinite bus voltage ($\vec{v}_g$).

a. According to Ref. [3], when a full network is reduced to a two-bus system, the resulting impedance is essentially inductive.

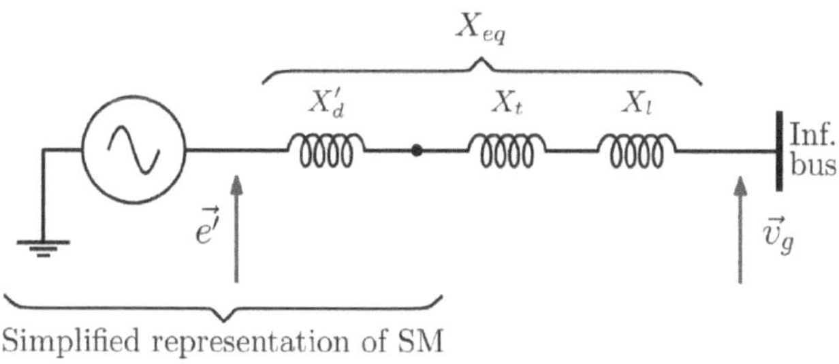

FIG. 2.4 Transient model of the synchronous machine ($E'X'H$, resistance of the stator neglected).

In the $E'X'H$ model, the damping factor (K_d) does not have the same meaning as for the detailed model. In the latter, K_d represents a damping purely related to the mechanics (i.e., mechanical losses) which can generally be neglected for all practical considerations [3], in the former, on the other hand, it represents a combination of different damping effects of the system, which is used to compensate the lack of damping due to simplifying hypotheses. According to Ref. [10], the equivalent value of K_d composing the damping provided by generators and loads is a function of the parameters of the synchronous machine and the grid. According to Ref. [3], a rigorous derivation of the damping power is long and complicated, and therefore, from simplifying assumptions, it is proposed a simplified expression of it, providing an average value of K_d. Refs. [1, 7] only consider the load damping (typical value 1 or 2%). Normally, for large systems, these parameters are estimated using Phasor Measurement Units (PMUs) at specific buses of the system [11].

$$\frac{d}{dt}\begin{bmatrix}\Delta\omega_r \\ \Delta\delta\end{bmatrix} = \underbrace{\begin{bmatrix}-\dfrac{K_d}{2H} & -\dfrac{K_s}{2H} \\ \omega_b & 0\end{bmatrix}}_{A}\begin{bmatrix}\Delta\omega_r \\ \Delta\delta\end{bmatrix} + \underbrace{\begin{bmatrix}\dfrac{1}{2\mathrm{H}} \\ 0\end{bmatrix}}_{B}\Delta P_m \tag{2.31}$$

where K_s is the synchronizing torque between the synchronous machine and the infinite bus, which is given in Eq. (2.32). The synchronizing torque is a restoring torque acting on the shaft of a synchronous machine to keep the synchronism when its angular frequency is different from the nominal frequency [7].

$$K_s = \frac{\partial P_e}{\partial \theta}\Big|_{(\theta,\theta_g)=(\theta_0,\theta_{g0})} = \frac{E'V_\infty}{X_{eq}}\cos\left(\theta_0 - \theta_{g0}\right) \tag{2.32}$$

As illustrated in Fig. 2.5, this system can also be displayed using a more tangible representation. Block diagrams are very flexible to represent small systems, providing visual interpretations of the propagation of actions through the parts of the system.

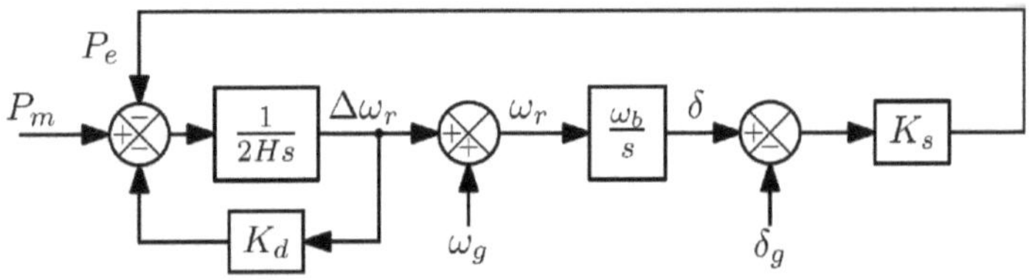

FIG. 2.5 Block diagram of the synchronous machine vs infinite bus system.

The characteristic polynomial of the system is found solving $\det(A-\lambda I)=0$:

$$\lambda^2+\frac{K_d}{2H}\lambda+\frac{\omega_b K_s}{2H}=0 \tag{2.33}$$

From Eq. (2.33), it is possible to determine the undamped natural frequency (ω_n) and the damping ratio (ξ) of the oscillatory mode of the system as a function of its parameters. It should be highlighted that, in time-domain simulations and eigenvalues analysis, the observed frequency is the damped frequency (ω_d), which is, as given in Eq. (2.34), a combination of ω_n and ξ.

$$\omega_d=\omega_n\sqrt{1-\xi^2},\text{with}:\omega_n=\sqrt{\frac{\omega_b K_s}{2H}}\text{ and }\xi=\frac{K_d}{2\sqrt{2H\omega_b K_s}} \tag{2.34}$$

2.4 Comparison between detailed and simplified models

This section is dedicated to the comparison between the $E'X'H$ model and various models for the synchronous machine.

Regarding the synchronous machine, despite not being realistic, initially, its regulators are not taken into account, and therefore, its inputs—mechanical power (P_m) and field voltage (v_{fd})—are constant and calculated from the power flow. The different elements needed for the operation of the synchronous machine (prime mover/governor, excitation system/voltage control, and PSS) are added progressively over this section. The parameters of the synchronous machine are given in Table 2.1.

The synchronous machine vs infinite bus of Fig. 2.6 is adopted as case study. The simplified representation of the system is similar to that from Fig. 2.4.

The parameters of the grid are given in Table 2.2. For this scenario, the adopted length for the transmission line is 25 km and, since it is a short line, its shunt susceptance is neglected [1]. The power flow of the system is presented in Table 2.3 and, from the given tables, it is possible to compute $\vec{e}\,'$ and, subsequently, K_s. The main parameters of the $E'X'H$ model, expressed using the *per*

TABLE 2.1 Dynamic model data for synchronous machines (SMs).

Transformers parameters	Value
Rated frequency (f_n, Hz)	60
Rated power ($S_{3\phi}$, MVA)	900
Rated voltage (V_n, kV)	20
Stator resistance (R_a, pu)	0.0025
Leakage reactance (X_l, pu)	0.02
Synchronous reactance d (X_d, pu)	1.80
Transient reactance d (X'_d, pu)	0.30
Subtransient reactance d (X''_d, pu)	0.25
Synchronous reactance q (X_q, pu)	1.70
Transient reactance q (X'_q, pu)	0.55
Subtransient reactance q (X''_q, pu)	0.55
Inertia coefficient (H, s)	6.5
Damping factor (K_d, pu)	0

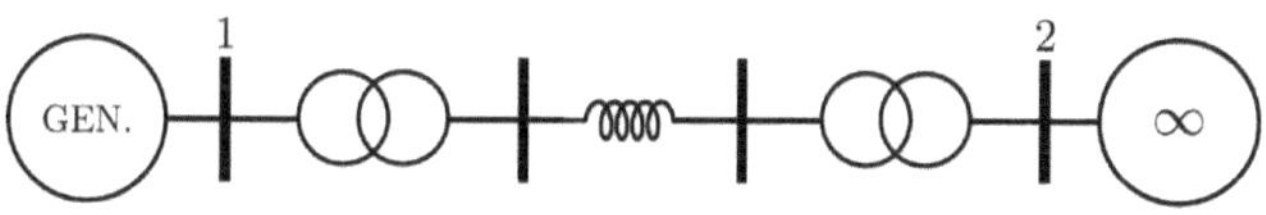

FIG. 2.6 Synchronous machine connected to an infinite bus.

unit base of the synchronous machine, are: $X_{eq} = 0.825$ pu, and $\vec{e}' = 1.07\angle 0.64$ pu. Therefore, applying Eq. (2.31), $K_s = 1.04$ pu torque/rad.

Using a similar method to that introduced in Ref. [11], the value of K_d of the $E'X'H$ model is estimated to reasonably represent the damping of the system. In the current example, $K_d = 7.78$ pu torque/pu speed.

Fig. 2.7 illustrates a comparison of the behaviors of frequency oscillations obtained with the electromagnetic and the $E'X'H$ model of the synchronous machine after a phase shift on the voltage of the infinite bus.

The dynamics obtained with both models are akin for the first swing and, after that, the mismatch between them increases over the time. This behavior is stated in the specialized literature and it is due to the aforementioned hypotheses related to the simplified model [1, 3, 4].

TABLE 2.2 Dynamic model data for transformers and transmission lines.

Transformers parameters	Value
Rated frequency (f_n, Hz)	60
Rated power ($S_{3\phi}$, MVA)	900
Rated voltage (V_n, kV)	20/230
Winding resistance (R_t, pu)	0
Winding reactance (X_t, pu)	0.15
Π-lines parameters	**Value**
Direct resistance (r_d, mΩ/km)	52.90
Direct reactance (x_d, mΩ/km)	529.0
Direct susceptance (b_d, μS/km)	3.31

TABLE 2.3 Power flow of the system.

Bus no.	V (pu)	θ (rad)	P (MW)	Q (MVAr)
1	1.0	0.417	700.00	117.19
2	1.0	0	−687.41	176.65

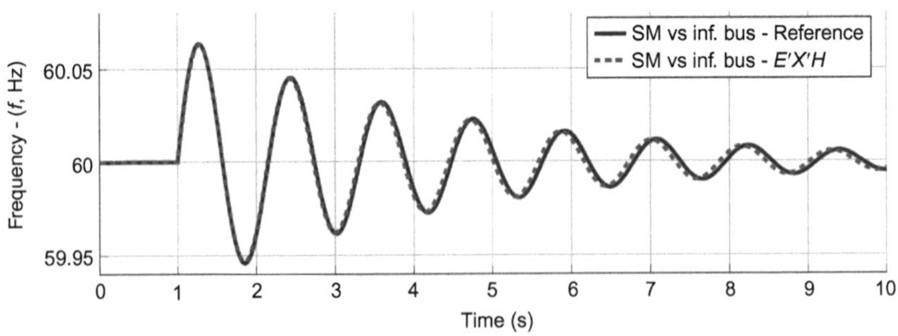

FIG. 2.7 Frequency oscillation behavior of the synchronous machine vs infinite bus system after a phase shift on the voltage of infinite bus ($\Delta\theta_g = \pi/40$ rad).

The eigenvalues of the system are found from the linearized representation of the system. The participation factor matrix [1, 3] is employed to identify the electromechanical eigenvalues of the system using the detailed model of the synchronous machine. The same is not necessary for the system using the $E'X'H$ model, since the system has only one pair of eigenvalues, which are the electromechanical ones. The comparison between the electromechanical eigenvalues of the synchronous machine considering both models is given in Table 2.4.

Fig. 2.7 and Table 2.4 provide substantial information about the consistency of the $E'X'H$ model with the electromagnetic model to represent the dominant behavior of the frequency of a synchronous machine without regulators.

2.4.1 Prime movers and governor

When operating in generator mode, the kinetic power developed on the shaft of synchronous machines is converted into electric power. In these applications, the rotors of synchronous machines are driven by turbines which in turn are driven by the energy provided by a primary source (i.e., hydro, fossil, or nuclear). It is important to take notice that the dynamics of the conversion chain of each primary source is different [1, 12, 13]. In the following studies, the dynamics of steam valves and a three-stage steam turbine [12] will be taken into account.

To maintain the proper functioning of the grid, the mismatch between generation and consumption must constantly be adjusted to maintain the frequency of the grid the closest to its nominal value. For this purpose, Transmission System Operators (TSOs) require that the generating units contribute to a series of ancillary services to the grid. At the European level, there are three main services to support the frequency of the grid [14]:

- Frequency Containment Reserve (FCR) is the active power reserve available to compensate the mismatch between generation and consumption. This service does not restore the frequency of the grid to its nominal value

TABLE 2.4 Electromechanical eigenvalues of the synchronous machine.

	Electromagnetic model	$E'X'H$ model
Eigenvalues ($\sigma \pm j\omega_d$)	$-0.28 \pm j5.46$	$-0.30 \pm j5.48$
Damping ratio (ξ, %)	5.14	5.45
Damped frequency (f_d, Hz)	0.869	0.872

after the disturbance. This automatic service is ensured by all FCR providing units at the European level, although, not all connected units are qualified to provide FCR. The scheduled cross-frontier power flows are affected by FCR contribution.

- Frequency Restoration Reserve (FRR) is the reserve available to restore the frequency of the system to its nominal value. The FRR is an automatic centralized service handled by the TSO responsible for the area where the disturbance took place, and provided only by the generating units of this area, re-establishing the scheduled cross-frontier power flows.
- Replacement Reserve (RR) is the power reserve which is activated to restore or support the required level of FRR, preparing the system for additional imbalances. This service is manual and centralized by the local TSO.

These services have different timescales, FCR being the fastest and RR the slowest between the three. In France, these timescales are between 15 and 30 s, 133 and 800 s, and 30 min and 8 h, for FCR, FRR, and RR, respectively [15]. A comparison between the grid codes of some countries can be found in Ref. [16]. Due to the timescale analyzed in this chapter, only FCR is considered. The FCR—and other frequency related ancillary services—is realized by the speed governor, and implemented using a power frequency droop, where the operating point of mechanical power ($P_{m,0}$) is proportionally adjusted as a function of mismatch between the reference and the instantaneous angular frequency of the shaft of the synchronous machine, as given in Eq. (2.35).

$$P_m^* = P_{m,0} - K_p(\omega^* - \omega_r) \tag{2.35}$$

where P_m^* is the adjusted operating point of mechanical power of the synchronous machine, K_p is the power frequency characteristic of the governor and ω^* is the setpoint of angular frequency of the synchronous machine (which is, in general, equal to its nominal value, ω_0). In Europe, the grid code requirement for generators determines that K_p must be comprised between 8 and 50 pu power/pu frequency [17]. Typical values of K_p for steam turbine governors are between 16.66 and 25 pu power/pu frequency.

Fig. 2.8 illustrates the generic frequency control chain of synchronous machines adopted, the IEESGO [12]. It can be observed that the model presents both the control and the power structures. It can be observed that the governor is composed of a power frequency droop function and two low-pass filters representing its dynamics. Standard models of turbines for power system simulations are presented in Ref. [12]. In Fig. 2.8, the power structure represents the dynamics of steam valves and a three-stage steam turbine.

This frequency control chain can be implemented in both electromagnetic and *E'X'H* models of the synchronous machine. Fig. 2.9 illustrates the second application. For the same scenario and parameters given in Section 2.4, the

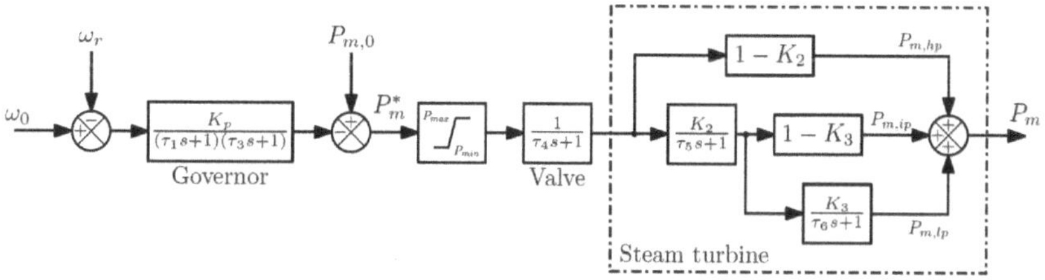

FIG. 2.8 IEESGO governor and steam turbine representation (simplified).

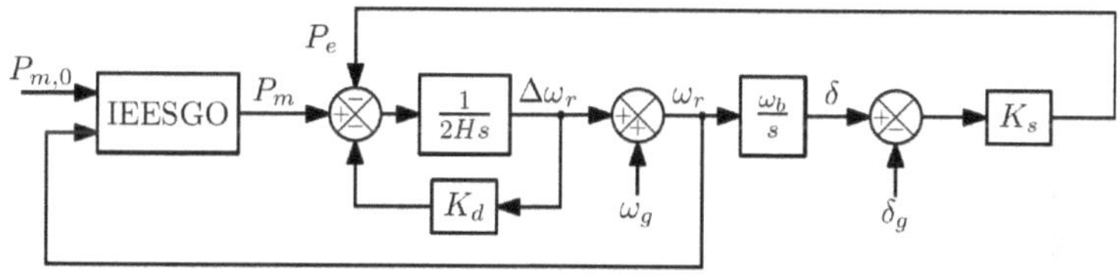

FIG. 2.9 Block diagram of the synchronous machine vs infinite bus system considering $E'X'H$ model and IEESGO.

impact of the frequency control chain on the electromechanical eigenvalues of the synchronous machine is compared in Table 2.5 considering both models of the synchronous machine. The parameters adopted for the control chain are given in Table 2.6. In the current case study, since the infinite bus imposes the frequency of the system, the synchronous machine, a priori, does not need to contribute to FCR, and therefore, the impact of prime movers and the governor can be separately observed.

TABLE 2.5 Impact of the frequency control chain on the electromechanical eigenvalues of the synchronous machine.

	Electromagnetic model		$E'X'H$ model	
	FCR off	*FCR on*	*FCR off*	*FCR on*
Eigenvalues ($\sigma \pm j\omega_d$)	$-0.28 \pm j5.46$	$-0.25 \pm j5.63$	$-0.30 \pm j5.48$	$-0.28 \pm j5.67$
Damping ratio (ξ, %)	5.14	4.49	5.45	4.93
Damped frequency (f_d, Hz)	0.869	0.896	0.872	0.902

TABLE 2.6 Dynamic model data for governor and steam turbine model IEESGO.

IEESGO parameters	Value
Power frequency characteristic (K_p, pu)	−0.25
Lag time constant (τ_1, s)	0.1
Lag time constant (τ_3, s)	0.2
Steam flow time constant (τ_4, s)	0.05
Reheater time constant (τ_5, s)	7.0
IP-LP time constant (τ_6, s)	0.4
Reheater power fraction (K_2, pu)	0.7
IP-LP power fraction (K_3, pu)	0.4

When just prime movers are considered (FCR off), the behavior of frequency oscillations is the same as that of the system without the prime movers (refer to Table 2.4).

On the contrary, when the FCR activated, it is observed a slight negative damping effect on the damping of the electromechanical eigenvalues, which is also stated by Refs. [12, 18]. This small impact can be explained by the fact that the frequency oscillations are filtered by the control chain, especially by the intermediate stage of the turbine which presents a slow dynamics—dynamics of the reheater—reducing the bandwidth of the feedback signal. Therefore, the impact of the frequency oscillations on the closed-loop system is minor. Moreover, it should be highlighted that, in both cases, when equipped with the frequency control chain, the $E'X'H$ model is still representative of the electromagnetic model of the synchronous machine.

2.4.2 Excitation system

In generator mode, the electromechanical energy conversion can only be ensured in the presence of an existing field flux (ψ_{fd}) which, in turn, is established when a dc field voltage (v_{fd}) is applied on the terminals of the field winding of the rotor of the synchronous machine. The excitation system supplies and automatically adjusts the field voltage of the synchronous machine to maintain the output voltage of the stator within the acceptable range [1, 19].

Excitation systems are classified in three categories based on the excitation power source [1, 19, 20]:

- The dc excitation system uses dc generators as the source of excitation power, injecting the field current via slip rings. The exciter may be driven by the shaft of the synchronous machine or by a motor. The exciter itself can be self or separately excited. The application of dc excitation system is continuously being replaced by ac or static excitation systems.
- The ac excitation system uses alternators as the source of excitation power, which usually shares the same shaft of the main generator. The conversion of the ac voltage of the exciter to dc voltage of the field is ensured by controlled or noncontrolled rectifiers. The power electronics switches can be static or rotating, where, in the former, the field current is injected via slip rings. The latter is known as brushless excitation system.
- The static excitation system does not possess any turning device. The excitation system is composed of static power electronics switches and provides the field current to the synchronous machine via slip rings. The excitation power is supplied to the rectifiers by the synchronous machine itself.

Voltage ancillary services to the grid are also required by the TSOs to maintain the voltages of the grid buses in acceptable values, and therefore, to maintain the proper functioning of the grid. At the European level, the regulations can be found in [14]. In France, there are two levels of voltage control required of the generating units by the French TSO [15]:

- The primary voltage control is an automatic service to control the local quantities of voltage and/or reactive power. There are three different types of primary voltage control: the supply of a fixed amount of reactive power; an adaptive output voltage following a reactive power voltage droop function; or the control of output voltage of the synchronous machine following a fixed operating point. The latter is the most used for synchronous machines.
- The secondary voltage control is an automatic centralized service deployed to expand the efficacy of the primary control and expand it to more global level. This service is coordinated by the TSO to control either the reactive level of a "geographical" area of the system, or the reference voltage of this area.

A third level of voltage control can be used to adapt, manually, the output voltage of the generating unit to respect the contractual reactive power voltage diagram. The timescale of primary and secondary voltage control are between hundreds of milliseconds and a few seconds, and 10 and 30 s, respectively. As for frequency ancillary services, only primary voltage control is considered in this chapter.

Fig. 2.10 illustrates the simplified block diagrams of three typical excitation systems DC1C, the AC1C and the ST1C, presented in Ref. [20]. All three models are composed of an Automatic Voltage Regulator (AVR) which, in turn, is composed of a voltage sensor and a low-pass filter, a main voltage regulator,

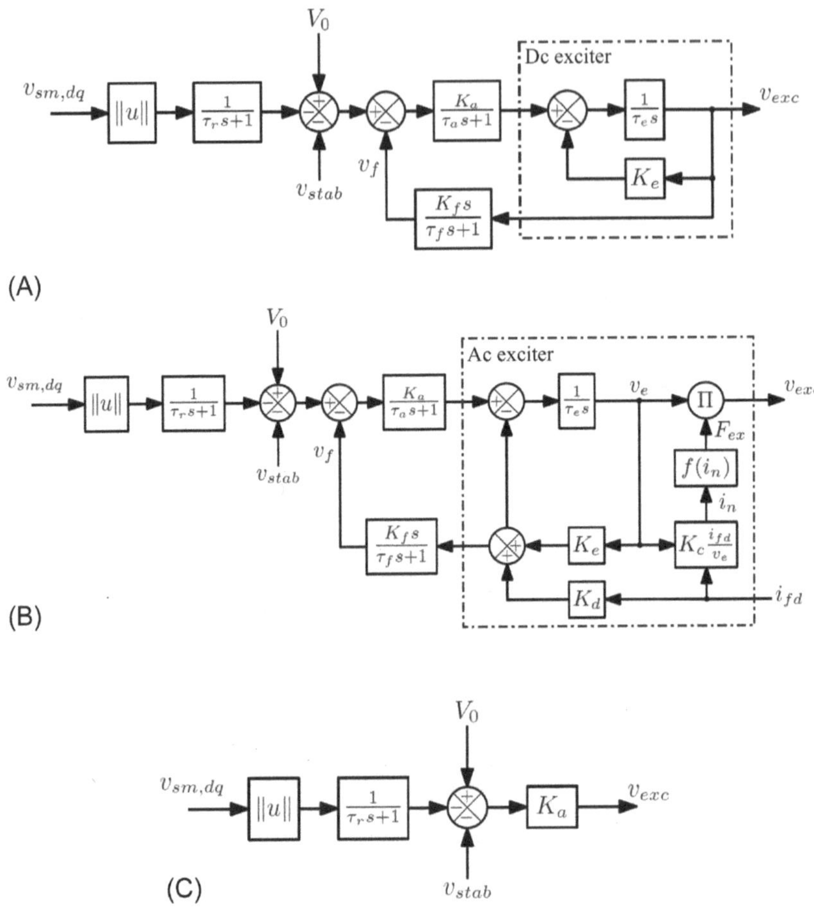

FIG. 2.10 Block diagram of dc, ac and static excitation systems (simplified). (A) DC1C, (B) AC1C, and (C) ST1C.

and an input for a stabilization signal (v_{stab}) provided by a power system stabilizer, which is discussed further in this section. The dc and ac excitation systems also present the dynamics of the exciter and a supplementary feedback stabilization signal. Considering the ST1C, since the dynamics of the static exciter are very fast, they are not represented in the block diagram of the model. More details about these models can be found in [1, 20].

From Fig. 2.10, it can be observed that the output of the excitation system is the excitation voltage (v_{exc}), yet in the introduction of this section, it was explained that the excitation system supplies the field voltage (v_{fd}). Indeed, in volts, these quantities are the same, although, in the *per unit* system, they are usually expressed using two different *per unit* bases to better interface the model of the excitation system with that of the synchronous machine (both the variables of the field and stator terminals) [1]. The relationship between v_{exc} and v_{fd} is expressed in Eq. (2.36) and developed in Ref. [1].

$$v_{exc} = \frac{L_{ad}}{R_{fd}} v_{fd} \tag{2.36}$$

In the following analysis, the idea is to observe how much the electromechanical eigenvalues estimated with the $E'X'H$ model get further from those found with the electromagnetic model of the synchronous machine equipped with the excitation system, evaluating then, how qualitative are the tendencies obtained with the former.

The adopted parameters for the different excitation systems are given in Table 2.7. The impact of the excitation system on the electromechanical eigenvalues of the synchronous machine is given in Table 2.8. It should be highlighted that, according to the exciter type, the impact on the damping ratio of the oscillations can be more or less pronounced. In the case of the AC1C, the electromechanical eigenvalues are not impacted by the exciter. On the contrary, for the ST1C, the system becomes unstable. The impact on the frequency oscillations are slightly negative when the synchronous machine is equipped with the DC1C excitation system. According to Ref. [3], if the AVR or the exciter has a large time constant, the AVR will not react during the transient state and the

TABLE 2.7 Dynamic model data for ac rotating excitation system model AC1C, dc rotating excitation system model DC1C and static excitation system model ST1C.

Parameters	DC1C	AC1C	ST1C
Filter time constant (τ_r, s)	0.01	0.01	0.01
Main regulator gain (K_a, pu)	20	78.14	200
Main regulator time constant (τ_a, s)	0.055	0.013	–
Regulator stabilizing gain (K_f, pu)	0.125	0.143	–
Reg. stabilizing time constant (τ_f, s)	1.8	1	–
Exciter gain (K_e, pu)	1	1	–
Exciter time constant (τ_e, s)	0.36	1.75	–
Demagnetizing factor (K_d, s)	–	0.17	–
Rectifier loading factor (K_c, s)	–	0.12	–

TABLE 2.8 Impact of the excitation system on the electromechanical eigenvalues of the synchronous machine.

Exciter type	$E'X'H$ model	DC1C	AC1C	ST1C
Eigenvalues ($\sigma \pm j\omega_d$)	$-0.28 \pm j5.67$	$-0.23 \pm j5.59$	$-0.25 \pm j5.62$	$0.004 \pm j6.12$
Damping ratio (ξ, %)	4.93	4.09	4.50	−0.07
Damped frequency (f_d, Hz)	0.902	0.890	0.894	0.974

regulated and unregulated systems behave akin. On the contrary, fast-acting AVRs with large gains may result in a negative damping, leading to an unstable system [2, 3], consequently requiring additional stabilization systems.

2.4.3 The power system stabilizer

The electromechanical oscillations of the synchronous machine are negatively impacted by the addition of its regulators [3, 12, 19], especially by the fast AVR and static exciter, where the system is even unstable. Despite the usual absence of specific legislations, recommendations or guidelines about the minimum required damping for these oscillations, 5% is often considered as the minimum accepted value [2, 21]. To enhance the damping of these oscillations, the Power System Stabilizer (PSS), a supplementary control unit, is added to the regulators of the synchronous machine [1–3, 20].

In the transmission system, because the ratio between resistance and reactance (R/X) is low, strong links between frequency and active power, and between voltage and reactive power are observed. Otherwise, cross-links between these quantities are fairly decoupled. Therefore, ideally, the PSS could easily mitigate the electromechanical oscillations by applying power oscillations of the same amplitude in the opposite direction (ac component) to counteract the former oscillations. According to [3], PSSs can be applied to the governor or to the excitation system, however, despite what could be indicated by common sense, in most cases, the PSS may not be effective when applied to the governor, because, as mentioned previously, the frequency bandwidth of the frequency control chain is very limited due to the slow dynamics of the turbine. Therefore, despite the weak link between power and voltage, the only practical solution is to apply the PSS on the AVR-excitation system.

The PSSs are usually based on the angular frequency of the synchronous machine, the active power, or both signals, and their main elements are the signal sensor and low-pass filter, a proportional gain, a washout filter to insulate the ac component of the oscillations, and phase compensations to match the stabilization signal to the ideal phase of the electromechanical oscillations [1–3, 20]. Fig. 2.11 illustrates the block diagram of the PSS1A, which is the adopted model of PSSs in this chapter. The tuning of the parameters of a PSS is a laborious task, and it is usually accomplished using optimization routines. The methodology and/or examples can be found in [1–3, 22]. The adopted parameters are given in Table 2.9.

The analysis of the impact of the PSS on the electromechanical eigenvalues of the system is performed and given in Table 2.10. The implementation of the PSS enhances the damping of the electromechanical oscillations for the three excitation systems, overcoming the aforementioned threshold of 5% of damping ratio. From the analysis of Table 2.10, it is observed that the performance of the PSS is more pronounced as the excitation system is faster, which is related to the bandwidth of the exciter.

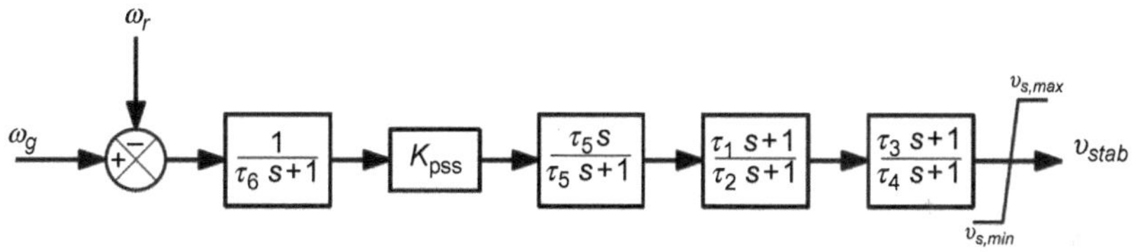

FIG. 2.11 Block diagram of power system stabilizer PSS1A (simplified).

TABLE 2.9 Dynamic model data for power system stabilizer model PSS1A.

PSS1A parameters	Value
Filter time constant (τ_6, s)	0.01
PSS gain (K_{pss}, pu)	variable
Washout filter (τ_5, s)	10
First lead time constant (τ_1, pu)	0.05
First lag time constant (τ_2, pu)	0.02
Second lead time constant (τ_3, pu)	3.0
Second lag time constant (τ_4, pu)	5.4

TABLE 2.10 Impact of the power system stabilizer on the electromechanical eigenvalues of the synchronous machine.

		$E'X'H$ model	DC1C	AC1C	ST1C
PSS off	Eigenvalues ($\sigma \pm j\omega_d$)	$-0.28 \pm j5.67$	$-0.23 \pm j5.59$	$-0.25 \pm j5.62$	$0.004 \pm j6.12$
	Damping ratio (ξ, %)	4.93	4.09	4.50	−0.07
	Damped frequency (f_d, Hz)	0.902	0.890	0.894	0.974
PSS on	PSS gain (K_{pss}, pu)	–	100	100	20
	Eigenvalues ($\sigma \pm j\omega_d$)	–	$-0.60 \pm j5.32$	$-0.47 \pm j5.41$	$-1.04 \pm j6.62$
	Damping ratio (ξ, %)	–	11.28	8.75	15.55
	Damped frequency (f_d, Hz)	–	0.847	0.861	1.054

2.5 Conclusion

This chapter has presented the main elements which constitute the classical power conversion chain based on a synchronous machine. The impact of these different elements on the electromechanical eigenvalues of the system has been analyzed. Due to the slow behavior of the turbine, the governor has very little effect on the dominant electromechanical poles of the system. The integration of the excitation may have a larger effect, especially with the static excitation which is the fastest among the three types of excitation. It should be highlighted that even if the excitation system and PSS are implemented, it is still possible to adjust the parameters of the $E'X'H$ model to provide a good estimation of the frequency oscillations. It can be concluded that the $E'X'H$ model of the synchronous machine is capable of providing qualitative tendencies about the behavior of the frequency oscillations of the system.

However, it has to be underlined that the parameters of this model may diverge from their initial physical meanings. For example, Fig. 2.12 illustrates the frequency oscillation behavior of the synchronous machine vs infinite bus system using both representations of the synchronous machine, where the

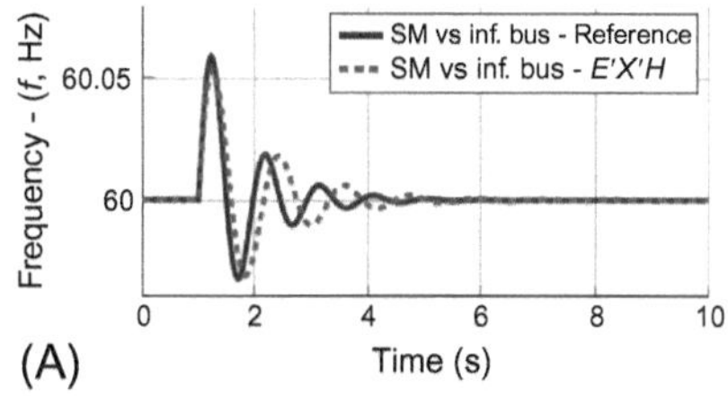

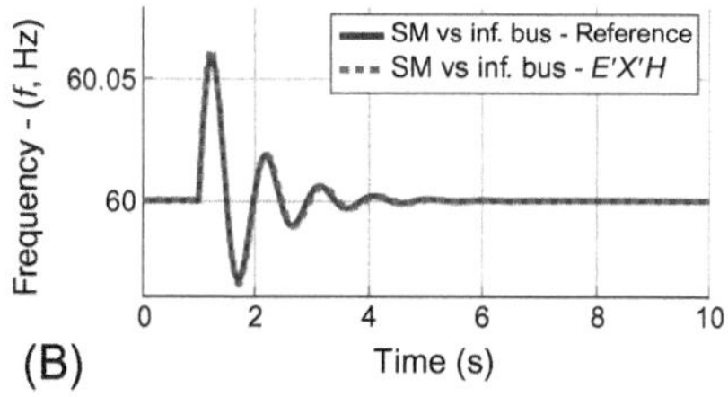

FIG. 2.12 Frequency oscillation behavior of the synchronous machine vs infinite bus Detailed model of synchronous machine equipped with IEESGO (FCR on), ST1C and PSS1A. Simplified model equipped with IEESGO (FCR on). (A) $K_s = 1.04$ pu and $K_d = 24.4$ pu. (B) $K_s = 1.44$ pu and $K_d = 31.1$ pu.

electromagnetic model is equipped with the IEESGO, ST1C and PSS, whereas the simplified model is only equipped with the governor. Using a method similar to that presented in [11], from the analysis of Eqs. (2.33), (2.34), the value of K_d of the $E'X'H$ model has be modified in order to estimate the damping of the frequency oscillations observed with the electromagnetic model, and if both K_d and K_s are adjusted, it is even possible to match the frequency of these oscillations.

References

[1] P. Kundur, Power System Stability and Control, second ed., McGraw-Hill, ISBN: 007035958X, 1994, p. 1201.

[2] G. Rogers, Power System Oscillations. Springer US, Boston, MA, ISBN: 978-1-4613-7059-8; 2000, pp. 1–332, https://doi.org/10.1007/978-1-4615-4561-3.

[3] J. Machowski, J.W. Bialek, J.R. Bumby, Power System Dynamics: Stability and Control, second ed., John Wiley & Sons, West Sussex, ISBN: 9780470725580, 2008, p. 660.

[4] IEEE Power Engineering Society, IEEE Guide for Synchronous Generator Modeling Practices and Applications in Power System Stability Analyses, vol. 2002, November, 2003, p. 81. ISBN: V, https://doi.org/10.1109/IEEESTD.2003.94408.

[5] R.H. Park, Two-reaction theory of synchronous machines generalized method of analysis-part I. Trans. Am. Inst. Electr. Eng. 0096-386048 (3) (1929) 716–727, https://doi.org/10.1109/T-AIEE.1929.5055275.

[6] P. Dandeno, R. Hauth, R. Schulz, Effects of synchronous machine modeling in large scale system studies. IEEE Trans. Power Syst. 0018-9510PAS-92 (2) (1973) 574–582, https://doi.org/10.1109/TPAS.1973.293760.

[7] M. Eremia, M. Shahidehpour, Handbook of Electrical Power System Dynamics Modeling, Stability, and Control, ISBN: 9781118497173, 2013, pp. 1–968, https://doi.org/10.1002/9781118516072.

[8] A. Blondel, Application de la méthode des deux réactions à l'étude des phénomènes oscillatoires des alternateurs accouplés, Rév. Gén. Électr. XIII (Février) (1923) 235–251.

[9] IEEE Power and Energy Society, IEEE guide for test procedures for synchronous machines. in: IEEE Std 115-2009 (Revision of IEEE Std 115-1995), 2009, p. 244, https://doi.org/10.1109/IEEESTD.2010.5953453.

[10] M.A. Swidan, Study of Damping Power in Interconnected Power Systems, Iowa State University, Digital Repository, Ames, 1964.
[11] M. Shiroei, B. Mohammadi-Ivatloo, M. Parniani, Low-order dynamic equivalent estimation of power systems using data of phasor measurement units. Int. J. Electr. Power Energy Syst. 0142061574 (2016) 134–141, https://doi.org/10.1016/j.ijepes.2015.07.015.
[12] IEEE Power and Energy Society, Dynamic Models for Turbine-Governors in Power System Studies, IEEE, 2013, p. 117. Tech. Rep.
[13] I. Boldea, Synchronous Generators, Taylor & Francis Group, ISBN: 9780849357251, 2006, p. 425.
[14] The European Commission, Commission Regulation (EU) 2017/1485 of 2 August 2017 establishing a guideline on electricity transmission system operation, Off. J. Eur. Union (2) (2017) 120.
[15] RTE, Documentation technique de référence. Paris, 2017, p. 2123.
[16] C. Roberts, Review of International Grid Codes, Lawrence Berkeley National Laboratory, 2018, p. 73. Tech. Rep. February.
[17] European Commission, Commission Regulation (Eu) 2016/631 of 14 April 2016 Establishing a Network Code on Requirements for Grid Connection of Generators, OJ L 112, 27.4.2016, 2016, pp. 1–68.
[18] CIGRE Task Force 38.01.07 on Power System Oscillations, Analysis and Control of Power System Oscillations, CIGRE Technical Brochure, December, 1996.
[19] P. Wetzer, Machines Synchrones Excitation, Techniques de l'Ingénieur, 1997, 11 no. D3545 v1.
[20] IEEE Recommended Practice for Excitation System Models for Power System Stability Studies. IEEE Std 421.5-2016 (Revision of IEEE Std 421.5-2005), 2016, pp. 1–207, https://doi.org/10.1109/IEEESTD.2016.7553421.
[21] E. Grebe, J. Kabouris, S.L. Barba, W. Sattinger, W. Winter, Low frequency oscillations in the interconnected system of Continental Europe. in: IEEE PES General Meeting, IEEE, ISBN: 978-1-4244-6549-1 2010, pp. 1–7, https://doi.org/10.1109/PES.2010.5589932.
[22] P.W. Sauer, M.A. Pai, Power System Dynamics and Stability. Illinois, ISBN: 978-1588746733, 2006, p. 361. ISBN: 13.

Chapter 3

Classical grid control: Frequency and voltage stability

Ettore Bompard[a], **Andrea Mazza**[a], **and Lucian Toma**[b]
[a]*Department of Energy "Galileo Ferraris", Politecnico di Torino, Torino, Italy,* [b]*Department of Electrical Power Systems, University Politehnica of Bucharest, Bucharest, Romania*

3.1 Power system states

The national/regional grid codes stipulate that the state quantities must remain within acceptable ranges for any disconnection of an element (line, transformer, generator, etc.). This is known as N-1 security criteria. Consecutive outages may occur in a power system and thus it is highly recommended that the power system be designed to withstand double disconnections, known as the N-2 security criterion. Unfortunately, due to the large investment requirements, most of the power systems may have problems to comply with the N-2 security criterion [1].

The operating conditions vary continuously, and the power system moves from one state to another, as suggestively indicated in Fig. 3.1. Transition to one state or another depends on the random events that may occur or on the decision taken by the system operator. In NORMAL state, all parameters are within acceptable ranges and the power system is stable and secure. Furthermore, from this state, disconnection of any element can bring no harm to the power system. However, significant changes such as large load increase or extreme weather conditions make the system vulnerable to disconnection on an element and the power system may enter in the ALERT state. Fig. 3.1 shows also a classification of possible states of the power system depending on the event that may occur.

When the power system enters in the alert state, immediate corrective actions must be taken to restore the normal operation. The restoration process may take shorter or longer time depending on the dynamics of the corrective actions. If during this transition a contingency occurs, the system can enter in an EMERGENCY state, in which there is a large number of bus voltage limits violations or exceeding of branch ampacity. In this state, ultimate (extreme) remedial actions can still be taken and system restoring to a normal operation

Converter-Based Dynamics and Control of Modern Power Systems
https://doi.org/10.1016/B978-0-12-818491-2.00003-1

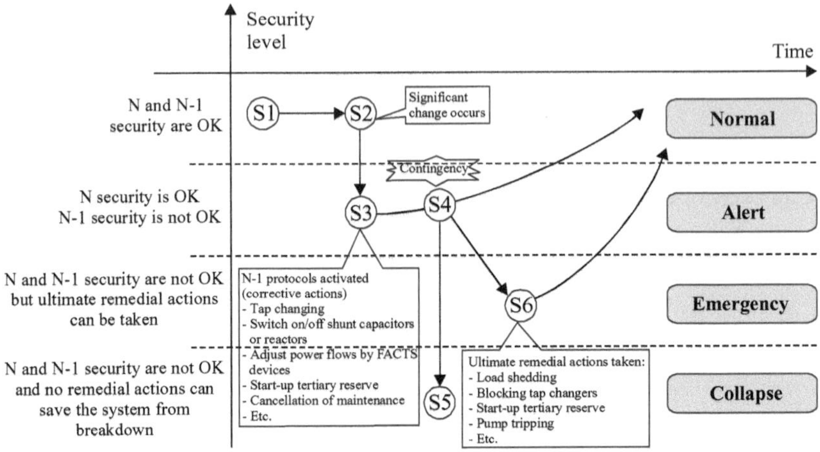

FIG. 3.1 Power system states [1, 2]. *(From S.S. Venkata, M. Eremia, L. Toma, Background of power system stability, in: M. Eremia, M. Shahidehpour (Eds.), Handbook of Electrical Power System Dynamics, 2013. https://doi.org/10.1002/9781118516072.ch8, based on UCTE Policy 3: Operational security, March 2009.)*

is possible. If the contingency is too severe, the power system may become instable and finally COLLAPSEs.

Violation of an operating constraint may not necessarily mean that the power system becomes instable. However, due to high loading, the voltage in the system nodes can become too low and the loads demand more current which, in time, may overload the transmission lines. This will finally jeopardize the electrical network integrity, limits the transmission capacity and lead to instability conditions.

3.2 Frequency control and stability in power systems

3.2.1 General aspects

To maintain the power system frequency to a reference value, the total generated active power has to be equal to the total consumed active power at every instant of time. In reality, the balance between generation and load is permanently perturbed by load variations, by the imprecision of real-time generation control or occasionally by the unscheduled/scheduled disconnection of a generator, a transmission line, etc. [3]. The increase in the total generation to a value greater than the total load, including the exports/imports, causes a frequency increase above the nominal value, while a total generated active power smaller than the total load causes the frequency to decrease below the nominal value. The frequency variations denote a rapid process so that its correction requires high performances for generators. The power system perturbations leading to frequency variations occur at very short time intervals and therefore corrective actions for power balancing have to be taken continuously. Interconnected operation of power system represents an advantage regarding frequency control since the frequency variations are directly influenced by the power imbalance and the instantaneous load. Thus, the bigger the instantaneous generation/load is with respect to the produced imbalance, the smaller is the frequency variation. On the other hand, due to the physical laws that govern the current flow, any powers imbalance in a system has an immediate consequence on the power flows on the interconnection lines. For this reason, besides frequency measurement, the frequency control involves power flow measurement on the interconnection lines.

In a synchronously interconnected power system, the frequency presents three main characteristics [4]:

- uniformity: in power systems based mainly on mechanical inertia, all generators swing together, in synchronism, at any instant, which makes the frequency to be identical in every point of the system; however, while the classical power plants are replaced by inverter-based generation units, this definition is no longer valid, leading to the need for a new definition.
- coherency: all generation units in the power system must be correlated in such a way that they produce a synchronized voltage waveform.
- quasistability: the frequency must be maintained around the nominal value (50 Hz in Europe or 60 Hz in United States, Canada, etc.) because the power system components are designed to operate optimally at this value. The generating units are equipped with protection systems that disconnect them from the power grid if the frequency deviation is greater than values that ensure the safe operation (Fig. 3.2). On the other hand, operation of the loads at frequencies different from the nominal value results in productivity loss, equipment aging, erroneous operation of some electrical devices based on time measurement, maloperation of some protection systems, etc.

Since the frequency is a global quantity that characterizes the whole power system, its control falls within the responsibility of the system operator. The frequency is an inertial quantity, and accurate control of it assumes its real-time monitoring as well as of the power flow on the interconnection lines. Balancing between generation and load requires continuous availability of some power reserves that can be (un)loaded in due time. These reserves involve some costs, and for their minimization, taking into consideration the need for assurance of an appropriate security level of the system, within particular control areas, the system operator should formulate an optimization problem with a multiobjective function.

3.2.2 Hierarchical frequency control

The frequency control is a process performed by hierarchical or semihierarchical coordination, in time, of the regulation resources. This section presents a general overview on the utilization, coordinated or noncoordinated, automatic

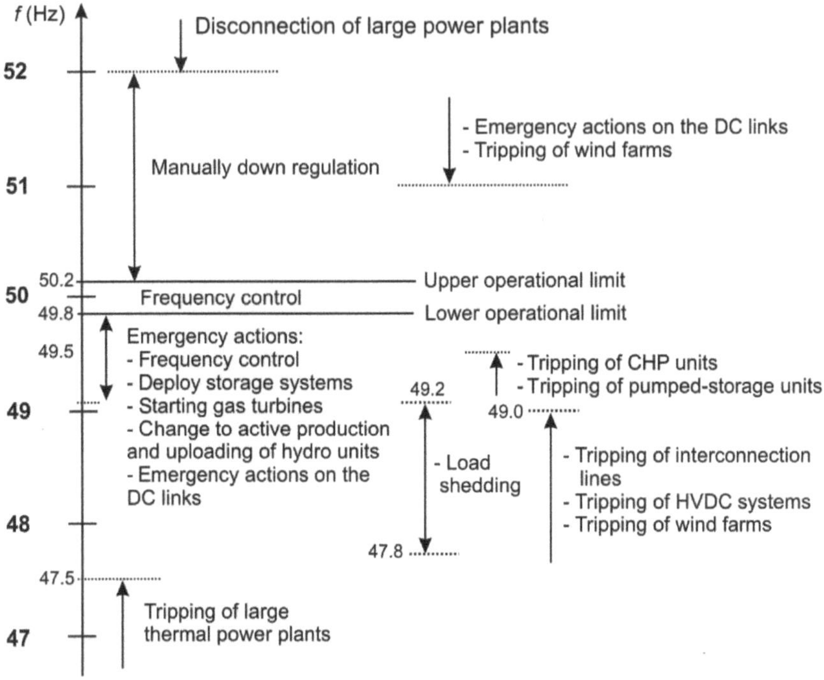

FIG. 3.2 Tripping thresholds to frequency variations [5, 6].

or manually, of the active power reserves when the frequency is subjected to variations caused by imbalances between generation and load.

The terminology related to frequency control is rich in definitions in various countries, but they are often referring to the same procedures or functions. There are, however, some differences that are given by several peculiarities of the different power systems, especially in geographically isolated countries or regions. The frequency control is performed mainly on three levels (Fig. 3.3), that is, primary level (frequency containment), secondary level (automatic frequency restoration), and tertiary level (manual frequency restoration).

Fig. 3.3 illustrates the origin of the active power control signals received by a qualified synchronous generator. Note that a generator can participate in the frequency control only based on previous qualification tests. The total power generation of a generator consists of a load reference, that is, P_{load}, which is the unit commitment in the energy market decided by the power plant owner and accepted by the system operator, and the power signals resulted from the participation to network ancillary services. Therefore, the generator, or more generally the energy resource, changes the power production slower or faster as a response to the frequency control orders or to economic optimization decisions, by automatic and manual orders. The ancillary services involve either upward or downward regulation depending on the negative or positive deviation

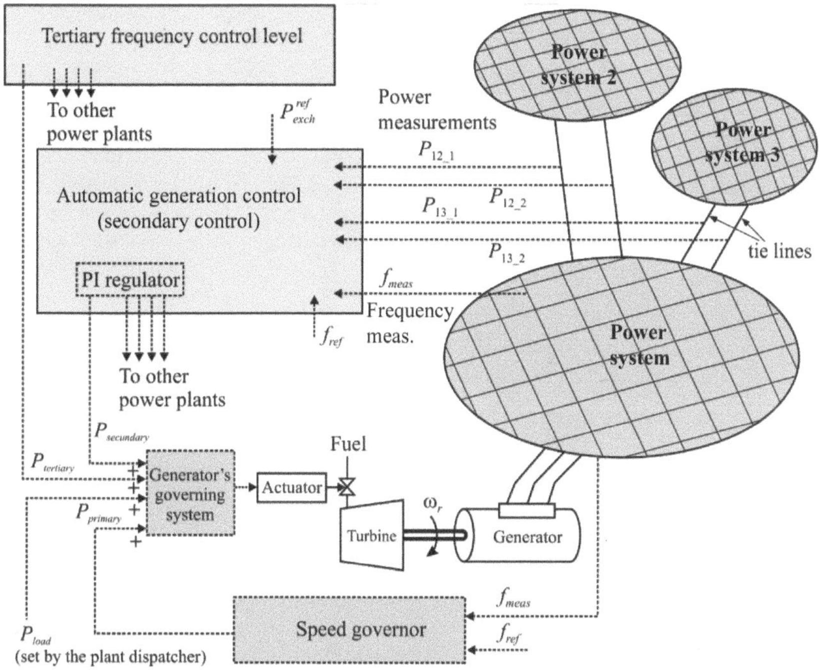

FIG. 3.3 The hierarchical frequency control scheme.

of the frequency, which means that the total power generation can be either higher or lower than the load reference.

An overview of the characteristics of the frequency control levels is presented in Table 3.1.

The synchronous operation of interconnected power systems assumes a common contribution from all the generators. For this reason, the policy for frequency control in the related interconnected areas contains common rules. These rules define the technical characteristics for the operation of all the energy resources connected to the grid, as well as for the centralized controllers.

3.2.3 The linearized electromechanical model of a synchronous generator

For the analysis of a power system from frequency dynamics point of view, the electromechanical model of a synchronous generator, expressed as a first-order differential equation, in per unit, can be used [8]:

$$
\begin{aligned}
&2H\frac{d\Delta\omega_r}{\mathrm{d}t}+D\Delta\omega_r=\Delta T_m-\Delta T_e\approx\Delta P_m-\Delta P_e\\
&\frac{d\Delta\delta}{\mathrm{d}t}=\omega_0\Delta\omega_r
\end{aligned}
\tag{3.1}
$$

TABLE 3.1 Characteristics of the three frequency control levels [7].

	Primary control	Secondary control	Tertiary control
Why is this control used?	To stabilize the frequency any time power unbalance is experienced	To restore the frequency and the interchange flows to their scheduled value	To restore the secondary control reserve, to manage eventual congestions, and to bring back the frequency and the interchange programs to their target if the secondary control reserve is not enough
How is this control achieved?	Automatically		Manually
Where is this control performed?	Locally/ decentralized	Centralized, from the dispatching center	
Who sends the control signal to the source of reserve?	Local sensor	Automatic generation controllers (AGC) located on the TSO premises	Generators, loads or other TSOs
Time activation	2–3 s from the perturbation detection	Continuously, the control order is sent every 1–5 s	Varies from system to system
Full activation	Usually up to 30 s	Usually up to 15 min	Usually between 15 min and 7 h
Resources used	Partially loaded generation units, storage systems, loads, fast/slow starting generation units, changes in exchange programs		

where T_m is the turbine mechanical torque; P_m is the mechanical power of the turbine-generator; T_e is the electrical torque; P_e is the electrical power of the generator; H is the inertia constant, with $2H = M$; M is the mechanical starting time; ω_r is the rotor speed; ω_0 is equal to $2\pi f$, in electrical radians per second; δ is the rotor angle, in radians; D is the self-regulation coefficient of the load; t is the time, in seconds; and Δ stands for the change in the variable.

When a power unbalance occurs in the system and the system frequency deviates from the reference value, each generator experiences an accelerating or decelerating torque given by $T_a = T_m - T_e$.

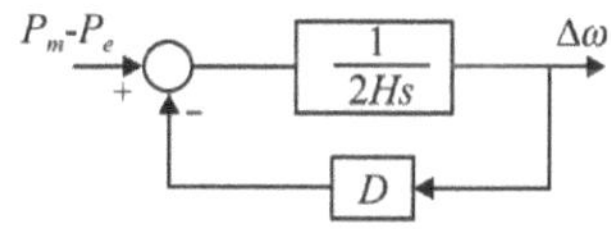

FIG. 3.4 Block diagram of the system dynamics, with load damping.

The term D in usually the same in all synchronous areas, and assumed to be 1%/Hz, which means that a load decrease of 1% occurs in case of a frequency drop of 1 Hz. Hence $D=1$ in Eq. (3.1) if load damping is considered [2]. The block diagram of the system dynamics and load damping is shown in Fig. 3.4.

Simplifying the block diagram from Fig. 3.4, a single forward block is given by the function $1/(2Hs+D)$.

3.2.4 Inertia frequency response

In the classical power systems, the mechanical inertia constant, denoted by H, represents the time (in seconds) a synchronous generator can provide the rated power using the kinetic energy, E_{kin}, stored in the rotating mass, and is given by [9]:

$$H=\frac{E_{kin}}{S_b}=\frac{1}{S_b}\underbrace{\left(\frac{1}{2}J\omega_{0m}^2\right)}_{E_{kin}}=\frac{J(2\pi f_m)^2}{2S_b}\left[\frac{\text{MW}\cdot\text{s}}{\text{MVA}}\right] \tag{3.2}$$

where J is the combined moment of inertia of the generator and turbine [kg × m^2], ω_{0m} is the nominal speed of rotation [mech rad/s], f_m is the rotating frequency of the machine and S_b is the base power, usually taken equal to the rated power of the generator.

In the case of a frequency disturbance, the inertial response of the synchronous machine can be described by the change in the rotational speed or the rotational frequency, that is,

$$\frac{\mathrm{d}(E_{kin})}{\mathrm{d}t}=\frac{\mathrm{d}\left(J\omega_{0m}^2/2\right)}{\mathrm{d}t}=J(2\pi)^2 f_m\frac{\mathrm{d}f_m}{\mathrm{d}t}=\frac{2HS_b}{f_m}\frac{\mathrm{d}f_m}{\mathrm{d}t} \tag{3.3}$$

The rate of change of frequency (rocof), which defines the variation of frequency with time, is therefore inversely proportional to the inertia constant, that is,

$$\frac{\partial f}{\partial t}\cong\frac{\Delta P=(P_m-P_e)}{2H} \tag{3.4}$$

Thus, the greater the mechanical inertia the smaller the frequency acceleration, which in turn will result in smaller frequency deviations. Minor attention was paid until recently to this quantity because of the appropriate amount of kinetic energy available in classical power plants. Examples of inertia constant values specific to various power plants are provided in Table 3.2.

The renewable energy sources connected to the electrical network via power converters have zero mechanical inertia, except for the doubly fed induction generator (DFIG) which has small inertia because its stator is synchronously connected to the electrical network.

Example: The Romanian Power System (RPS) operates synchronously with the ENTSO-E network of the Continental Europe since 2004. The Cernavoda Nuclear Power Plant (CNPP) consists of 2 units 700 MW rated each, and, in some operating conditions, account for up to 40% of the total mechanical inertia available in RPS. CNPP is located at the edge of the ENTSO-E system. Two events involving the disconnection of one nuclear unit, during which the frequency was recorded by several PMU devices at remote locations in RPS, can reveal the deployment of the kinetic energy and the natural time duration to recover the frequency. The first event is denoted by CNPP_ev1 (normal line in Fig. 3.5): occurred in 2017, one unit was in scheduled maintenance, during which the other units was suddenly tripped. The second event is denoted by CNPP_ev2 (dashed line in Fig. 3.5): occurred in 2018, both units were in operation, and one unit was suddenly tripped. The graphical representation in Fig. 3.5 of the frequency during the CNPP_ev2 event is at scale on the frequency axis, while on the time axis only the time span corresponds to the time interval illustrated.

The frequency drop recorded during CNPP_ev2 on the 400 kV busbar of the CNPP was half of that recorded during CNPP_ev1. The frequency recorded in the remote buses, which are located electrically and geographically closer to the ENTSO-E Interconnected Network of the Continental Europe, has experienced much smaller variations; this is because of the strong mechanical inertia available in both Western side of RPS and the ENTSO-E.

As the two events were not planned, and the disconnection was almost instantaneous, the inertia response was the only contribution captured in this

TABLE 3.2 Values of the inertia constant.

Type of plant	H (s)
Nuclear	6–7
Coal and natural gas	3–6
Large hydro	2–4
Small hydro	1
Synch. condenser	1
Wind	0–1

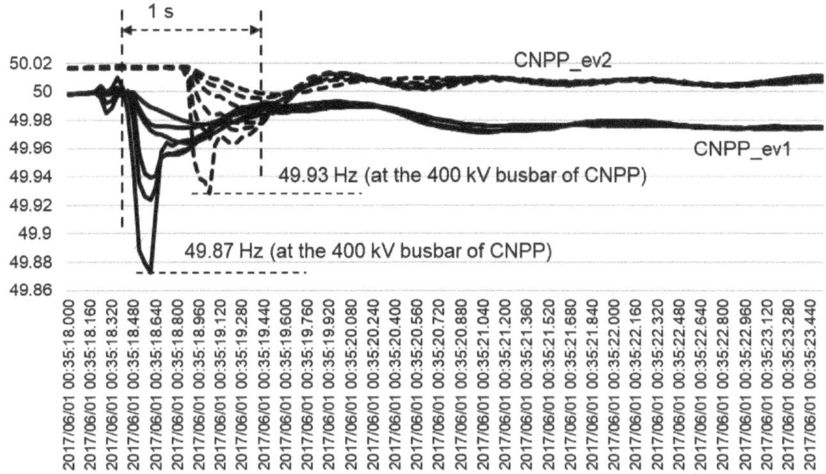

FIG. 3.5 Frequency variation during CNPP_ev1 and CNPP_ev2 events in the Romanian power system.

PMU based frequency records. In both cases, the main inertial response was observed during one second after the perturbation inception, within which the frequency was restored around a quasistationary value. The total transient period was around 5–6 s, specific to the mechanical inertial response.

3.2.5 Primary frequency control

The primary frequency control, called also frequency containment control (FCC) in ENTSO-E, is performed automatically and in a decentralized way (independently) by all the power resources designed and qualified for this purpose. It aims at stabilizing the frequency around an acceptable quasistationary value after a disturbance that causes the frequency to deviate outside the predefined limits. The disturbance consists in the unbalance between generation and load demand, caused by the change in load, by the change in generation, or by the inadvertent change in the power flow on the interconnection lines.

In ENTSO-E, the total primary frequency reserve that must be available within the system, called also frequency containment reserve (FCR), should cover the simultaneous loss of the largest two power plants or the loss or a line section or a busbar. Therefore 3000 MW must be deployed by a frequency deviation of 200 mHz [10]. This reserve is allocated to each control area/block or interconnection partner proportionally to their maximum load. Other events are also considered, but they are considered extremely rare. The reserve allocated to each participant to the interconnection is then allocated to all its qualified resources using the principles defined by the grid codes.

While some power systems rely on classical large synchronous generators to provide primary frequency control, with the advancement in the power electronic based generation technology and the information and communication technology (ICT): (1) renewable energy sources (mainly wind and photovoltaic); (2) energy storage systems; and (3) controllable loads are today included in the primary frequency control level. For economic and social reasons, the last option is taken in emergency conditions, by means of the load shedding automation.

Any power imbalance occurring in the power system results in the frequency deviation and implicitly in the change in the rotor speed of all generators. A speed governor is employed in the case of synchronous generators to adjust the energy flow (steam, water, gas) to the turbine (Fig. 3.6) and hence to adjust the active power output of the generator every time the rotor speed deviation is greater than a threshold.

There are three main types of governors employed to regulate the turbine-generator speed, that is, isochronous, speed droop, and compensated governor [2, 8]. While the speed-droop governor is appropriate in interconnected power systems, the isochronous governors are required in islands.

Speed-droop governor

In a speed-droop governor based frequency control scheme, a negative mechanical feedback is added as shown in Fig. 3.7. The governor and the turbine are simply represented by a delay transfer function.

The droop value, R, is the amount of speed (or frequency, when expressed in per unit) change that is necessary to cause the main prime mover control mechanism to move from fully closed to fully open. The droop, in per unit, can be calculated as [11]:

$$R = \frac{\Delta\omega}{\Delta P} = \frac{\Delta f}{\Delta P} \tag{3.5}$$

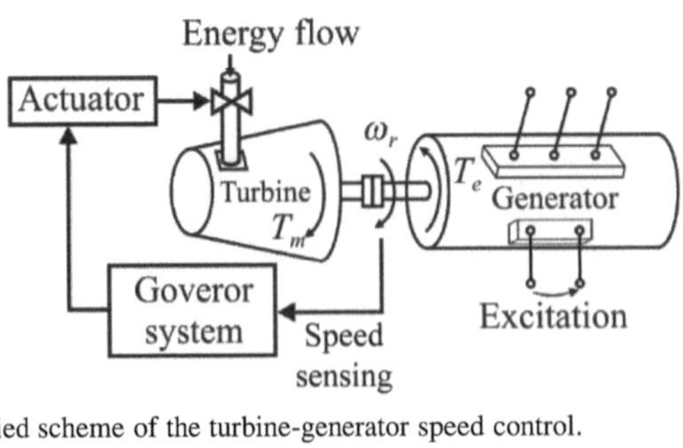

FIG. 3.6 Simplified scheme of the turbine-generator speed control.

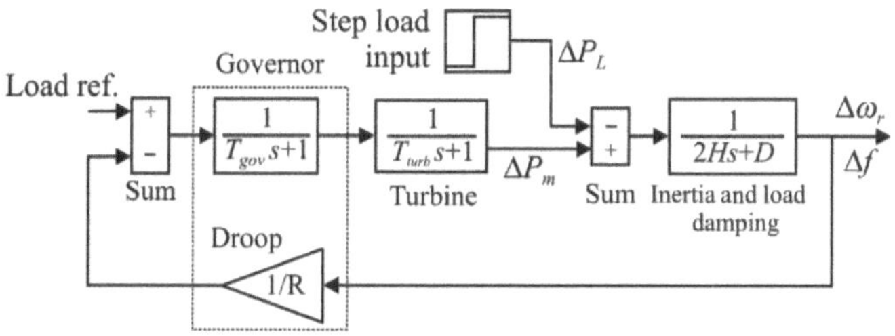

FIG. 3.7 Droop control model of a single machine supplying load in islanded operation.

The resulted change in power is $\Delta P = \Delta\omega/R$. In other words, the droop represents an amplification of the change in speed ($\Delta\omega_r$) or frequency ($\Delta f = f - f_0$) necessary to achieve the required power output adjustment.

The droop can be also calculated, in per unit, as the ratio between the change in frequency from no-load to full-load operation and the reference frequency [8].

$$R = \frac{\omega_{NL} - \omega_{FL}}{\omega_0} = \frac{f_{NL} - f_{FL}}{f_0}\ (\text{p.u.}) \tag{3.6}$$

The droop is usually set between 3% and 5%, depending on power system characteristics. A droop value of 5% means that a 5% frequency variation causes the generated power to change by 100%. Let us consider the regulation of a generation unit with 5%, which is set to achieve the reference frequency ($f_0 = 1.0$ p.u.) at 50% (0.5 p.u.) loading (Fig. 3.8). This unit operates at zero load with a 2.5% overfrequency and at 100% load at 2.5% underfrequency.

Exercise: Let us consider the following parameters for the frequency control scheme illustrated in Fig. 3.8: $T_{gov} = 0.2$ s; $T_{turb} = 0.5$ s; $H = 5$ s; $D = 1$ p. u.; $R = 0.05$ p. u. A step change in the load, $\Delta P_L = 0.01$ p.u., results in the frequency variation as shown in Fig. 3.9.

In a speed-droop frequency control, the frequency recovers to a quasisteady-state value. The steady-state frequency deviation is determined by:

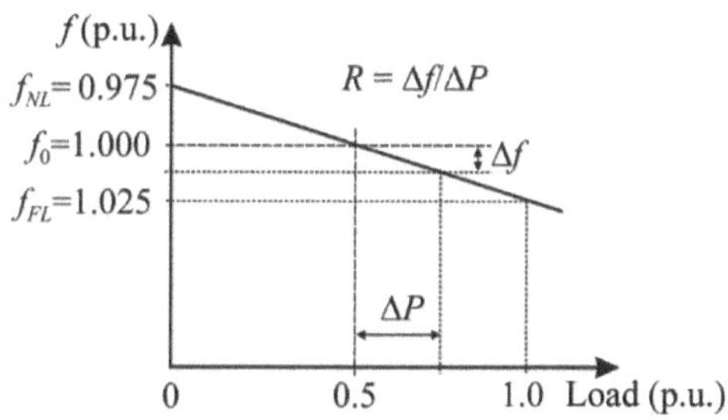

FIG. 3.8 Droop characteristic of s speed governor.

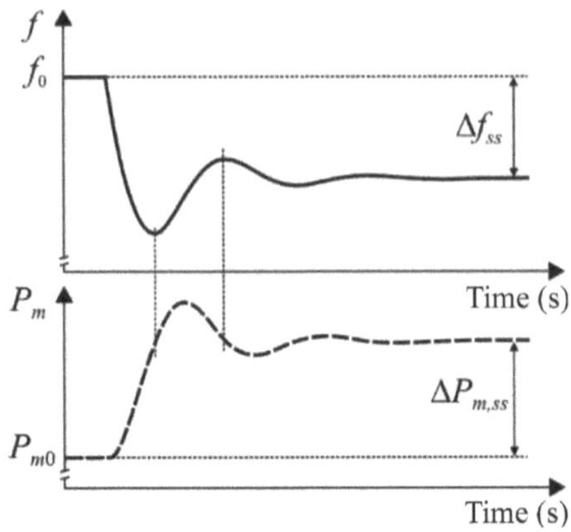

FIG. 3.9 Frequency and active power variation for speed-droop governor.

$$\frac{\Delta\omega}{\Delta P} = -\frac{1}{1/R+D} \Rightarrow \Delta\omega = \Delta f_{ss} = -\frac{\Delta P}{1/R+D}$$

then, by substituting the above-mentioned values, we obtain the following:

$$\Delta f_{ss} = -\frac{0.01}{1/0.05+1} = -4.762\times 10^{-4}\text{p.u.}$$

If the governor is tuned to be "isochronous" (i.e., zero droop), it will keep opening the valve until the frequency is restored to the original (reference) value. This type of tuning is used on small, isolated power systems, but would result in excess governor movement on large, interconnected systems, thus becoming unstable.

3.2.6 Secondary frequency control

The secondary control level, called most recently automatic Frequency Restoration Control (aFRC) in ENTSO-E, aims at restoring the power system frequency to the reference value by balancing the generation and load within the control area (usually a national power system). The Automatic Generation Controller (AGC) is a software application, component of the TSO's Energy Management System (EMS), which determines the amount of power unbalance within the control area by monitoring the frequency and the interchange power flows on the interconnection lines, and sends control orders via telecommunication connections to the qualified generation units to provide balancing power. The specific time cycle for data acquisition, calculation, and provision of automated instructions is 4–10 s [12].

As the unbalance can be either positive or negative, the secondary reserve (automatic frequency restoration reserve—aFRR) is established for both upward and downward regulation directions, both forming the secondary reserve band (SRB). The amount of secondary power reserve deployed is also intended to restore the primary reserve. Since the participation to the secondary control involves a visible effort from producers, the procurement of this reserve by the system operator is subjected to commercial activity.

The power plants integrated in the secondary frequency control are selected based on qualification tests, during which they must be capable of deploying power according to the ramping up or ramping down requirements, expressed in MW/min, for the required time period. The most appropriate generation units employed for this service are the hydraulic power plants and the gas-fired combined-cycle power plants. However, nuclear power plants and classical thermal power plants (coal-fired and oil-fired) are also used in power systems with small hydro-potential.

Even if not applied so far, the AGC controllers in neighbor countries can be coupled to share the frequency restoration reserves. Signals from one AGC to another can be sent to activate FRRs in the neighbor country. By this means, a power margin should be added to the scheduled exchanged power considered in the input to the AGC. This will allow developing inter-TSO markets for frequency control.

Fig. 3.10 illustrates two interconnected power systems provided with AGC controllers, with inputs from both internal measurements and external reserve demand from the neighbor power system.

The AGC calculates the area control error (ACE) resulted within the control area using the following expression (Fig. 3.10):

$$ACE = P_{exch} - P_{exch}^{ref} + K\left(f_{meas} - f_{ref}\right) \tag{3.7}$$

where P_{exch} is the total instantaneous power flow (exchanged) on the interconnection lines, in MW; P_{exch}^{ref} is the scheduled net interchange power, in MW; K is the K-Factor of the control area, in MW/Hz; f_{meas} is the measured frequency, in Hz; f_{ref} is the reference/nominal frequency, in Hz.

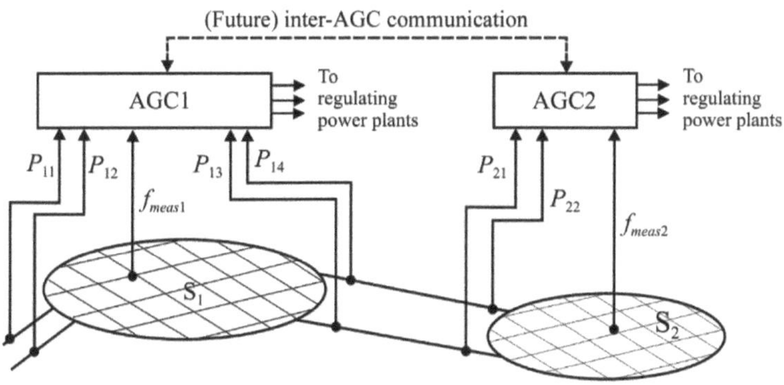

FIG. 3.10 Representation of the secondary frequency control.

Under balanced conditions, $P_{exch}^{ref} = P_{sch}$ and $f_{meas} = f_{ref}$, and thus the ACE is equal to zero.

The active power required for system balancing, ΔP_B, is determined using a proportional-integral (PI) regulator, and sometimes an integral (I) regulator, in accordance with the following relationship:

$$\Delta P_B = -\beta \cdot ACE - \frac{1}{T_r}\int ACE\,\mathrm{d}t \tag{3.8}$$

where β is the proportional factor (gain) of the secondary controller in the control area, and T_r is the integration time constant. The integral term ensures both the system frequency and power deviations to return to their set points within the required time (without additional control needed). The proportional factor must be carefully chosen to ensure the stability of the interconnected system, especially when hydroelectric plants are employed within the secondary control system. Redesigning the secondary controller function will be a challenging factor when the generation sector relies only on renewable energy sources.

The output of the PI regulation is limited by the scheduled secondary reserve band for upward (SRB_{up}) and downward (SRB_{down}) regulation (Fig. 3.11). The two bands are obtained by summing up the reserves available from all the control generation units. Fig. 3.12 illustrates a snapshot of the AGC operation in the Romanian power system, where the left side shows the power exchange, the right side shows the control plants (blue (gray in print version) is for hydro, yellow (light gray in print version) is for coal, and red (dark gray in print version) is for natural gas), and the central positions shows the band used (100% in this case).

An example of frequency restoration is illustrated in Fig. 3.13 [10]. The frequency is stabilized by the intervention of the inertial response and the primary response, then is recovered to the reference value by contribution of the secondary control.

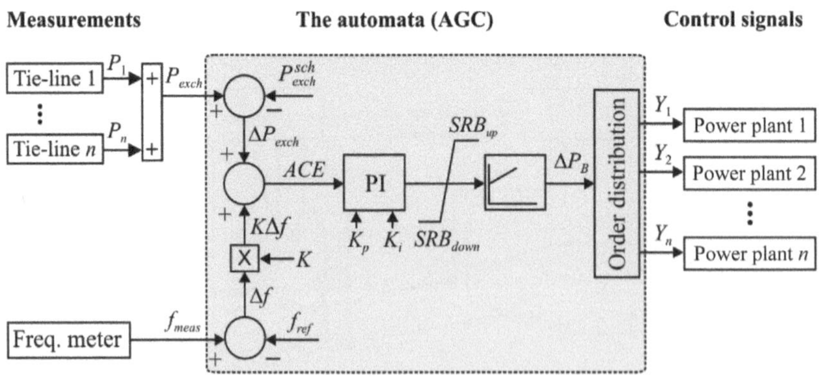

FIG. 3.11 Representation of the AGC scheme.

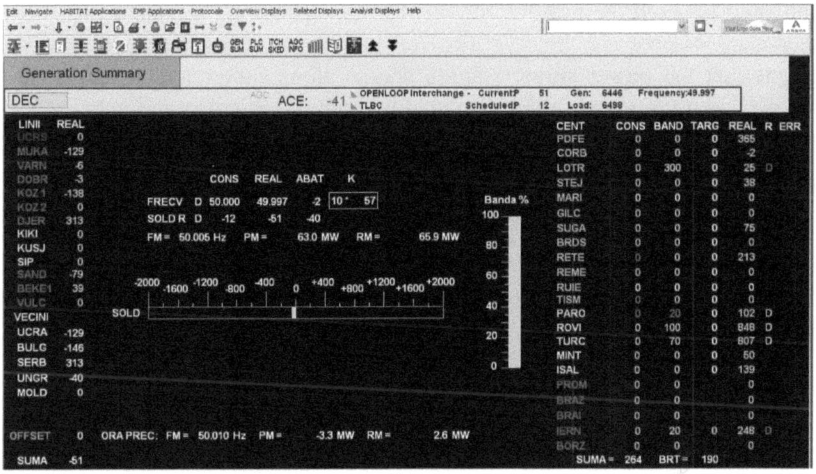

FIG. 3.12 Snapshot of AGC real-time parameters in the Romanian power system. *(Source: L. Toma, M. Sanduleac, S.A. Baltac, Review of relevance of current techniques to advanced frequency control, Deliverable 2.2, H2020 RESERVE Project No. 727481, September 2017.)*

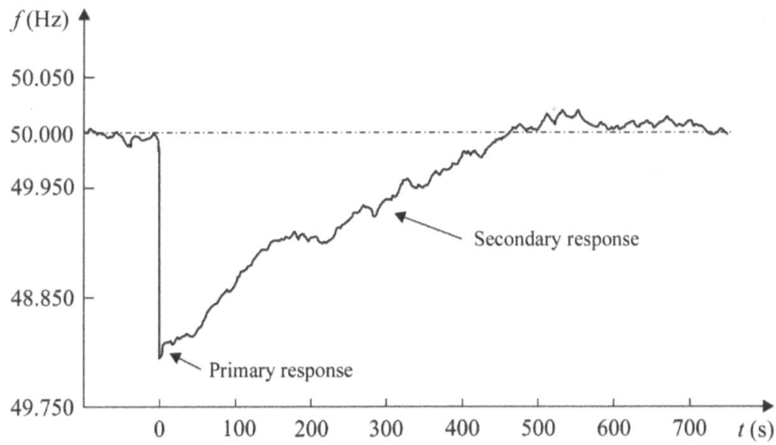

FIG. 3.13 Example of automatic frequency correction.

3.2.7 Tertiary frequency control

The tertiary control reserve, called also manual Frequency Restoration Reserve (mFRR) in ENTSO-E, is activated manually when requested by the system operator, and is used for the relief of the secondary reserve, following the loss of large generation or load power, or to correct forecast errors. As the power unbalances can be negative or positive, the tertiary reserve is determined and activated for both the upward and downward regulation. Extended use of the secondary reserve can involve also full use of the primary reserves, which means that the power system cannot handle alone subsequent power unbalances. Tertiary reserve can be provided by both the spinning and nonspinning generators.

3.2.8 Frequency stability

Frequency stability refers to the ability of a power system to maintain steady frequency following a severe system upset resulting in a significant unbalance between generation and load [13]. The unbalance between the total generated active power and the total load may result in frequency deviation with respect to the reference value. In the case of slow unbalances, the frequency is the same in every point of the network, while sudden unbalances result in unequal frequency values during the transient period. The severity of a perturbation depends first on the amount of mechanical inertia available in the synchronous generators with respect to the total instantaneous load, and secondly on the capability of power electronic based equipment to contribute to frequency stabilization. Small mechanical inertia may result in an excessively large frequency deviation after a sudden power unbalance, which in the long run may result in frequency instability.

The national grid codes require that appropriate power reserve be maintained available in the power system for both upward and downward regulation to face the inadvertent disconnection of either the largest power plant or the largest load center. However, danger may come from faults that can affect the network integrity. Disconnection of one transmission line or one transformer may cause overloading of other network branches and finally weakening the transfer capacity on some network corridors which in the long run create severe local unbalance between generation and load. The Italian blackout in 2003 [14] and the UCTE network splitting in 2006 [15] originated from disconnection of one transmission line, and the first phenomena were related to frequency problems. The reason is that in large interconnected power systems, large power unbalance may cause weakening of the coupling between generators and groups of generators may run at different frequencies leading to network splitting into islands. In such cases the question is whether the frequency in the islands can reach a steady value. It depends on the coordination of control and protection equipment. To prevent the collapse of the islands, underfrequency load shedding relays are activated to maintain the balance between generation and load [1].

3.2.9 The European network codes

Frequency control is a multilevel multiple time frame activity which requires the equitable participation of all partner countries. The energy transition, which refers to the replacement of all the fossil fuel-fired power plants with renewable energy sources and other clean energy units, involves a significant change in the global characteristics of the energy sources. To support this transition, common rules have been adopted by ENTSO-E and included in the European Union's legislation in April 2016 [5]. There are three families of network codes:

- Requirements for connection to the electrical network,
- Requirements for system operation, and
- Market rules.

Several codes have been defined in these families, of which some have already been issued and already adopted in national legislation, while others are under elaboration. The first step in the harmonization of the network codes was the adoption of the requirements for connection to the electrical network of the loads, generators, and HVDC systems.

Of particular interest for frequency control is the "Requirements for Generators—RfG." Following the increased penetration of renewable energy sources and the distributed generation, a classification of the generation units was done into four types of generation units, called power-generating modules (PGM). A PGM can be either a synchronous power generation module (SPGM) or a power park module (PPM), the latter being a generation unit connected to the electrical network via power electronic equipment. They are classified on the basis of the voltage level at the point of common coupling ($U_{n,PCC}$) and the maximum generation capacity ($P_{g,max}$) of the, that is,

(a) Type A: $U_{n,PCC} < 110\,\text{kV}$ and $P_{g,max} \geq 0.8\,\text{kW}$;
(b) Type B: $U_{n,PCC} < 110\,\text{kV}$ and $P_{g,max}$ at or above a threshold (Table 3.3);
(c) Type C: $U_{n,PCC} < 110\,\text{kV}$ and $P_{g,max}$ at or above a threshold (Table 3.3);
(d) Type D: $U_{n,PCC} \geq 110\,\text{kV}$ and $P_{g,max}$ at or above a threshold (Table 3.3).

Replacement of synchronous generators with power electronic-based generation sources requires new rules as regards the active power response to

TABLE 3.3 Limits for thresholds for type B, C and D power-generating modules.

Synchronous area	Type B (MW)	Type C (MW)	Type D (MW)
Continental Europe	1	50	75
Great Britain	1	50	75
Nordic	1.5	10	30
Ireland and Northern Ireland	0.1	5	10
Baltic	0.5	10	15

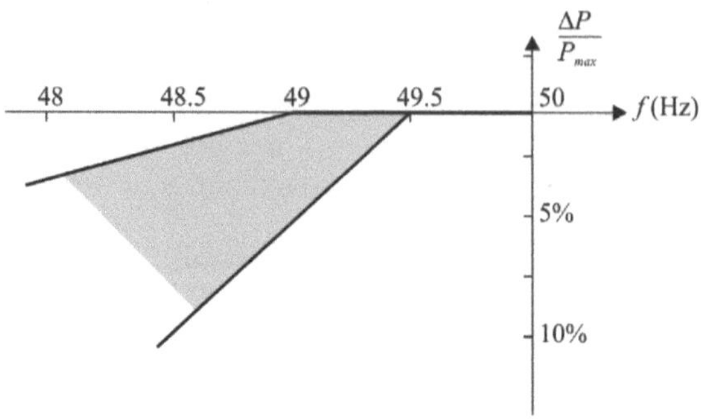

FIG. 3.14 Maximum power capability reduction with falling frequency, according to [5].

frequency deviation (Fig. 3.14). Customized values can be adopted by each country in terms of their particularities. These rules are necessary for PPMs to set the operation of the power converters when creating the voltage waveforms.

The RfG adopted at the European level is a necessary first step to deal with the changing characteristics of the power system as a whole. However, some rules may be seen as too restrictive. For instance, the need to control nano-generation with an installed capacity of 0.8 kW by the control signal from the system operator would be a barrier in creating nano-grids. Additionally, the inexistence of network codes specially designed for energy storage systems and virtual power plants makes incomplete the current family of ENTSO-E's network codes.

3.3 Voltage control and stability in power systems

3.3.1 General aspects

The concept of voltage stability refers to the ability of the electrical system to maintain the nodal voltages within the allowable ranges both in normal conditions and in contingencies ones [16].

Unlike the frequency, which is a global parameter maintained at the desired value through the balance between generated and consumed active powers, the voltage is a local variable that differs from a node to another due to the voltage drops that occur along the grid branches.

The main purpose of the electrical networks is to transfer the active power from the production sources to the consumers, whereas reactive power requirement can be covered through local sources. However, the transfer of the active power is possible only by maintaining the voltage level of the network nodes between certain limits.

The voltage instability is strictly related to the inability of the system to meet the reactive power demand [8]. This means that the amount of reactive power injected in the node differs from the *desired* nodal reactive power. Due to the difficulty to transport reactive power in systems with $X \gg R$ (such as the transmission system), the local voltage problems can evolve towards an *instability problem* that can interest the entire system leading to a *voltage collapse* [17]. The system can meet a voltage instability condition due to a *disturbance cause*, which may lead to a noncontrollable decline of the voltage values, such as:

1. *Load increase*, due to local reactive power deficit
2. *Incidents*, such as the disconnection of either some generation sources providing reactive power support, or *transmission lines* increasing the loading of the network
3. *Nonproper operation of On-Load Tap Changers* (OLTCs), which are part of the HV/MV transformers and act as devices to regulate the voltage at distribution level

3.3.2 Issues in the transmission of the reactive power

To understand the difficulties of transmitting reactive power along a transmission line, the simple power system shown in Fig. 3.15 is taken as an example. The model aims to represent a lossless long transmission line. As hypothesis, the voltages at the two sides are controlled.

The active and reactive power at the sending and receiving ends of the transmission line can be obtained as follows:

$$\underline{S}_r = \underline{V}_r \underline{I}^* = P_r + jQ_r \Rightarrow \left\{ \begin{array}{l} P_r = \dfrac{E_s V_r}{X} \sin\delta = P_{max} \sin\delta \\ Q_r = BV_r^2 + \dfrac{E_s V_r \cos\delta - V_r^2}{X} \end{array} \right\} \tag{3.9}$$

$$\underline{S}_s = \underline{E}_s \underline{I}^* - jBE_s^2 = P_s + jQ_s \Rightarrow \left\{ \begin{array}{l} P_s = \dfrac{E_s V_r}{X} \sin\delta = P_{max} \sin\delta \\ Q_s = -BE_s^2 + \dfrac{E_s^2 - E_s V_r \cos\delta}{X} \end{array} \right\} \tag{3.10}$$

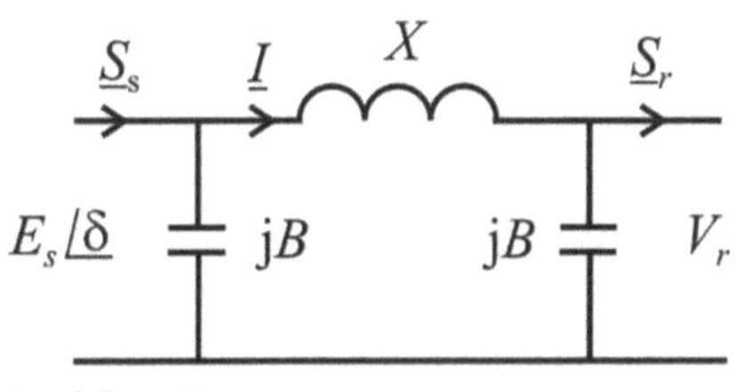

FIG. 3.15 Simple network branch.

From Eqs. (3.9), (3.10) it is possible to see that the values of active power at the sending and receiving ends are the same, because the considered line is without losses, that is,

$$P_s = P_r = \mathrm{P} \tag{3.11}$$

By considering that the voltages are equal in magnitude (i.e., $E_s = V_r = V$), the reactive powers at the sending and receiving ends are:

$$Q_s = V^2\left(\frac{1-\cos\delta}{X} - B\right) \tag{3.12}$$

$$Q_r = -V^2\left(\frac{1-\cos\delta}{X} - B\right) \tag{3.13}$$

From Eqs. (3.12), (3.13), it is evident that $Q_s + Q_r = 0$: if P is low then δ is small and $Q_s - Q_r < 0$, that means that the line *produces* reactive power. On the other hand, if P is high, δ is larger and $Q_s - Q_r > 0$ (in particular $Q_s > 0$ and $Q_r < 0$) and the line *absorbs* reactive power. So, the presence of the same voltage magnitude does not allow the transmission of reactive power from one side to the other. It is worth to note that the maximum active power transmitted is reached with $\delta = \pi/2$ and, with equal voltage magnitude on both sides, is equal to:

$$P_{max} = \frac{V^2}{X} = S_c \tag{3.14}$$

Let us now make another example, by considering $\delta = 0$, but different values of the voltage magnitude at the two sides: in this case the transmitted value of active power is zero. By starting from the formulation of the reactive power related to the receiving end shown in Eq. (3.9), it can be rewritten as:

$$Q_r = BV_r^2 + \frac{E_s V_r \cos\delta - V_r^2}{X} = \left(\frac{BX - 1}{X}\right)V_r^2 + \frac{E_s V_r}{X} \tag{3.15}$$

The formulation (3.15) represents a parabola, with a maximum as shown in Fig. 3.16.

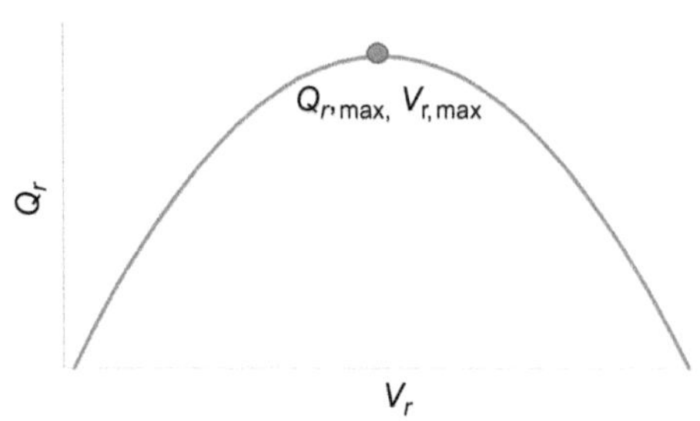

FIG. 3.16 Qualitative shape of the formulation (3.15).

The two coordinates referring to the maximum and point of maximum of the parabola are:

$$V_{r,\max} = \frac{E_s}{2(1 - BX)} \tag{3.16}$$

$$Q_{r,\max} = \frac{E_s^2}{4X} = \frac{S_c}{4} \tag{3.17}$$

Comparing Eqs. (3.14), (3.17) it is obvious that even without the active power transmission, the maximum reactive power transmittable is 25% of the maximum active power transmittable. Furthermore, despite this more restrictive limit about the reactive power transmission, the voltage at the receiving end is lower because, considering that $BX \ll 1$, we have $V_{r,\max} \approx \frac{E_s}{2}$.

3.3.3 Classification of the voltage stability problems

As clarified in the introduction, voltage problems are strictly related to the lack of reactive power. However, the voltage issues, even having in common the same cause, can be classified according to the magnitude of the cause, that can be *small disturbances* or *large disturbances*:

- *Small-disturbance* causes refer to the incremental changes in the system load [8]. The characteristics of load and control systems in a time instant are required to evaluate the stability at any time instant to small disturbances. Due to the steady-state origin of the basic processes affecting the small-disturbance stability, static analysis is sufficient to determine the stability margins. At a given operation point, if an increase of the injection of reactive power at a given node leads to an increase of voltage in the same node (and this is true for all the nodes of the system), the system is stable with respect to small disturbances, that is, the *V-Q* sensitivity is positive.
- *Large-disturbance* causes are system faults, loss of generation or circuit contingencies [8]. The ability of the system to handle with this kind of problems strictly depends on the interaction among controls and protection, as well as by the system-load characteristics. The study of the stability of the system in cases of large-disturbance events needs the analysis of the dynamic behavior of the system over a relatively long time horizon (from some seconds to tens of minutes) and thus long-term dynamic simulations are required. Once the disturbance happens, the control systems have to act in such a way that, in steady state, the voltage values of every node result acceptable.

According to [18] and to the real cases of voltage instability happened in the past (as reported in Appendix F of [16]), it is possible to individuate two time domains for voltage instability:

- *Transient stability*, which involves *large disturbances* and *loads* with rapid response such as induction motors or HVDC converters.
- *Longer term*, usually involves factors as load increase, load recovery after faults, reactive losses in power lines due to high power transfer, loss of reactive supply [17].

In the following sections, an analysis of the mechanism of the voltage collapse will be shown.

3.3.4 PV and VQ curves

Let us consider the system shown in Fig. 3.17.

The parameters of the system are listed below:

- System base: $V_b = 132\,\text{kV}$, $S_b = 100\,\text{MVA}$
- Line: $r_l = 0.1\,\Omega/\text{km}$, $x_l = 0.4\,\Omega/\text{km}$, $l = 40\,\text{km}$, number of lines in parallel: 1
- Transformer: $S_t = 63\,\text{MVA}$, $v_{cc,\ \%} = 14.66\%$,$P_{Cu} = 175\,\text{kW}$, tap changer $\pm 12 \cdot 1.5\%$, number of transformer in parallel: 2, $V_1/V_{20} = 132/22\,\text{kV}$
- Load: $S_r = 90 + j55\,\text{MVA}$

Solving the circuit, the series parameters (including both the line and the transformer parameters) expressed in p.u. of the system are $R = 0.0252\,\text{p. u.}$ $X = 0.2082\,\text{p. u.}$

The following quadratic equation, linking together voltage, active and reactive power can be obtained [16]:

$$V_r^4 + \left[2S_r(R\cos\delta + X\sin\delta) - V_r^2\right]V_r^2 + Z^2S_r^2 = 0 \tag{3.18}$$

This is a quadratic equation of order two with respect to the variable V_r^2. This equation has two different solutions only if its discriminant is higher or equal to zero, that means:

$$\left[2S(R\cos\delta + X\sin\delta) - E_s^2\right]^2 - 4Z^2S_r^2 \geq 0 \tag{3.19}$$

Considering Eq. (3.11) as true, and by using the notation shown in [2], the two voltage solutions can be written as follows:

$$V_{r,A} = \sqrt{\frac{\alpha + \sqrt{\Delta}}{2}};\ V_{rB} = \sqrt{\frac{\alpha - \sqrt{\Delta}}{2}} \tag{3.20}$$

where $\alpha = E_s^2 - 2(RP_r + XQ_r)$ and $\Delta = [2(RP_r + XQ_r) - E_s^2]^2 - 4Z^2(P_r^2 + Q_r^2)$.

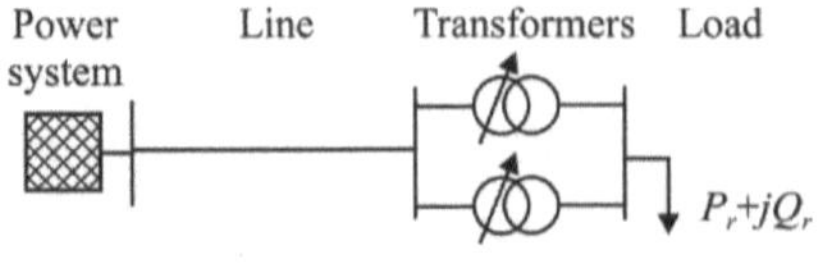

FIG. 3.17 Schematic of the case study.

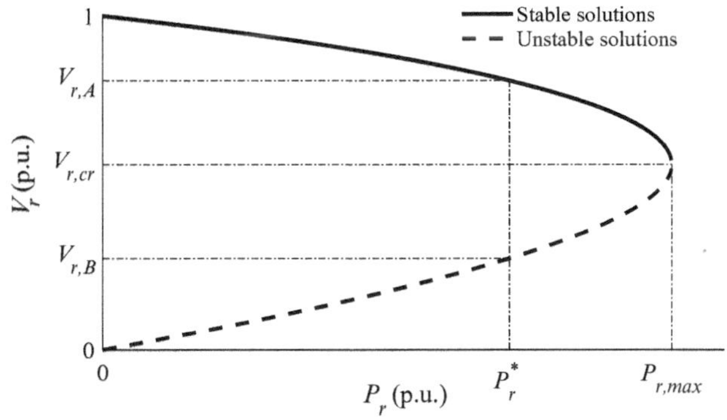

FIG. 3.18 PV characteristic ("nose curve").

Two voltage values are possible for P_r, that is, $V_{r,A}=0.8060$p. u. and $V_{r,B}=0.2742$p. u. Due to the extremely low value of voltage, it is evident that the set of solution at which $V_{r,B}$ belongs is the locus of the *unstable solution*, whereas the upper part of the curve represents the set of *stable solutions* (Fig. 3.18).

If $\Delta=0$ is considered in the expressions (3.20), identical solutions $V_{r,A}$ and $V_{r,B}$ are obtained, representing the *critical point*:

$$V_{r,A}=V_{r,B}=V_{r,cr}=\sqrt{\frac{\alpha}{2}}=\sqrt{\frac{E_s^2-2\left(RP_{r,\max}+XQ_{r,\max}\right)}{2}} \tag{3.21}$$

In this example, the value of critical voltage is $V_{r,cr}=0.5670$p. u.

The *maximum active power* $P_{r,max}$ and the *maximum reactive power* $Q_{r,max}$ corresponds to the critical voltage, that is,

$$P_{r,\max}=\frac{E_s^2\cos\varphi}{2(R\cos\varphi+X\sin\varphi+Z)} \tag{3.22}$$

$$Q_{r,\max}=\frac{E_s^2\sin\varphi}{2(R\cos\varphi+X\sin\varphi+Z)} \tag{3.23}$$

In this example, the value of maximum active power is $P_{r,max}=1.3682$p. u., whereas the maximum reactive power is $Q_{r,\,max}=0.6922$p. u.

3.3.4.1 Voltage sensitivities to active and reactive powers variation

By applying differential operator at Eq. (3.18), the sensitivities of the system to the variation of active and reactive power variations are obtained.

The calculation is simplified by neglecting the phase difference between the sending and receiving node voltages (i.e., $\delta\cong 0$). The relationship between the complex values of the voltages at the two ends of line is:

$$\underline{E}_s=V_r+\frac{RP_r+XQ_r}{Vr}+j\frac{XP_r-RQ_r}{V_r} \tag{3.24}$$

Because the hypothesis $\delta \cong 0$, the imaginary part is zero, then:

$$V_r^2 - E_s V_r + RP_r + XQ_r = 0 \tag{3.25}$$

Thus, $V_{r,A}$ and $V_{r,B}$ can be obtained as follows:

$$V_{r,A} \cong \frac{E_s + \sqrt{E_s^2 - 4(RP_r + XQ_r)}}{2} \tag{3.26}$$

$$V_{r,B} \cong \frac{E_s - \sqrt{E_s^2 - 4(RP_r + XQ_r)}}{2} \tag{3.27}$$

The sensitivities of voltage to the active and reactive powers in the A and B points can be calculated starting from the expressions (3.26), (3.27) [2, 19]:

$$\left.\frac{\partial V_{r,A}}{\partial P_r}\right|_{Q_r=\text{const.}} = \frac{-R}{\sqrt{E_s^2 - 4(RP_r + XQ_r)}} < 0 \tag{3.28}$$

$$\left.\frac{\partial V_{r,A}}{\partial Q_r}\right|_{P_r=\text{const.}} = \frac{-X}{\sqrt{E_s^2 - 4(RP_r + XQ_r)}} < 0 \tag{3.29}$$

$$\left.\frac{\partial V_{r,B}}{\partial P_r}\right|_{Q_r=\text{const.}} = \frac{R}{\sqrt{E_s^2 - 4(RP_r + XQ_r)}} > 0 \tag{3.30}$$

$$\left.\frac{\partial V_{r,B}}{\partial Q_r}\right|_{P_r=\text{const.}} = \frac{X}{\sqrt{E_s^2 - 4(RP_r + XQ_r)}} > 0 \tag{3.31}$$

From the sensitivities above, the point A is stable because is controllable (an increase of load leads to a decrease of the voltage), whereas the point B results in uncontrollable operating point, thus unstable.

The values obtained for this case study are

$$\left.\frac{\partial V_{r,A}}{\partial P_r}\right|_{Q_r=\text{const.}} = -\left.\frac{\partial V_{r,B}}{\partial P_r}\right|_{Q_r=\text{const.}} = -0.0374$$
$$\left.\frac{\partial V_{r,A}}{\partial Q_r}\right|_{P_r=\text{const.}} = -\left.\frac{\partial V_{r,B}}{\partial Q_r}\right|_{P_r=\text{const.}} = -0.3098$$

3.3.4.2 *Effect of the power factor seen from the line*

The influence of the reactive power compensation is shown in Fig. 3.19.

The value of maximum active power increases with the passing from inductive to capacitive power factor, and the same happens for the critical voltages, as summarized in Table 3.4.

Another useful representation of the effect of reactive power compensation is shown in Ref. [2]. In this case, the plot aims to put in relation the value of the

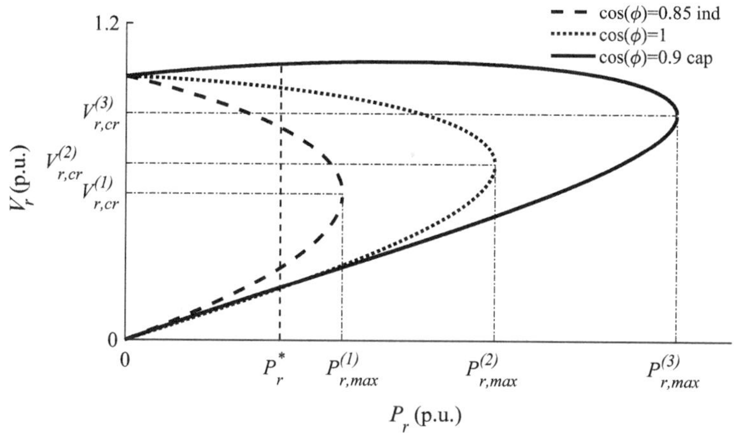

FIG. 3.19 Effect of the variation of the power factor.

TABLE 3.4 Variation of the notable points for the nose curve by applying a reactive power compensation.

Cases	$\cos\varphi$	$P_{r,\max}$	$V_{r,cr}$	$V_{r,A}$	$V_{r,B}$
Case 1	0.85 ind	1.3682	0.5670	0.8060	0.2742
Case 2	1	2.1292	0.6682	0.9570	0.1972
Case 3	0.9 cap	3.1783	0.8605	1.0469	0.2003

reactive compensation and the voltage at the receiving end. The curves have two different derivatives: the left-part of the curves represents unstable operation points, because the increase of the reactive compensation Q_c would lead to a worsening of the voltage values, that is, $\frac{dQ_c}{dV_r} < 0$. The right-side of the curves have $\frac{dQ_c}{dV_r} > 0$, which means that an increase of the reactive power compensation leads to an improvement of the voltage values (Fig. 3.20).

It is worth to note that in case of relatively low value of active power P_r, the system can be operated also without reactive power compensation (i.e., $Q_c = 0$). The higher is the load power, the higher is the value of Q_c required to maintain a controlled value of voltage. As extreme case, it has been chosen to shown the behavior for $P_r = 10$ p. u., to demonstrate that with high values of active power there is no possibility to provide such active power without a reactive power compensation.

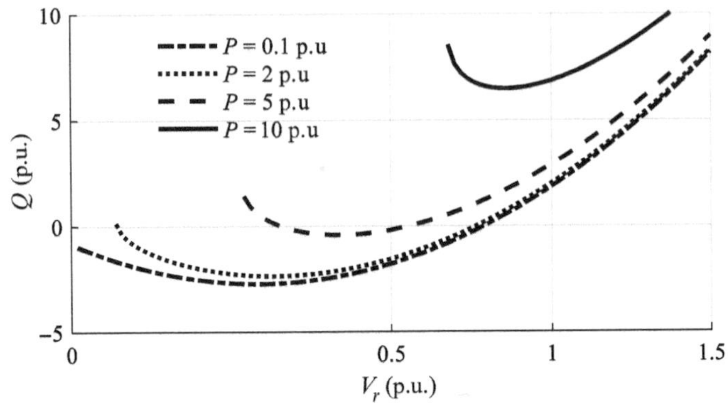

FIG. 3.20 Q characteristics by considering negligible value of R.

3.3.5 Voltage regulation

Voltage regulation is an activity of the network operators and is carried out through a series of support actions that combine the hierarchical use of some control devices.

The voltage adjustment can be achieved:

(1) By acting on the *parameters* of the electrical network, through *branch-based control strategies*, with the aim to modify the parameters of the branches of the electrical network. This is reached through OLTC-equipped transformers, the connection/disconnection of line circuits or transformers, as well as by employing static devices connected in series, which compensate the parameters of the electrical lines.

(2) By varying the *reactive power generation/absorption*, through *bus-based control strategies*, aiming at maintaining the system voltage by injecting reactive power. This goal is reached by employing generators, synchronous compensators, static compensators and compensation coils. Synchronous generators, synchronous compensators and static compensators may dynamically contribute to voltage regulation because they can produce/absorb reactive power continuously as compared to static reactor and capacitor banks.

The strategy of using reactive power devices depends on the load level of the network. On one hand, when the power grid is *heavily loaded*, the reactive power devices needs to *provide* reactive power because the network is predominantly inductive. On the other hand, when the network is *lightly loaded*, the reactive power devices have to absorb reactive power because the network is predominantly capacitive.

Voltage control is performed both by automatic systems and manual actions. In addition to the primary setting (which is automatic), the voltage adjustment is mostly done by manual actions, with the exception of France, Italy and Belgium, where the secondary voltage control was implemented several years ago [18].

The "manual" voltage control procedures usually consist in dispatching the reactive (forecast) powers available in the generating units, setting the reference voltages in the power plants, connecting/disconnecting capacitor banks or compensating reactors, setting the supply voltages at the transformers through OLTCs or FACTS devices. These conventional procedures used to solve the problems related to the voltage regulation in the electrical network do not include the real-time decisions of the system operator based on the monitoring of the system and the management of alarms by orders sent by remote controls.

In the following subsections, a brief description of the main characteristics of the devices for reactive power compensation is shown, with some remarks regarding the technical implementation.

3.3.5.1 Synchronous generators: Capability curve

The main electrical quantities that characterize a synchronous generator are: (i) nominal power (usually expressed as apparent power, in MVA), (ii) nominal voltage, and (iii) nominal power factor. When the generator is working in conditions that differ from the nominal ones, the feasible operation region is delimited by the so-called *capability curve*. This chart defines the limits for the $\{P,Q\}$ values that the machine can provide to the network [18], and an example in shown in Fig. 3.21.

Due to the complexity of the processes that take place in the synchronous machine, in order to build the generator's capability curve, the following assumptions are made:

- the stator resistance is neglected;
- the magnetization characteristic is linear;

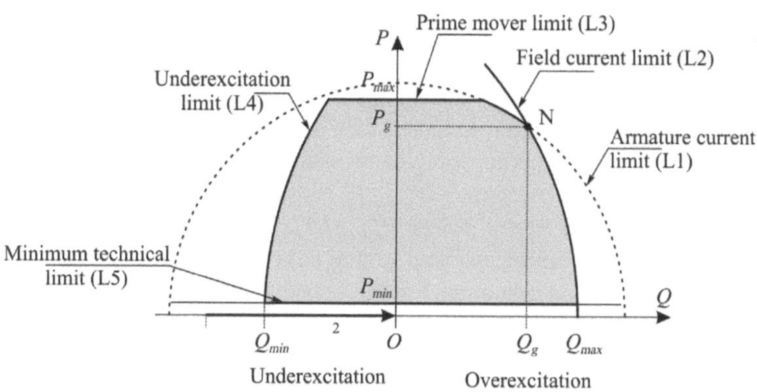

FIG. 3.21 Example of capability chart for a synchronous generator.

- the power losses due to the Joule effect in the stator winding are neglected, as well as the power losses in the stator iron;
- the synchronous reactance is kept constant.

Based on these assumptions, the following operating limits of the synchronous generator are defined:

(1) *Stator current limit* (*L1*), imposed by the stator winding heating limit. This limit is a circle with the origin at point O and radius equal to the apparent nominal power S_n, which represents the geometric locus of the points given by the relation $S_n = P_n^2 + Q_n^2$. Given that the generated apparent power S must not exceed the nominal value S_n, the operating points must be inside the limit L1. For a given value of the stator current greater than the limit value, the generator can safely operate for a short time, depending on how much the limit value is exceeded.

(2) *Limit of rotor current* (*L2*). The maximum value of the excitation current is imposed by the heating limit of the rotor windings. This imposes also a limit value of the generator internal voltage, which is equal to the nominal value. Also, for reasons of safe operation at mechanical torque shocks, a minimum value of the excitation current is required. The limit curve of the rotor current is a circle with the center in O′ and radius equal to the nominal voltage E_n. The position of point O′ is given by the generator reactance value, which means that the capability chart differs depending on the type of generator (i.e., salient pole machines or turbogenerator). As can be seen in Fig. 3.21, for an active power lower than the nominal active power P_n, the rotor current limit is more restrictive than the stator current limit. The nominal operating point of the generator is at the intersection of the two limits L1 and L2, where the generator is used to the maximum in terms of the generated apparent power.

(3) *Maximum limit of the mechanical power* $\boldsymbol{P_{max}}$ (*L3*), imposed by the maximum torque of the turbine. Since in general the mechanical power of the turbine is lower than the maximum electrical power of the generator, this limit is a horizontal line drawn at a value of the active power lower than the nominal power P_n of the generator.

(4) *Limit of the subexcitation limiter* (*L4*), imposed by the automatic control systems of the generator, to have a safe operation in inductive mode. This limit is related to problems of dynamic stability of the generator that limits the angle $\boldsymbol{\delta}$ of the machine.

(5) *Minimum limit of active power* $\boldsymbol{P_{min}}$ (*L5*). If the generator has a power lower than P_{min} the mechanical torque developed by the turbine will be insufficient to keep it running. For this reason, a minimum value of the generated active power is required.

The generator capability chart is voltage dependent: thus, the over and under excitation limits change dynamically. More precisely, if the generator voltage increases, the overexcitation limit reduces the deliverable reactive powers, while the underexcitation limit increases the absorbable reactive powers.

When the generator increases the produced reactive power, a reduction of the produced active power is required, according to the capability chart limits.

3.3.5.2 Shunt reactors

In general, the shunt reactors are used to compensate the capacitive effect of the power lines, limiting the increase of the voltage in case of low- or no-load operation. They are used on high voltage lines with lengths greater than 150–200 km.

For information purposes regarding the importance played by the value of the capacitive reactive power in these conditions, let us consider a line operated at 380 kV, with length 100 km and capacitance per length unit equal to 12 nF/km (typical for a two conductor configuration line) [20]. At no load, the line generates a capacitive reactive power almost equal to 55 Mvar, that is,

$$Q_c = \omega C V^2 = \omega \cdot 12 \cdot 10^{-9} \cdot 100 \cdot \left(380 \cdot 10^3\right)^2 = 55\,\text{Mvar}$$

Denoting by Q_L the reactive power absorbed by the reactors, the *level of compensation* of the line is defined as:

$$K_t = \frac{Q_L}{Q_c} \cdot 100 \tag{3.32}$$

The usual value of K_t for long transmission system lies in the range 50%–70%.

The reactors cannot be distributed as the capacitance of the lines and can be connected directly to the power line or to the tertiary winding of a three winding transformers [2] (see Fig. 3.22).

3.3.5.3 Synchronous compensator

Synchronous compensators, called also synchronous condensers, are synchronous machines designed to operate at no mechanical load (without shaft-resistant torque, that is, the rotor is free to rotate without any constraints due

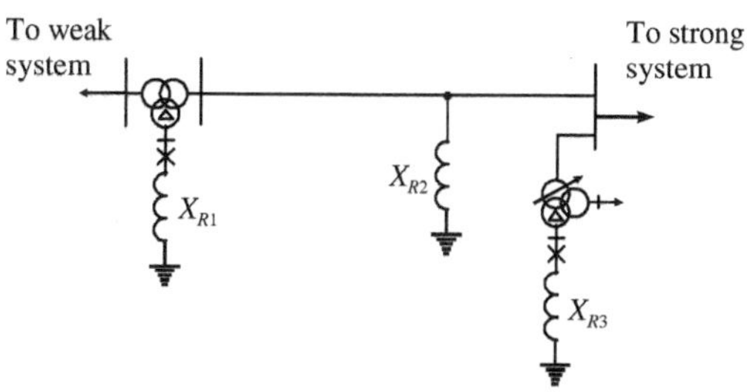

FIG. 3.22 Example of connection of shunt reactance direct to the line and through three-winding transformer in the electrical substation [8].

to load or turbines) and can produce or consume reactive power depending on the value of the excitation current. The synchronous compensator is an equipment specially designed to continuously control the voltage level in a network area, both during the normal operation (by contributing to improve the static stability and thus to increase the active power transport capacity), as well as in transient operation (by contributing to maintain voltage stability in the event of major disturbances in the electrical network). In evaluating the operating costs of the synchronous compensator, the losses of active power must be taken into account, which include Joule losses, mechanical losses, and, in case of rotating excitation system, also the excitation system losses.

The characteristic of the synchronous compensator is shown in Fig. 3.23: it is evident that the synchronous compensator is able to operate both as capacitor and reactor, depending on the value of the excitation current I_e.

Usually, the synchronous compensators are salient pole machines, designed to normally operate in underexcitation conditions, allowing the absorption of 75% of the nominal power.

The synchronous compensators are self-regulating. In fact, by indicating with E the phase voltage at the machine terminals, with E_0 the internal voltage of the machine (induced by the rotor) and with X_d the reactance of the machine, the reactive power provided by the synchronous compensator can be expressed as:

$$Q_c = 3EI = 3E\frac{(E_0 - E)}{X_d} \tag{3.33}$$

By considering the value E_0 constant, it results that the value of Q_c increases (decreases) when the voltage at the machine terminals decreases (increases).

3.3.5.4 Static var compensators

The requirements regarding the reactive power on an electric line vary according to the load from an inductive to a capacitive one. Static var compensators (SVCs) are ideal equipment to compensate the reactive power demanded by large loads and to dynamically limit overvoltages in the event of load

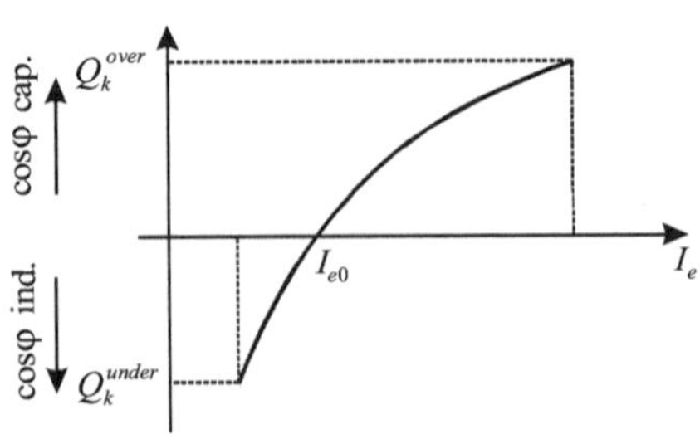

FIG. 3.23 Characteristic of the synchronous compensator [2].

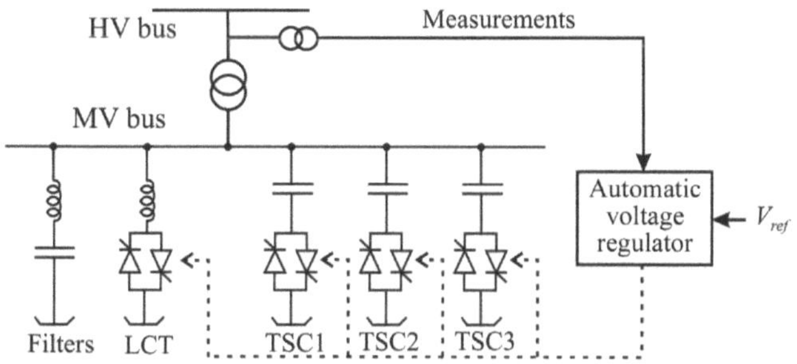

FIG. 3.24 Schematic of SVC system [21].

disconnection. They are also used in HVDC link conversion stations, as well as when reactive power flow control is required.

The main aim of using the SVC is to control the voltage level in a node and, consequently, to increase the transfer capacity of an electrical line [21]. In order to maintain an active power transit on the transmission line when disturbances or contingencies occur in the system, the reactive power at the output (capacitive or inductive) of the SVC is modified to control the voltage in a certain node of the electrical network.

Fig. 3.24 shows the schematic of an SVC device: it consists of one reactor and several shunt capacitor banks, which are controlled very quickly with the help of power electronics. Therefore, the reactive power can be produced or absorbed dynamically, thus controlling the voltage level very quickly.

The maximum value that an SVC device can produce depends on the size of the capacitor banks, while the maximum power that can be absorbed depends on the size of the reactor, assuming that the voltage of the connection node is equal to the nominal voltage. An SVC device can produce any value of reactive power within the capability range, being very simply modeled as a synchronous compensator [2, 22, 23]. However, when the reactive power capability limits are reached, SVC behaves like the fixed shunt compensation equipment.

The STATic COMpensator (STATCOM) is another reactive power static controller that is similar to the synchronous compensator in terms of performance, but with better dynamics, lower CAPEX and OPEX and no inertia. It acts on the voltage and current waveforms in such a way that the desired value of reactive power is provided. It works basically as a controllable voltage source. This implies that capacitor banks and reactors are no longer needed, allowing a more compact design.

3.3.5.5 Shunt and series capacitors

Shunt capacitors compensate locally for the reactive power absorbed by loads and are widespread in the electrical network. The main advantages are low cost, easy installation, and easy maintenance. The main disadvantage of the shunt capacitors is that the reactive power produced by them is proportional to the square of the voltage and therefore, at low voltage values, the reactive power produced by them is reduced.

The series capacitors are connected in series with the line conductors to compensate for their inductance. They reduce the inductive reactance of the electrical line, and thus change the voltage along the line and increase the maximum power that can be transferred through the line by reducing the reactive power losses. Unlike the shunt capacitor, the series capacitors are self-regulating, because the provided reactive power refers to the current flowing in the line: if the voltage at the receiving end decreases (increases), the current value increases (decreases) and so the reactive power provided increases (decreases).

3.3.5.6 OLTC-equipped transformers

A representation of OLTC-equipped transformers is shown in Fig. 3.25. Transformers with OLTC are typical of the HV/MV substation. At lower level, that is, for MV/LV distribution transformers, the tap changer is usually manual, and it is rarely modified.

By changing the tap position by means of the OLTC, the transformers have the capability to adjust the voltage at the secondary winding terminals. Voltage modification can be carried out over time periods longer than the one required by other devices such as the synchronous compensators or SVC devices. The transformers are connected in series with the power lines, and the main voltage is changed in cascade.

3.3.5.7 Summary of the features of the compensation devices

A summary of the characteristics of the devices briefly presented above is shown in Table 3.5.

3.3.6 Hierarchical reactive power regulation

The reactive voltage-power control is indispensable in both normal operation and disturbed conditions. In normal operation, the role of voltage regulation

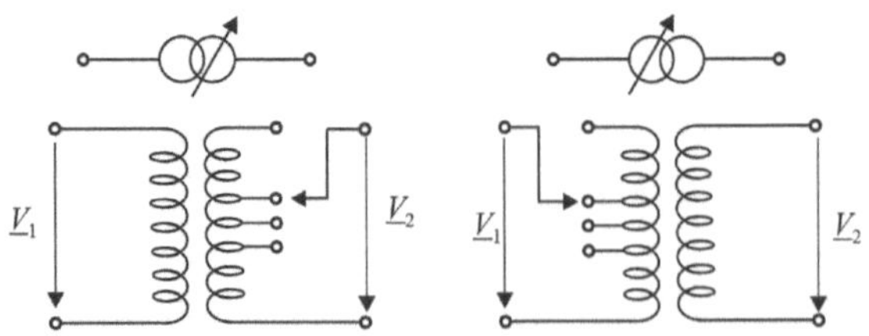

FIG. 3.25 Representation of the OLTC-equipped transformers, operating on (A) secondary and (B) primary winding.

TABLE 3.5 Summary of the characteristics of the devices.

Device	Response time	Voltage regulation effectiveness
Synchronous generators	Fast	Very good, provides extra capacity in short term
Synchronous compensators	Fast	Excellent, provides extra capacity in short term
Shunt capacitors	Slow, steps	Small, effect proportional to $1/V^2$
Series capacitors	Fast	Very good, problems with harmonics
SVC	Fast	Small, decreases with V^2
STATCOM	Fast	Good, decreases with V

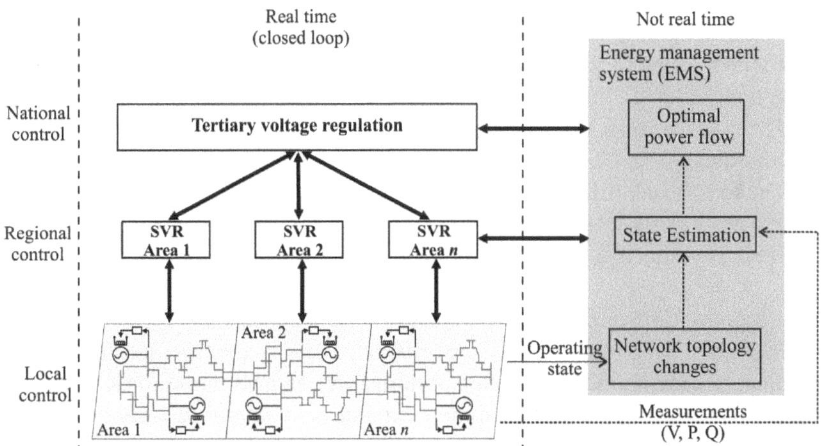

FIG. 3.26 Representation of the hierarchical voltage regulation.

is to supply consumers with electricity at the required quality. In case of a disturbance, the actions of voltage control have the role of preventing the phenomenon of dynamic degradation of the voltage and avoiding extensive damage caused by its fall.

The most complex structure of organization of the reactive voltage-power control system generally comprises the following hierarchical steps: primary, secondary and tertiary [21] (Fig. 3.26).

3.3.6.1 Primary voltage regulation

The primary voltage regulation (Fig. 3.26) consists in operating the local voltage regulation equipment (synchronous generators, synchronous compensators, static compensators) in the event of a disturbance. The control actions are based on local measures and aim to bring the voltage in the connection node to the set value. The response time is between several milliseconds and one minute.

The primary control involves the independent intervention of the control equipment, acting directly on the voltage regulators, without involving reactive power measurements/calculations. The excitation voltage is modified so that the voltage of the controlled node is maintained as close as possible to the reference value.

3.3.6.2 Secondary and tertiary voltage regulation

The principle of secondary voltage control consists in the control of voltages inside an area of the electrical network called "control area". By employing control devices located in the respective area of the network in a coordinated way, the voltages are maintained within admissible limits. Voltage control is done online, in a closed loop, and assumes that interactions with neighboring areas are minimized. In a control area there is a large number of load nodes in which the evolution of voltage is representative for the evolution of voltage in the other load nodes of the area; thus, the voltage measurements will be made only in these nodes called "pilot nodes."

The response time of this type of adjustment is between one minute and several minutes.

The tertiary control represents the global coordination action in the voltage values. It consists in determining the optimal voltage settings at the control devices. The aim is to maintain the reactive power flow between the various areas of the power system and to increase the stability limits of the system. The response time of this type of adjustment is of the order of tens of minutes.

In some power systems, only the tertiary control is adopted, seen as a static problem of reactive power-voltage optimization, treated in the open loop.

In this situation, the most common purpose is to identify the steady operating state with minimum power losses. Minimization of power losses modifies reactive power at generators to maintain voltage at appropriate levels across all system nodes.

References

[1] UCTE, UCTE Policy 3: Operational Security, UCTE, March 2009.

[2] M. Eremia, M. Shahidehpour (Eds.), Handbook of Electrical Power System Dynamics: Modeling, Stability, and Control, Power Engineering Series, Wiley & IEEE Press, Hoboken, NJ, March 2013.

[3] D. Kirschen, G. Strabac, Fundamentals of Power System Economics, John Wiley & Sons, Ltd, England, 2005.

[4] Y. Harmand, C. Nebas-Hamoudia, B. Larripa, B. Neupont, Le mecanisme d'ajustement. Comment assurer l'équilibre production-consommation de l'électricité dans un marche ouvert a la concurrence?, REE, 2005. no. 6/7, June/July.
[5] European Commission, Commission Regulation (EU) 2016/631 of 14 April 2016 Establishing a Network Code on Requirements for Grid Connection of Generators, 2016.
[6] ENTSO-E, Technical Background for the Low Frequency Demand Disconnection Requirements, November, 2014.
[7] Y. Rebours, D. Kirschen, A Survey of Definitions and Specifications of Reserve Services, UMIST, 2005. Release 2.
[8] P. Kundur, Power System Stability and Control, McGraw-Hill, New York, 1994.
[9] T.S. Borsche, G. Andersson, A. Ulbig, Impact of low rotational inertia on power system stability and operation, IFAC Proc. Vol. 47 (2014) 7290–7297.
[10] UCTE, Load-Frequency Control and Performance, UCTE, 2004.
[11] WECC Control Work Group, WECC Tutorial on Speed Governors, February 1998. Revised 2002.
[12] ENTSO-E, Impact of Merit Order Activation of Automatic Frequency Restoration Reserves and Harmonised Full Activation Times, ENTSO-E, 2015.
[13] P. Kundur, J. Paserba, V. Ajjarapu, G. Andersson, A. Bose, C. Canizares, N. Hatziargyriou, D. Hill, A. Stankovic, C. Taylor, T. Van Cutsem, V. Vittal, Definition and classification of power system stability, IEEE/CIGRE Joint Task Force on Stability terms and definitions, IEEE Trans. Power Syst. 19 (2) (2004) 1387–1401.
[14] A. Berizzi, The Italian 2003 blackout, in: IEEE-PES General Meeting, Denver, USA, 6–10 June, 2004.
[15] UCTE, System Disturbance on 4 November 2006. Final Report, Union for the Co-Ordination of Transmission of Electricity, currently ENTSO-E, UCTE, November 2007.
[16] C.W. Taylor, N.J. Balu, D. Maratukulan, Power System Stability, McGraw-Hill, 1994.
[17] G. Andersson, Power System Analysis, Lecture 227-0526-00, ITET ETH Zurich, September 2012.
[18] CIGRE, 1993. "Technical Brochure 75—Modelling of Voltage Collapse Including Dynamic Phenomena",Ed. Carson W. Taylor.
[19] M. Eremia, et al., Electric Power Systems. Volume I. Electric Networks, Publishing House of the Romanian Academy, Bucharest, 2006. ISBN 973-27-1324-0.
[20] F. Iliceto, Chapter 5—Regolazione della tensione, in: Impianti elettrici, vol. 1, Patron Editore, 1981 (in Italian).
[21] C. Bulac, C. Diaconu, M. Eremia, B. Otomega, I. Pop, L. Toma, Power transfer capacity enhancement using SVC, in: 2009 IEEE Bucharest PowerTech, Bucureşti, 28 iunie—2 iulie, 2009.
[22] M. Eremia, J. Trecat, A. Germond, Réseaux électriques, in: Aspects actuels, Technical Publisher, Bucharest, 2000 (in French).
[23] I. Pisica, C. Bulac, L. Toma, M. Eremia, Optimal SVC placement in electric power systems using a genetic algorithms based method, in: 2009 IEEE Bucharest PowerTech, Bucharest, 28 June—2 July, 2009.

Chapter 4

Modal analysis

Lucian Toma[a], Ettore Bompard[b], and Andrea Mazza[b]
[a]*Department of Electrical Power Systems, University Politehnica of Bucharest, Bucharest, Romania,* [b]*Department of Energy "Galileo Ferraris", Politecnico di Torino, Torino, Italy*

4.1 Linearization of dynamic equations

The modal analysis is a methodology to study the stability of a *dynamic system* by calculating the eigenvalues and the eigenvectors of the state matrix of the system [1]. The state of the system represents the minimum amount of information required to study the evolution of the system from a defined time step t, without knowing what happened before t. In particular, the state matrix allows to describe the evolution of the state variables either when the system is *linear*, or it has been *linearized*. In the latter case, the state matrix allows evaluating the system evolution when a small input variation is applied to the system operating at its equilibrium point.

Mathematically, a dynamic system can be defined as follows:

$$\dot{\mathbf{x}} = \mathbf{f}(\mathbf{x}, \mathbf{u}, t) \tag{4.1}$$

where $\mathbf{f}$ represents the set of functions describing the system, $\mathbf{x}$ and $\mathbf{u}$ are the vectors of the state variable and inputs, respectively, and t is the time. The vector $\dot{\mathbf{x}}$ represents the time derivatives of the state variables.

In the particular case of the power system, the system is *autonomous*, which means that $\dot{\mathbf{x}}$ does not explicitly depend on the time [2], i.e.:

$$\dot{\mathbf{x}} = \mathbf{f}(\mathbf{x}, \mathbf{u}) \tag{4.2}$$

At the equilibrium point, denoted as $\{\mathbf{x}_0, \mathbf{u}_0\}$, Eq. (4.2) becomes:

$$\mathbf{0} = \mathbf{f}(\mathbf{x}_0, \mathbf{u}_0) \tag{4.3}$$

In the case of small perturbations applied to the system (i.e., applying an input variation $\mathbf{u} = \mathbf{u}_0 + \Delta\mathbf{u}$ or by varying the state variable vector $\mathbf{x} = \mathbf{x}_0 + \Delta\mathbf{x}$), Eq. (4.2) can be written as:

$$\dot{\mathbf{x}} + \Delta\dot{\mathbf{x}} = \mathbf{f}(\mathbf{x} + \Delta\mathbf{x}, \mathbf{u} + \Delta\mathbf{u}) \tag{4.4}$$

Converter-Based Dynamics and Control of Modern Power Systems
https://doi.org/10.1016/B978-0-12-818491-2.00004-3

Since the perturbation is small, it is possible to write the expression with its Taylor's expansion, by neglecting the terms of superior order [1]:

$$\dot{\mathbf{x}} + \Delta\dot{\mathbf{x}} = \mathbf{f}(\mathbf{x}_0, \mathbf{u}_0) + \mathbf{A}\Delta\mathbf{x} + \mathbf{B}\Delta\mathbf{u} \rightarrow \Delta\dot{\mathbf{x}} = \mathbf{A}\Delta\mathbf{x} + \mathbf{B}\Delta\mathbf{u} \tag{4.5}$$

that means:

$$\Delta\dot{\mathbf{x}} = \mathbf{A}\Delta\mathbf{x} + \mathbf{B}\Delta\mathbf{u} \tag{4.6}$$

with the matrices **A** and **B** containing the derivatives of the functions f_i concerning the state variables and the inputs, i.e.:

$$\mathbf{A} = \begin{bmatrix} \frac{\delta f_1}{\delta x_1} & \cdots & \frac{\delta f_1}{\delta x_n} \\ \vdots & \ddots & \vdots \\ \frac{\delta f_n}{\delta x_1} & \cdots & \frac{\delta f_n}{\delta x_n} \end{bmatrix} \tag{4.7}$$

$$\mathbf{B} = \begin{bmatrix} \frac{\delta f_1}{\delta u_1} & \cdots & \frac{\delta f_1}{\delta u_r} \\ \vdots & \ddots & \vdots \\ \frac{\delta f_n}{\delta u_1} & \cdots & \frac{\delta f_n}{\delta u_r} \end{bmatrix} \tag{4.8}$$

It is worth noting that $\dim\{\mathbf{A}\} = n \times n$ and $\dim\{\mathbf{A}\} = n \times r$.

The eigenvalues of the matrix **A** (i.e., the Jacobian matrix of the system) provides information regarding the *stability* of the system, according to the Lyapunov stability criterion [3], i.e.:

- Eigenvalues with *negative* real part indicate that the system is *asymptotically stable*;
- Eigenvalues with *positive* real part indicate that the system is *unstable*;
- Eigenvalues with real part equal to zero do not provide any information.

It is worth noting that the elements of matrix **A** are *partial derivatives* evaluated at the equilibrium point of the system.

4.2 Eigenvalues and eigenvectors

The word "eigen" is German and means "characteristic of" or "peculiar to." The terms "characteristic values" and "characteristic vectors" are preferred in mathematics.

The scalar λ is an *eigenvalue* of **A** if and only if $(\mathbf{A} - \lambda\mathbf{I})$ it is singular, which means that it is not invertible. Therefore:

$$\det(\mathbf{A} - \lambda\mathbf{I}) = 0 \tag{4.9}$$

Expression of (4.9) gives the *characteristic equation*, whereas the eigenvalues are the solutions of this equation, which are determined by specific calculations.

When $\mathbf{A}$ is singular, $\lambda = 0$ is one of the eigenvalues.

Eigenvalues may be real or complex. If $\mathbf{A}$ is real, complex eigenvalues always occur in conjugate pairs. Similar matrices have identical eigenvalues. It can also be shown that a matrix and its transpose have the same eigenvalue [1].

A square matrix $\mathbf{A}$ has an eigenvalue λ_i associated with a nonzero eigenvector $\boldsymbol{\phi}_i$, called *right eigenvector*, if

$$\mathbf{A} \cdot \boldsymbol{\phi}_i = \lambda_i \cdot \boldsymbol{\phi}_i \tag{4.10}$$

for some vector $\boldsymbol{\phi}_i \neq \mathbf{0}$. The zero vector is never an eigenvector, by definition.

To find an eigenvector $\boldsymbol{\phi}_i$ corresponding to an eigenvalue λ_i, one may solve

$$(\mathbf{A} - \lambda_i \mathbf{I})\boldsymbol{\phi}_i = 0 \tag{4.11}$$

where $\phi_i = [\phi_{1i} \ \phi_{2i} \ \cdots \ \phi_{ni}]^T$

Similarly, a vector $\boldsymbol{\psi}_i$ which satisfies the relationship

$$\boldsymbol{\psi}_i \cdot \mathbf{A} = \lambda_i \cdot \boldsymbol{\psi}_i \tag{4.12}$$

is called *left eigenvector*. After normalization, it can be calculated from

$$\boldsymbol{\psi}_i \cdot \boldsymbol{\phi}_i = 1 \tag{4.13}$$

where $\boldsymbol{\psi}_i$ is the left eigenvector corresponding to the eigenvalue λ_i, and has the form $\boldsymbol{\psi}_i = [\psi_{1i} \ \psi_{2i} \ \cdots \ \psi_{ni}]$.

The right eigenvector $\boldsymbol{\phi}_i$ gives the *mode shape* of the ith mode, i.e., the relative activity of the original state variables when the ith mode is excited (perturbed). The left eigenvector $\boldsymbol{\psi}_i$ gives the *mode composition* of the ith mode, i.e., what weighted composition of original variables is needed to construct the mode.

The participation factor $P_{ki} = \phi_{ik}\psi_{ki}$ measures the participation of the kth variable x_k in the ith mode. The participation factors are dimensionless and hence invariant under changes on scale of the variables. In the general form, the matrix of participation factors is given as:

$$\mathbf{P} = \begin{bmatrix} \phi_{11}\psi_{11} & \phi_{12}\psi_{12} & \cdots & \phi_{1n}\psi_{1n} \\ \phi_{21}\psi_{21} & \phi_{22}\psi_{22} & \cdots & \phi_{2n}\psi_{2n} \\ \vdots & \vdots & \ddots & \vdots \\ \phi_{n1}\psi_{n1} & \phi_{n2}\psi_{n2} & \cdots & \phi_{nn}\psi_{nn} \end{bmatrix} \tag{4.14}$$

4.3 Time response of the linear systems

The time response of a set of variables $\mathbf{x}$ due to nonzero initial conditions, $\Delta\mathbf{x}_0 \neq 0$, is given by

$$\Delta\mathbf{x}(t) = [\boldsymbol{\phi}_1 \ \ \boldsymbol{\phi}_2 \ \ \dots \ \ \boldsymbol{\phi}_n] \begin{bmatrix} C_1 e^{\lambda_1 t} \\ C_2 e^{\lambda_2 t} \\ \vdots \\ C_n e^{\lambda_n t} \end{bmatrix} \tag{4.15}$$

where

$$\begin{bmatrix} C_1 \\ C_2 \\ \vdots \\ C_n \end{bmatrix} = \begin{bmatrix} \boldsymbol{\psi}_1 \\ \boldsymbol{\psi}_2 \\ \vdots \\ \boldsymbol{\psi}_n \end{bmatrix} \Delta\mathbf{x}_0 \tag{4.16}$$

In equation form, variable Δx_i of the vector $\Delta\mathbf{x}$ is

$$\Delta x_i(t) = \phi_{i1} C_1 e^{\lambda_1 t} + \dots + \phi_{in} C_n e^{\lambda_n t} \tag{4.17}$$

4.4 The modal analysis applied to small-signal rotor angle stability

4.4.1 Aspects of small-signal rotor angle stability

Rotor angle oscillations may appear in the power system any time also because of the weak electrical connection between generators or between generators and loads due to long distances (large reactances), due to uncoordinated fast voltage regulators or other types of controls. In large interconnected systems, local generators are swinging together but, in some conditions, they can oscillate against other groups of generators. If not damped very fast, the rotor angle oscillations may cause damages to the power plants and other equipment in the power system [1, 4].

Small-disturbance rotor stability is concerned with the ability of the power system to maintain synchronism under small disturbances, like small variations in loads and generation [5]. The small disturbances are those changes occurring in the power system for which the rotor angle presents an almost linear variation allowing linearization of the system equations around the equilibrium point without encountering errors. The process following the disturbance occurrence depends on several factors, including the initial operating conditions, the transmission system strength, and the excitation systems performance. Depending on the generators and the states involved in the unstable oscillations, the stability of the following types of oscillations is of concern [1, 6, 7]:

- *Local modes of oscillations* are experienced by one generator or one power plant that oscillates against the rest of the power system, at frequencies of oscillation in the range 1–2 Hz, depending on the generator characteristics and the power system operating conditions [1, 4]. These oscillations are typically damped out utilizing power system stabilizers (PSS).
- *Interarea modes of oscillations* are associated with groups of generators or groups of power plants that oscillate against the other synchronous machines from the power system at frequencies of oscillations in the range 0.1–0.8 Hz [8, 9]. They are usually observed in areas of a power system interconnected by weak tie lines (lower voltage and/or large reactances). When these oscillations become unstable, the specific protection systems isolate the unstable area by disconnecting the tie lines. The PSS exists on many generators, but they are not effective in damping out interarea oscillations. To overcome this inconvenient, phasor measurement units (PMU) are today employed in wide areas together with flexible alternate current transmission system (FACTS) devices.
- *Control mode oscillations* are due to poorly tuned control systems attached to field exciters, speed governors, high voltage direct current (HVDC) links, static VAR compensators (SVC), etc. [1, 4];
- *Torsional modes oscillations* are associated with the torsional vibrations induced in the generator-turbine rotational components due to sporadic changes either on the turbine side or on the power grid side. The torques applied on the rotor shaft in opposite directions create twisting of the shaft. Such phenomena can cause shaft break or turbine blade failures especially at the low-pressure cylinder of the thermal power plants. A particular case of torsional mode is the subsynchronous resonance (SSR) which is due to the interaction between series capacitors and nearby power plants with long and flexible shafts [1].

4.4.2 The Single-Machine Infinite Bus (SMIB) system

For educational purposes, let us consider the simple model of an equivalent generator supplying power to an infinite power system via an equivalent network, represented only by reactances (Fig. 4.1). The generator is represented by the classical model, which consists of a constant e.m.f., E_g', behind the transient reactance, X_d'. To neglect the rotor saliency ($X_d'=X_q'$), the model of a turbo-generator is considered. The network is represented by the reactance X_{net}, and the infinite bus is represented by a constant voltage E_S.

Let δ be the angle by which the voltage E_g' leads the infinite bus voltage E_S. By neglecting the generator electrical resistance, the air-gap power is equal to

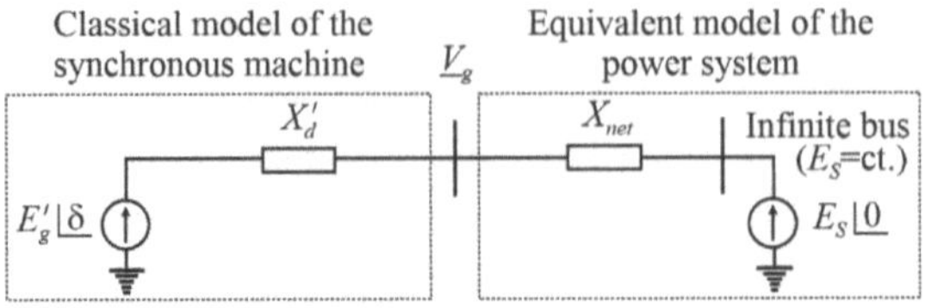

FIG. 4.1 Equivalent circuit of the SMIB system.

the output power, which is also the power transferred from the machine to the infinite power system. Hence,

$$P_e = \frac{E'_g \cdot E_S}{X'_d + X_{net}} \sin\delta \tag{4.18}$$

In the per unit, the air-gap torque is equal to the air-gap power, i.e., $T_e = P_e$.

Any change in the machine output results in a change in δ. As the power output oscillates, the angle δ oscillates too. Fig. 4.2 illustrates the P-δ characteristic of the SMIB system. The initial operating point is found at the intersection between the output power characteristic and the initial mechanical power (P_{m0}) characteristic, to which it corresponds the angle δ_0. The P-δ characteristic is a strongly nonlinear path of the operating point. Linearization of the equations governing the SMIB system and thus the application of the modal analysis can be applied with acceptable errors only if the operating point follows a linear path on the P-δ characteristic. In other words, the linear region of the P-δ characteristic is the region that overlaps on the tangent to the characteristic. For instance, the change of the P_{m0} to P_{m1} may follow a linear path.

Linearizing the air-gap torque about the initial operating point ($\delta = \delta_0$) yields

$$\Delta T_e = \frac{\partial T_e}{\partial \delta}\Delta\delta = \frac{E'_g \cdot E_S}{X'_d + X_{net}} \cos\delta_0 \cdot \Delta\delta = K_S \cdot \Delta\delta \tag{4.19}$$

where K_S is the synchronizing torque coefficient.

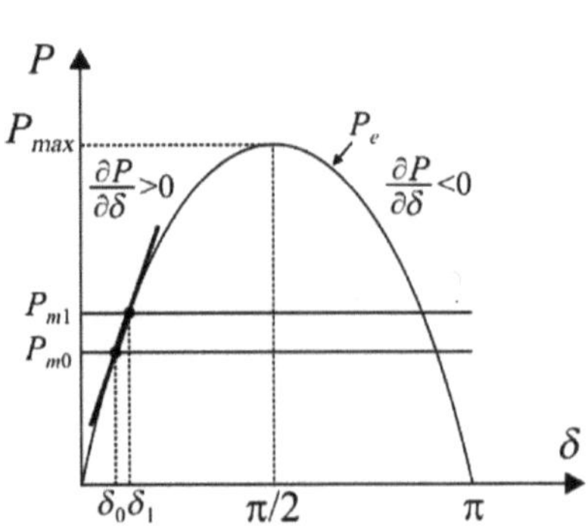

FIG. 4.2 The P-δ characteristic [6].

The swing equation expressed in per unit, as two first order differential equations, is [1]

$$\begin{cases} 2H\dfrac{d\Delta\omega_r}{dt} = T_m - T_e - K_D\Delta\omega_r \\ \dfrac{d\Delta\delta}{dt} = \omega_0\Delta\omega_r \end{cases} \tag{4.20}$$

where: T_m is the turbine mechanical torque; H is the inertia constant, in MW·s/MVA; ω_r is the rotor speed; ω_0 is equal to $2\pi f$, in electrical radians per second; $K_D = D$ is the damping torque coefficient of the machine; t is the time, in seconds.

Linearization of equation (4.20) and substituting for ΔT_e yields

$$\begin{cases} 2H\dfrac{d\Delta\omega_r}{dt} = \Delta T_m - K_S\Delta\delta - K_D\Delta\omega_r \\ \dfrac{d\Delta\delta}{dt} = \omega_0\Delta\omega_r \end{cases} \tag{4.21}$$

The block diagram from Fig. 4.3 can be built using the two differential equations from Eq. (4.21).

The system stability is conditioned by the existence of both components of torque for each synchronous generator. Insufficient synchronizing torque will lead to *aperiodic* or *nonoscillatory* instability whereas insufficient damping torque results in *oscillatory instability*.

According to Ref. [8], the synchronizing torque is the torque that acts on the shaft of a synchronous machine when the rotational speed of the rotor deviates from the synchronous speed that keeps the machine in synchronism. In other words, it is the torque that brings the rotor speed back to the synchronous speed. This can be done by the contribution of the speed governor, the excitation system, and other internal control loops attached to the generator. External actions, e.g., from FACTS devices, can also contribute to the improvement of the synchronizing torque.

The power system is characterized by relatively small damping torque, which is mainly provided by the damper windings of the synchronous machines and to some load types. The damping capability is approximated by a damping term, which is included in the swing equation. However, in the attempt to improve the synchronizing torque, the fast-acting excitation systems can

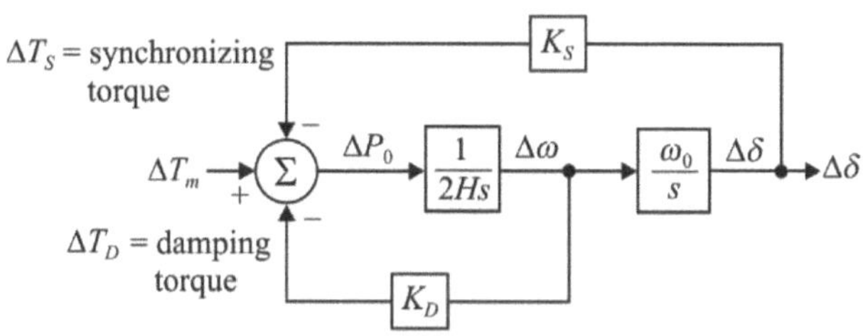

FIG. 4.3 A linearized model of a generator in a SMIB system.

significantly weaken the damping torque thereby causing oscillatory instability. This effect is referred to as an *artificial negative damping*. This problem can be corrected by using an additional signal to the excitation system loop called the *power system stabilizer* [1].

4.4.3 Eigenvalues, eigenvectors, and participation factors applied to small signal stability

Under constant mechanical power, P_m=ct. yields $\Delta P_m=0$, and the system of equations (4.21) becomes:

$$\begin{cases} \dfrac{d\Delta\omega_r}{dt} = -\dfrac{K_D}{2H}\Delta\omega_r - \dfrac{K_S}{2H}\Delta\delta \\ \dfrac{d\Delta\delta}{dt} = \omega_0\Delta\omega_r \end{cases} \tag{4.22}$$

or in matrix form:

$$\underbrace{\begin{bmatrix} \Delta\dot{\omega}_r \\ \Delta\dot{\delta} \end{bmatrix}}_{\Delta\dot{\mathbf{x}}} = \underbrace{\begin{bmatrix} -\dfrac{K_D}{2H} & -\dfrac{K_S}{2H} \\ \omega_0 & 0 \end{bmatrix}}_{\mathbf{A}} \cdot \underbrace{\begin{bmatrix} \Delta\omega_r \\ \Delta\delta \end{bmatrix}}_{\Delta\mathbf{x}} \tag{4.23}$$

where **A** is the state matrix, **x** is the vector of state variables and $\Delta\dot{\mathbf{x}}$ is the vector of changes of the time derivatives of the state variables.

The characteristic equation of the state matrix is:

$$\det(\mathbf{A}-\lambda\mathbf{I}) = \det\begin{bmatrix} -\dfrac{K_D}{2H}-\lambda & -\dfrac{K_S}{2H} \\ \omega_0 & -\lambda \end{bmatrix} = \lambda^2 + \frac{K_D}{2H}\lambda + \frac{K_S}{2H}\omega_0 = 0 \tag{4.24}$$

The eigenvalues are the solutions of the state matrix, i.e.,

$$\lambda_{1,2} = -\frac{K_D}{4H} \pm j\sqrt{\frac{K_S\omega_0}{2H} - \left(\frac{K_D}{4H}\right)^2} \equiv \sigma \pm j\omega \tag{4.25}$$

The real term, σ, of the eigenvalues is the *damping* of the corresponding mode, expressed in Neper/second (Np/s), and the imaginary term, $\omega=2\pi f$, is the *frequency of oscillation*, in radian/second (rad/s). Knowledge of σ is important as it indicates if the system is stable. The system is stable if $\sigma<0$, and unstable if $\sigma>0$. The magnitude of σ provides a measure of the margin of stability.

If damping is missing, i.e., $K_D=0$, for $K_S>0$, then eigenvalues are:

$$\lambda_{1,2} = \pm j\sqrt{\frac{K_S\omega_0}{2H}} \equiv \pm j\omega_n \tag{4.26}$$

where ω_n is the undamped natural frequency, in rad/s.

For the form provided in equation (4.24), the damping ratio is:

$$\xi = -\frac{\sigma}{\sqrt{\sigma^2+\omega^2}} = \frac{K_D}{4H}\frac{1}{\sqrt{\frac{K_S\omega_0}{2H}}} \tag{4.27}$$

Replacing the expressions of ξ and ω_n in the characteristic equations from (4.25) yields:

$$\lambda^2 + 2\xi\omega_n\lambda + \omega_n^2 = 0 \tag{4.28}$$

and:

$$\lambda_{1,2} = -\xi\omega_n \pm \sqrt{\xi^2\omega_n^2 - \omega_n^2} = \underbrace{-\xi\omega_n}_{\sigma} \pm \underbrace{j\omega_n\sqrt{1-\xi^2}}_{\omega} \tag{4.29}$$

Fig. 4.4 illustrates the representation of the eigenvalues in the complex plane given by ω and σ.

In Fig. 4.4, the damping angle θ can be calculated as

$$\theta = \cos^{-1}\xi = \tan^{-1}\left(\frac{\sqrt{1-\xi^2}}{\xi}\right) \tag{4.30}$$

Fig. 4.5 illustrates the modes of oscillations for various locations of the eigenvalues in the complex plane. The sign of ξ is the opposite of the sign of σ. Therefore, a negative sign of σ ensures damped oscillatory modes, whereas a positive sign of σ provides unstable oscillations. Zero value real parts of the eigenvalues provide undamped oscillations. A real eigenvalue ($\omega = 0$) corresponds to a nonoscillatory mode; negative value results in damped aperiodic mode, whereas positive value results in aperiodic instability.

The transition of a pair of complex conjugate eigenvalues from the left semiplane to the right semiplane, following a perturbation occurring in the power system, corresponds to a *Hopf bifurcation*.

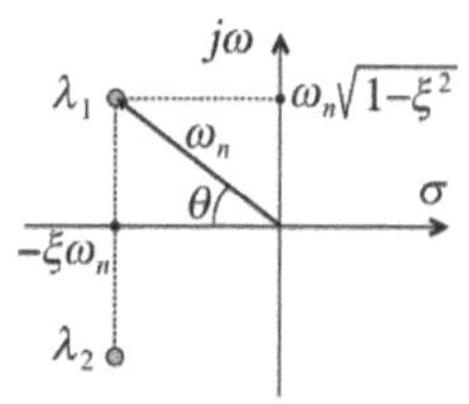

FIG. 4.4 Eigenvalue representation in the complex space.

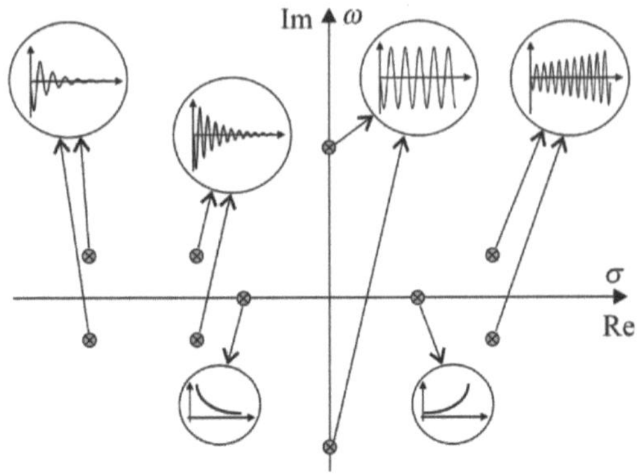

FIG. 4.5 Representation of the modes of oscillation in the complex plane.

4.4.4 Application 1: Single machine infinite bus system

Let us consider a thermal power plant consisting of two synchronous generators supplying power to an infinite power system through their unit transformers and a double circuit transmission line (Fig. 4.6).

The parameters of the system are:

- Line: $x_0 = 0.4\,\Omega/\text{km}$, $L = 80\,\text{km}$, $V_n = 220\,\text{kV}$, double circuit;
- Each step-up transformer has a reactance $X_T = 0.14\,\text{p.u.}$ on $S_{nt} = 2 \times 250 = 500\,\text{MVA}$ and $V_b = 220\,\text{kV}$ base, and has an off-nominal ratio of 1.0;
- The power plant is modeled by an equivalent generator, with the transient reactance $X_d' = 0.27\,\text{p.u.}$, the inertia constant $H = 6.1\,\text{MW s/MVA}$, and damping $D = 2\,\text{p.u.}$, on $S_{ng} = 2 \times 235 = 470\,\text{MVA}$ and $V_b = 220\,\text{kV}$ base.

4.4.4.1 *Parameters calculation*

- The line reactance is calculated in per unit

$$X_L = \frac{1}{n_L} x_0 L \frac{1}{Z_b\left(= V_b^2/S_b\right)} = \frac{1}{2} 0.4 \cdot 80 \frac{1}{484} = 0.0331\,\text{p.u.}$$

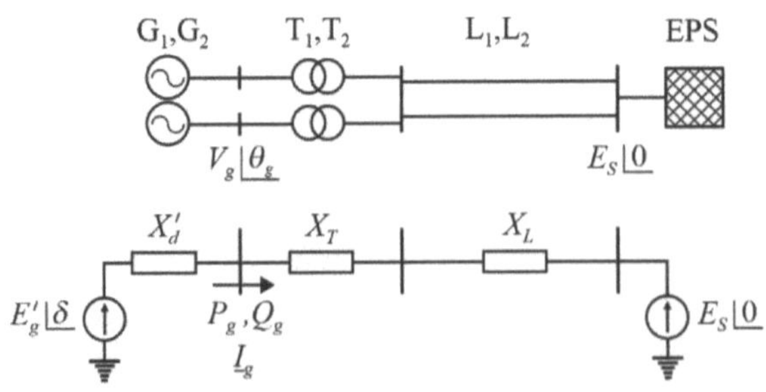

FIG. 4.6 One-line diagram and equivalent diagram of the SMIB system.

- The equivalent generator and the transformers are referred to a common base

$$X'_d = 0.27 \frac{S_b}{n_g \cdot S_{ng}} = 0.27 \frac{100}{2 \cdot 235} = 0.0574 \text{p.u.}$$

$$H = 6.1 \frac{n_g \cdot S_{ng}}{S_b} = 6.1 \frac{2 \cdot 235}{100} = 28.67 \text{MWs/MVA}$$

$$K_D = 2 \cdot \frac{n_g \cdot S_{ng}}{S_b} = 2 \frac{2 \cdot 235}{100} = 9.4 \text{p.u.}$$

$$X_T = 0.14 \frac{S_b}{n_T \cdot S_{nT}} = 0.14 \frac{100}{2 \cdot 250} = 0.028 \text{p.u.}$$

- The network reactance is

$$X_{net} = X_L + X_T = 0.0331 + 0.028 = 0.0611 \text{p.u.}$$

- The total system reactance is

$$X_{eq} = X'_d + X_{net} = 0.0574 + 0.0611 = 0.1185 \text{p.u.}$$

4.4.4.2 Calculation of the initial operating conditions

In steady-state, the power plant supplies active power $P_g = 300$ MW at a terminal voltage $V_g = 1.00$p.u., whereas the magnitude of the voltage at the infinite bus is $E_S = 1.00$p.u.

The voltage angle θ_g can be obtained using the expression of the active power transferred from the power plant to the infinite bus, i.e.,

$$P_g = \frac{V_g E_S}{X_{net}} \sin\theta \Rightarrow \theta = \text{asin} \frac{P_g X_{net}}{V_g E_S} = \text{asin} \frac{3 \cdot 0.0611}{1 \cdot 1} = 0.1842 \text{rad} = 10.55^\circ$$

and the generated reactive power is

$$Q_g = \frac{V_g^2}{X_{net}} - \frac{V_g E_S}{X_{net}} \cos\theta_g = \frac{1^2}{0.0611} - \frac{1 \cdot 1}{0.0611} \cos(0.1842) = 0.2771 \text{p.u.}$$

The complex power supplied by the power plant is

$$\underline{S}_g = P_g + jQ_g = (3.0 + j0.2771) \text{p.u.}$$

The complex voltage at the power plant bus is

$$\underline{V}_g = (0.9831 + j0.1832) \text{p.u.} \quad \text{or} \quad \underline{V}_g = V_g e^{j\theta_g} = 1.0 \cdot e^{j10.55^\circ} \text{p.u.}$$

and the current is

$$\underline{I}_g = \left(\frac{S_g}{\underline{V}_g}\right)^* = \frac{3.0 - j0.2771}{0.9831 - j0.1832} = (3.0000 - j0.2771)\,\text{p.u.}$$

The transient e.m.f. can now be calculated by

$$\underline{E}'_g = \underline{V}_g + jX'_d\underline{I}_g = (0.9672 + j0.3555)\,\text{p.u.}$$

or

$$\underline{E}' = E'\underline{|\delta_0} = 1.0304\underline{|20.183^\circ}$$

where δ_0 is the initial value of the rotor angle δ.

4.4.4.3 Modal analysis

The synchronizing torque coefficient can be calculated as

$$K_S = \frac{E'_g \cdot E_S}{X_{eq}}\cos(\delta_0) = \frac{1.0304 \cdot 1.0}{0.1185}\cos(20.183^\circ) = 8.1614$$

The state matrix is

$$\mathbf{A} = \begin{bmatrix} -K_D/2H & -K_S/2H \\ \omega_0 & 0 \end{bmatrix} = \begin{bmatrix} -0.1639 & -0.1423 \\ 314.16 & 0 \end{bmatrix}$$

the eigenvalues are

$$\lambda_{1,2} = -0.0820 \pm j6.6864 = \sigma \pm j\omega$$

and the damping ratio is

$$\xi = 0.0123$$

The oscillatory mode frequency is $\omega/2\pi = 6.6864/2\pi = 1.0642$ Hz. The oscillatory frequency of the machine versus the infinite bus is known as local mode, which is typically from 1 to 2 Hz [10].

The right eigenvectors are calculated from

$$\begin{bmatrix} -0.1639 - \lambda_i & -0.1423 \\ 314.16 & -\lambda_i \end{bmatrix}\begin{bmatrix} \phi_{1i} \\ \phi_{2i} \end{bmatrix} = 0$$

The right eigenvector corresponding to eigenvalue $\lambda_1 = -0.0820 + j6.6864$ is

$$\begin{bmatrix} \phi_{11} = -0.0003 + j0.0213 \\ \phi_{21} = 1 \end{bmatrix}$$

and the right eigenvector corresponding to eigenvalue $\lambda_2 = -0.0820 - j6.6864$ is

$$\begin{bmatrix} \phi_{12} = -0.0003 - j0.0213 \\ \phi_{22} = 1 \end{bmatrix}$$

The left eigenvectors are normalized so that $\boldsymbol{\psi}_i \cdot \boldsymbol{\phi}_i = 1$. Then, we get

$$\boldsymbol{\Psi} = \begin{bmatrix} \psi_{11} = -j23.4976 & \psi_{12} = 0.5001 - j0.0061 \\ \psi_{21} = j23.4976 & \psi_{22} = 0.5001 + j0.0061 \end{bmatrix}$$

The time response results from

$$\begin{bmatrix} \Delta\omega_r(t) \\ \Delta\delta(t) \end{bmatrix} = \begin{bmatrix} \phi_{11} & \phi_{12} \\ \phi_{21} & \phi_{22} \end{bmatrix} \begin{bmatrix} C_1 e^{\lambda_1 t} \\ C_2 e^{\lambda_2 t} \end{bmatrix}$$

Assuming that the angle varies by $\Delta\delta = 2° = 0.0349$ rad resulted from the change in the mechanical power by $\Delta P_m = 0.303$ p.u., while $\Delta\omega_r(0) = 0$, yields

$$\begin{bmatrix} C_1 \\ C_2 \end{bmatrix} = \begin{bmatrix} \psi_{11} & \psi_{12} \\ \psi_{21} & \psi_{22} \end{bmatrix} \begin{bmatrix} \Delta\omega_r(0) = 0 \\ \Delta\delta(0) = 0.0349 \end{bmatrix} = \begin{bmatrix} 0.0175 - j0.0002 \\ 0.0175 + j0.0002 \end{bmatrix}$$

Then, expressions of the time response of the rotor speed and the rotor angle become:

$$\Delta\omega_r(t) = \phi_{11} \cdot \psi_{12} \cdot \Delta\delta(0) \cdot e^{\lambda_1 t} + \phi_{12} \cdot \psi_{22} \cdot \Delta\delta(0) \cdot e^{\lambda_2 t}$$
$$\Delta\delta(t) = \phi_{21} \cdot \psi_{12} \cdot \Delta\delta(0) \cdot e^{\lambda_1 t} + \phi_{22} \cdot \psi_{22} \cdot \Delta\delta(0) \cdot e^{\lambda_2 t}$$

Fig. 4.7 illustrates the variation of ω and δ for the given SMIB data. The graph shows poorly damped variations. This is natural as they correspond to a small damping coefficient. An identical graph has been obtained on the scheme from Fig. 4.3, with the data resulted from the calculations mentioned previously.

To emphasize the importance of the damping coefficient K_D, three representative values have been chosen. Fig. 4.8 illustrates the angle deviation for the three values of the damping coefficient. The greater the damping coefficient, the faster the oscillations are damped. A zero-value damping coefficient results in undamped oscillations, whereas a negative damping coefficient results in system instability.

4.4.5 Application 2: Small-signal stability in the multimachine system

The analysis of a multimachine system is performed on the 4-machine network (Example 12.6 page 813 from [1]) modeled in Eurostag [11]. Fig. 4.9 illustrates the load flow data, where the bus voltages are in per unit, the voltage angles are in degrees, the power flows are in MW, and Mvar, the loads are also in MW and Mvar, and the shunt capacitors (C7 and C9) are in Mvar.

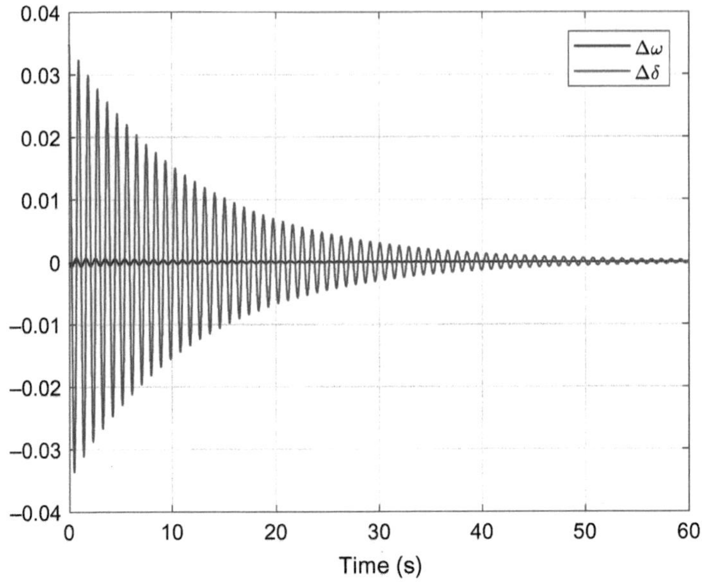

FIG. 4.7 Time response of ω and δ for $K_D = 2$ p.u.

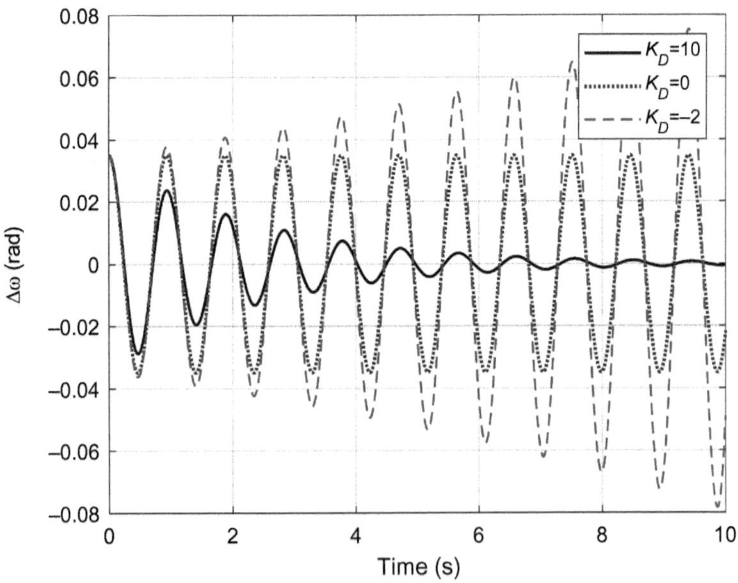

FIG. 4.8 Time response of ω for various values of K_D.

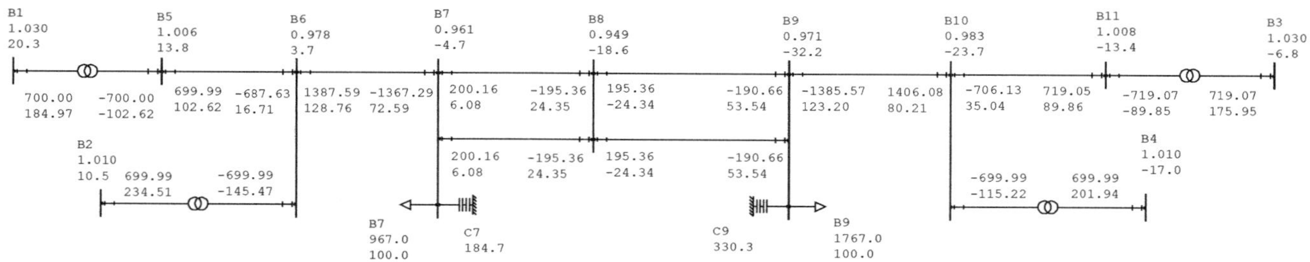

FIG. 4.9 Load flow data of the 4-machine network.

The automatic voltage regulator (AVR) and the PSS provided in [1] have been modeled in Eurostag (Fig. 4.10) to perform time-domain simulations and verify the stability behavior of the Kundur' 4-machines test network by visual inspection.

A perturbation consisting of a change in the load at bus B7 by 5% in both active and reactive powers are assumed at instant 1 s from the simulation inception. Fig. 4.11 illustrates the rotor speeds of the four machines for different modeling conditions. When no AVR and no PSS is considered, the system is unstable. When AVR only is considered, both local oscillation modes and interarea oscillation modes are observed. The addition of a PSS at the four machines cancels out the local oscillation modes, while the interarea oscillation modes are reduced to just a few oscillations.

Figs. 4.12 and 4.13 illustrate the variation of the terminal voltage and the output electrical torque, both in per unit, at machine G1. The PSS acts on the field voltage which results in a change in the terminal voltage that helps the output active power oscillations to better damp, eventually with much smaller amplitudes of the oscillations.

A PSS provides an additional input signal to the AVR to cancel out the local and interarea oscillations. While these simulations have been performed by considering the change in the rotor speed as an input signal to the PSS, as indicated in [1], other input signals can be also considered, e.g. the active power.

4.5 Application of modal analysis to voltage stability

The earlier concepts can be also applied to study voltage stability. The state variables of the power system are the angle and the magnitude of the bus voltages, indicted as $\boldsymbol{\theta}$ and $\mathbf{V}$, respectively. The network calculation is made through the *load flow calculation*, which represents the successive *equilibrium points* at which the system operates, which are characterized by $\Delta\dot{\boldsymbol{\theta}} = \Delta\,\dot{\mathbf{V}} = 0$. The load flow equations for a power system represent the link between the *input* variable (active and reactive power) and the voltage of the system [10]. One of the methodologies used for performing the load flow calculation is the Newton-Raphson method: this method aims to solve in iterative way the load flow equations and, for each time step, iteratively adjust the voltage angle and magnitude to reach the equality between the scheduled bus active and reactive powers and the calculated ones. Mathematically [6]:

$$\begin{bmatrix} \Delta\mathbf{P} \\ \Delta\mathbf{Q} \end{bmatrix} = \underbrace{\begin{bmatrix} \mathbf{J}_{P\theta} & \mathbf{J}_{PV} \\ \mathbf{J}_{Q\theta} & \mathbf{J}_{QV} \end{bmatrix}}_{\mathbf{A}} \begin{bmatrix} \Delta\boldsymbol{\theta} \\ \Delta\mathbf{V} \end{bmatrix} \tag{4.31}$$

It is worth to note that the matrix $\mathbf{A}$ in Eq. (4.31) is exactly the matrix represented in Eq. (4.6) and so it is possible to use it to evaluate the voltage stability of the system.

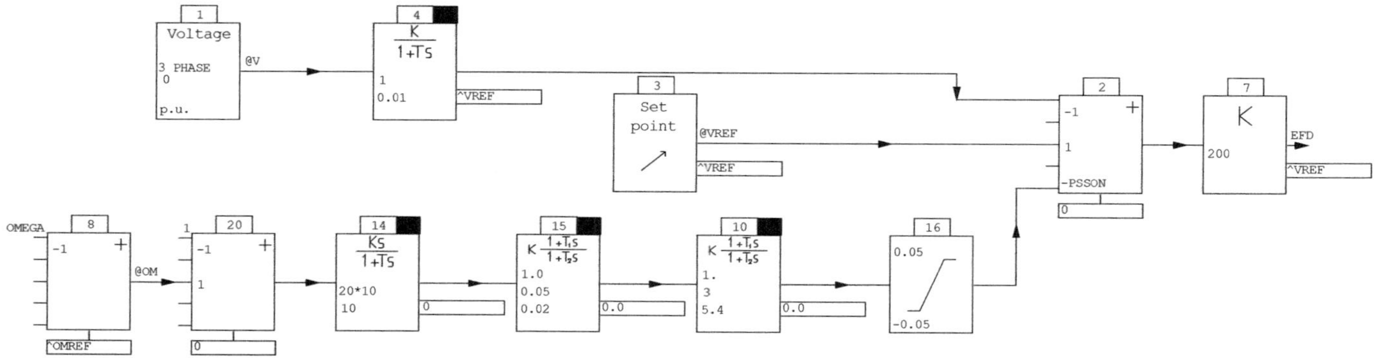

FIG. 4.10 Thyristor exciter with a high transient gain and power system stabilizer implementer in Eurostag.

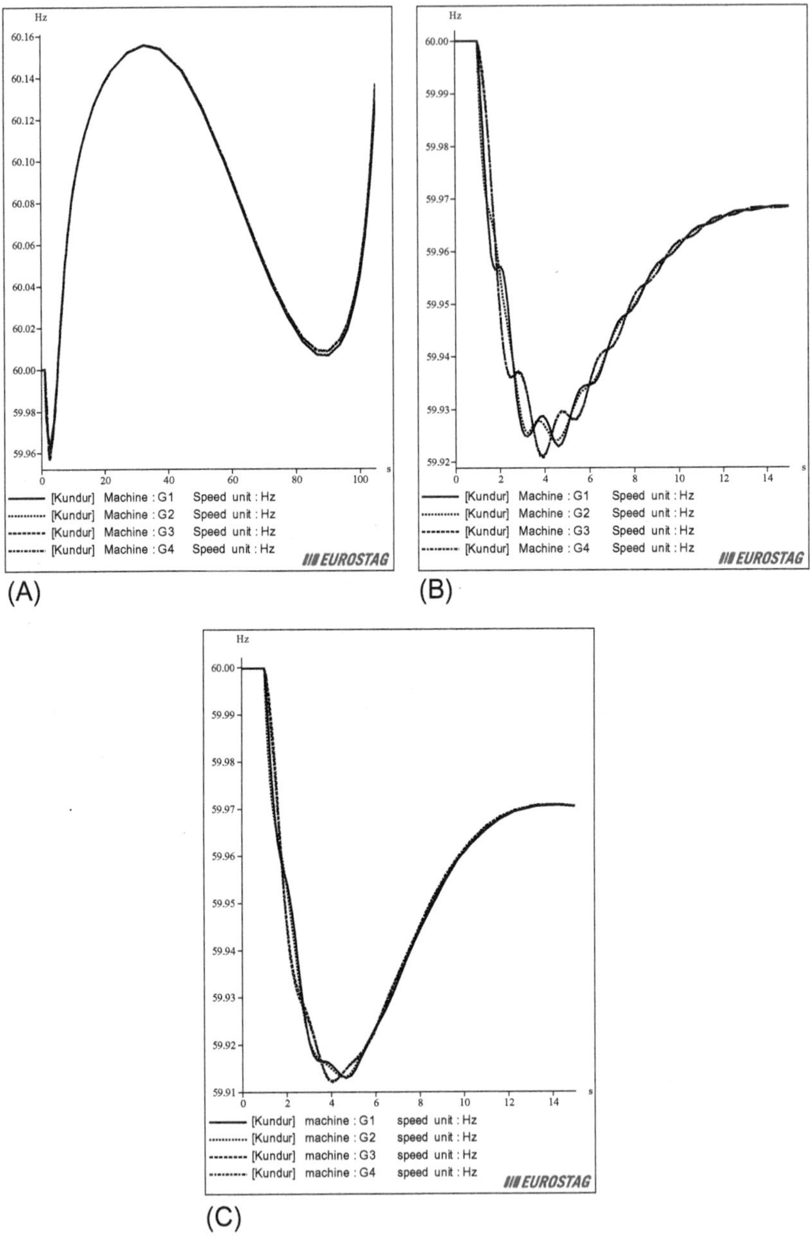

FIG. 4.11 Rotor speeds for (A) no AVR and no PSS; (B) with AVR and no PSS; (C) with AVR and with PSS.

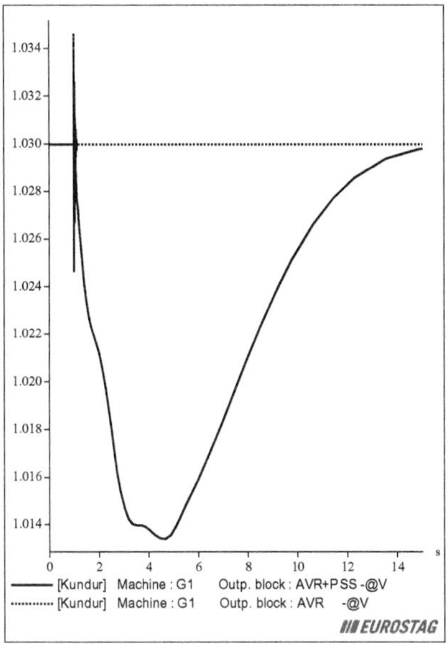

FIG. 4.12 Variation of terminal voltage at machine G1.

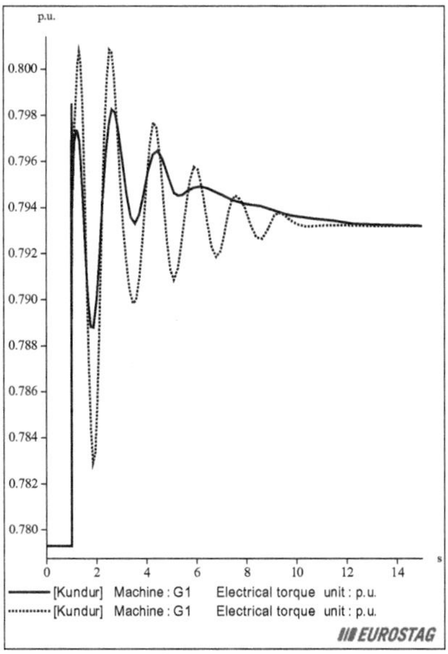

FIG. 4.13 Variation of output electrical torque at machine G1.

From the formulation (4.31), it is evident that the voltage stability is affected by both active and reactive power. However, the usual study approach uses the active power as a parameter, and evaluate the effect of the *reactive* power on the voltage (equivalent to the QV curves reported in Chapter 3). Thus, imposing $\Delta\mathbf{P}=0$, it is possible to write from the first equation of (4.31):

$$\Delta\boldsymbol{\theta} = -\mathbf{J}_{P\theta}^{-1}\mathbf{J}_{PV}\Delta\mathbf{V} \tag{4.32}$$

and substituting (4.32) in the second equation of (4.31), it is possible to obtain:

$$\Delta\mathbf{Q} = \mathbf{J}_r\Delta\mathbf{V} \tag{4.33}$$

with $\mathbf{J}_r = \mathbf{J}_{QV} - \mathbf{J}_{Q\theta}\mathbf{J}_{P\theta}^{-1}\mathbf{J}_{PV}$. Since the variable under study is the voltage, the formulation to be used is $\Delta\mathbf{V} = \mathbf{J}_r^{-1}\Delta\mathbf{Q}$.

The matrix $\mathbf{J}_r$ can be written as:

$$\mathbf{J}_r = \boldsymbol{\Phi}\boldsymbol{\Lambda}\boldsymbol{\Psi} \tag{4.34}$$

with $\boldsymbol{\Phi}$ and $\boldsymbol{\Psi}$ indicating the matrices containing the right and left eigenvectors of $\mathbf{J}_r$, respectively, and $\boldsymbol{\Lambda}$ is a diagonal matrix containing all the eigenvalues of $\mathbf{J}_r$.

Thus, by exploiting the property of the eigenvector matrices (i.e., $\boldsymbol{\Phi} = \boldsymbol{\Psi}^{-1}$), it is possible to get the voltage variation $\Delta\mathbf{V}$ as:

$$\Delta\mathbf{V} = \mathbf{J}_r^{-1}\Delta\mathbf{Q} = (\boldsymbol{\Phi}\boldsymbol{\Lambda}\boldsymbol{\Psi})^{-1}\Delta\mathbf{Q} = \boldsymbol{\Phi}\boldsymbol{\Lambda}^{-1}\boldsymbol{\Psi}\Delta\mathbf{Q} \tag{4.35}$$

which is equivalent to:

$$\boldsymbol{\Psi}\Delta\mathbf{V} = \boldsymbol{\Lambda}^{-1}\boldsymbol{\Psi}\Delta\mathbf{Q} \tag{4.36}$$

Eq. (4.36) can be rewritten as:

$$\mathbf{v} = \boldsymbol{\Lambda}^{-1}\mathbf{q} \tag{4.37}$$

Eq. (4.37) allows obtaining every *modal voltage variation* v_i as a function of its corresponding *modal reactive power variation* q_i as:

$$v_i = \frac{1}{\lambda_i}q_i \tag{4.38}$$

From Eq. (4.38), one can get that:

- The voltage can be properly regulated by acting on the reactive power when the eigenvalues are positive. Thus, the system is *stable*;
- With negative eigenvalues, acting on reactive power does not provide the requested correction action, and this means that the system is unstable;
- In the case of positive eigenvalues (i.e., stable system), small eigenvalues indicate that the system is close to the instability;
- Eigenvalues equal to zero indicate a collapse of the voltage because any change on reactive power makes an "infinite" variation of the voltages

4.5.1 Application 3: Voltage stability of a 3-bus system by modal analysis

Let us consider the 3-bus power system from Fig. 4.14.

The power system is composed of three HV buses: one representing an infinite power system (B1), one connecting a power plant (B2), and one supplying a load (B3). The parameters of the network branches are shown in Table 4.1, and the quantities attached to the network buses are shown in Table 4.2.

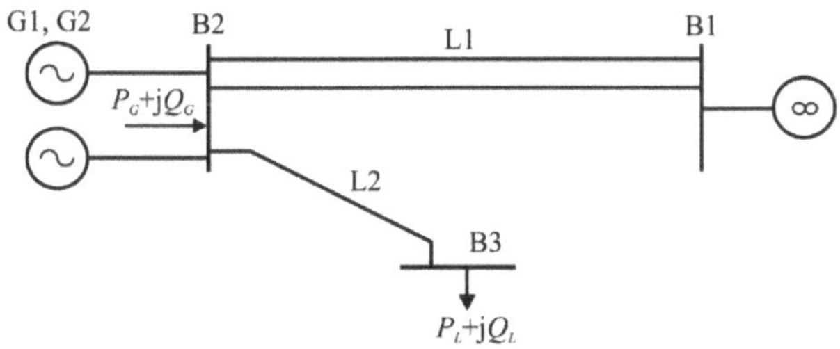

FIG. 4.14 Case study.

TABLE 4.1 Branch parameters.

Line	V_n (kV)	Length (km)	r_0 (Ω/km)	x_0 (Ω/km)	g_0 (μS/km)	b_0 (μS/km)
L1	110	35	0.1	0.40	0	2.5
L2	110	30	0.1	0.40	0	2.5

TABLE 4.2 Bus quantities.

Bus	Type	V_n (kV)	P_G (MW)	Q_G (Mvar)	P_L (MW)	Q_L (Mvar)	V_{sch} (kV)	Q_{min} (Mvar)	Q_{max} (Mvar)
B1	SL	132	–	–	0	0	V_n	–	–
B2	G	132	50	–	0	0	V_n	0	60
B3	L	132	0	0	40	30	–	–	–

Note: SL denotes the slack bus or Vθ bus; G denotes the generator bus or PV bus; L denotes the load bus or PQ bus.

The load flow calculation provides the values of the Jacobian matrix at the last iteration, i.e.,

$$\mathbf{JAC} = \begin{bmatrix} \mathbf{J}_{P\theta} & \mathbf{J}_{PV} \\ \mathbf{J}_{Q\theta} & \mathbf{J}_{QV} \end{bmatrix} = \left[\begin{array}{cc|c} 35.2333 & -7.8786 & -1.6422 \\ -7.7245 & 7.7245 & 1.4587 \\ \hline 2.2585 & -2.2585 & 7.1247 \end{array}\right]$$

The reduced Jacobian is a 1×1 matrix, with $\mathbf{J}_r = 7.5511$. The eigenvalue (λ) of the matrix is equal to the matrix itself. The V-Q sensitivity is equal to the inverse of the eigenvalue, i.e., $\mathrm{d}V/\mathrm{d}Q = 0.1324$. Both the eigenvalue and the V-Q sensitivity shows that the system is stable from a voltage point of view.

To emphasize the weakness conditions of the power system from the reactive power-voltage point of view, let us increase the load at bus B3 at values higher than the generation capacity of the power plant located at bus B2 so that the power plant is no longer capable of controlling the voltage at the scheduled value and bus B2 turns from a PV bus into a PQ bus. We will keep the same conditions for the power factor, which is PF = 0.80.

Bus B2 turns into a PQ bus when the load reaches the operating point formed by $P_L = 68$ MW and $Q_L = 51$ Mvar. The steady-state voltages at this operating point are $V_2 = 0.991$ p.u. and $V_3 = 0.9122$ p.u. Under these conditions, the Jacobian matrix is expanded from 3×3 to a 4×4 matrix, which results in a 2×2 sizes reduced Jacobian

$$\mathbf{J}_r = \begin{bmatrix} 38.2546 & -7.8446 \\ -8.1008 & 6.6378 \end{bmatrix}$$

Two eigenvalues result for the two load buses, i.e., eigenvalue λ_1 associated with bus B2 and eigenvalue λ_2 associated with bus B3. The diagonal matrix of eigenvalues is:

$$\mathbf{\Lambda} = \begin{bmatrix} \lambda_1 = 40.1508 & 0 \\ 0 & \lambda_2 = 4.7416 \end{bmatrix}$$

The matrix of right eigenvector associated with the two eigenvalues is:

$$\mathbf{\Phi} = \begin{bmatrix} \underbrace{\begin{bmatrix} 0.9720 \\ -0.2350 \end{bmatrix}}_{\lambda_1} & \underbrace{\begin{bmatrix} 0.2279 \\ 0.9737 \end{bmatrix}}_{\lambda_2} \end{bmatrix}$$

We now can calculate the voltage—reactive power sensitivities, i.e.,

$$\mathrm{d}V/\mathrm{d}Q = \left[\begin{array}{c|c} 0.0349 & 0.0412 \\ \hline 0.0426 & 0.2009 \end{array}\right]$$

Let us analyze the stability of bus B3. As the load increases, the eigenvalue decreases, and the $\mathrm{d}V/\mathrm{d}Q$ sensitivity increases. A very large value of the sensitivity shows that for a small change in the voltage, a very large effort in terms of reactive power is required. Smaller eigenvalue shows the proximity to the critical voltage.

References

[1] P. Kundur, Power System Stability and Control, McGraw-Hill, New York, 1994.

[2] T. Van Cutsem, C. Vournas, Modelling: system perspective, in: Voltage Stability of Electric Power Systems, Kluwer Academic Publisher, Power Electronics and Power Systems Series, 1998 (Chapter 6).

[3] A.M. Lyapunov, Stability of Motion, Academic Press, 1967.

[4] P. Kundur, J. Paserba, V. Ajjarapu, G. Andersson, A. Bose, C. Canizares, N. Hatziargyriou, D. Hill, A. Stankovic, C. Taylor, T. Van Cutsem, V. Vittal, Definition and classification of power system stability, IEEE/CIGRE Joint Task Force on stability terms and definitions, IEEE Trans. Power Syst. 19 (2) (2004) 1387–1401.

[5] The Great Soviet Encyclopedia, Russian Academy of Sciences, Moscow, 2004.

[6] M. Eremia, M. Shahidehpour (Eds.), Handbook of Electrical Power System Dynamics: Modeling, Stability, and Control, Wiley & IEEE Press, Power Engineering Series, Hoboken, NJ, March 2013.

[7] M. Klein, G.J. Rogers, P. Kundur, A fundamental study of inter-area oscillations in power systems, IEEE Trans. Power Syst. 6 (3) (1991) 914–921.

[8] P. Kundur, M. Klein, G.J. Rogers, M.S. Zywno, Application of power system stabilizers for enhancement of overall system stability, IEEE Trans. Power Syst. 4 (2) (May 1989) 614–626.

[9] Eurostag Used to Guide, Version 4.5.

[10] J.J. Grainger, W.D. Stevenson Jr., Power-flow calculations, in: Power System Analysis, McGraw-Hill Series in Electrical and Computer Engineering, 1994 (Chapter 9).

[11] J.H. Chow, J.J. Sanchez-Gasca, Power System Modeling, Computation, and Control, John Wiley & Sons, 2020.

Chapter 5

Dynamics of modern power systems

Claudia Battistelli and Antonello Monti
Institute for Automation of Complex Power Systems, RWTH Aachen University, Aachen, Germany

Abbreviation

AC	alternate current
ADN	active distribution network
CP	custom power
DC	direct current
DER	distributed energy resource
DFC	dynamic flow controller
DG	distributed generation
DSO	distribution system operator
D-STATCOM	distribution static compensator
DVR	dynamic voltage restorer
EM	energy management
FACTS	flexible AC transmission system
FCL	fault current limiter
GTO	gate turn-off
HV	high voltage
HVDC	high voltage direct current
ICT	information and communication technology
IGBT	insulated gate bipolar transistor
IGCT	integrated gate commutated thyristor
IPFC	interline power flow controller
LCC	line commutated converter
LV	low voltage
LVDC	low voltage direct current
MAS	multiagent system
MG	microgrid
MGCC	microgrid central controller

Converter-Based Dynamics and Control of Modern Power Systems
https://doi.org/10.1016/B978-0-12-818491-2.00005-5

MV	medium voltage
PE	power electronics
PWM	pulse width modulation
RDG	renewable distributed generation
RES	renewable energy resource
SCADA	supervisory control and data acquisition
SG	smart grid
SSCB	solid-state circuit breaker
SSSC	solid state synchronous compensator
ST	smart transformer
STATCOM	static compensator
SVC	static VAR compensator
TCPST	thyristor controlled phase shifting transformer
TCSC	thyristor controlled series capacitor
TS	transfer switch
TSO	transmission system operator
UPFC	unified power flow controller
UPQC	unified power quality controller
UPS	uninterruptible power source
VSC	voltage source converter

5.1 Introduction

Distributed generation (DG) with high levels of different renewable energy sources (RES), such as wind turbines or photovoltaic systems, will dominate the dynamics of future electric power grids. The ongoing transformation of the electric supply from large, centralized power plants based on nuclear or fossil fuels to smaller, decentralized sources based on renewable energy, constitutes an enormous challenge for the stable operation of the grid. Power electronic (PE) converters play a key role in the system, operating as controlled power interface devices, and allowing local control of both directions of flow, typically at notable distances from the main grid. RES installations, as well as energy storage systems, are connected to PE devices either individually or as a part of a new structure called *microgrid (MG)*.

In this new structure of modern power systems, it is of great importance to control the RESs' dynamics, as well as the operation and stability of the MG in islanding and grid-connected mode. Besides, it is paramount to create new links that can facilitate the power flow from point-to-point, to balance the power produced/demanded in the grid. As the rotating system inertia is reduced, frequency response issues become essential and the intermittent nature of RES mandates new ways of regulating voltage. To confront the many challenges, new solutions considering the fast and slow RES dynamics and the control capabilities of the PE interfaces are needed. In this respect, this chapter presents an overview of the PE-based technological solutions nowadays available for the control of dynamics in modern power systems.

The chapter is organized focusing the three levels of the electrical power grid—namely, transmission, distribution, and end-user/consumption point—and addressing the discussion of dynamics separately, for each of these levels. The advantage of this approach is that in this way, it is possible to identify the different issues affecting the electrical grid from the dynamic point of view, and accordingly it is possible to present the technological solutions available to resolve them in each case.

The chapter is structured as follows Section 5.2 introduces the reader to the topic of dynamics and stability in modern power system, discussing in detail how the introduction of new advanced technologies such as renewable power sources, energy storage, and related apparatus can have an impact on the grid at transmission, distribution, and end-user levels, as well as how PEs can be beneficial to correct this impact. Section 5.3 discusses the controllability of transmission systems via PEs, presenting two technological solutions: high voltage direct current (HVDC) and flexible alternate current transmission system (FACTS). Furthermore, Section 5.4 discusses the controllability of distribution systems via PEs, presenting the technological solutions of low voltage direct current (LVDC) and custom power (CP) devices. Finally, Section 5.5 presents the concept of the smart transformer (ST), which soon could help large commercial facilities use power more efficiently and more sustainably.

5.2 Dynamics and stability of modern power system

In the last decades, worldwide the electrical power systems have experienced substantial changes in the principles and philosophy of operation. Some changes are related to the need of resolving climate change, energy shortage, and environmental pollution (e.g., roll out and integration of RESs), or to the advancements in technologies (e.g., PEs, protection devices, digital processors, communication). Other changes are linked to the liberalization of electricity markets, with the establishment of competitive markets with many independent actors. With this drastic evolution, modern power systems become increasingly complex and dynamic.

The shift of electric power supply from large central generating units to smaller plants such as wind and photovoltaic farms, rooftop solar panels, and energy storage systems is fast evolving. Such plants, associated with controllable loads that are connected to a local distribution system or a hosting facility, are collectively known as "distributed energy resources," (DERs) [1]. This shift is complemented by the integration in the network of hundreds of PE devices creating new control points, by the appearance of new sensing and metering technologies (smart sensors and smart meters), by the roll-out of automation, and by the establishment of new services (such as demand-side management, e-vehicle charging, etc.). With these dramatic changes happening in the grid, the dynamics and stability of the power system becomes more complex and critical than ever, calling for new control strategies, technologies, and methods.

5.2.1 Towards the modern structure of the power system

The traditional power grid is substantially an interconnection of different elements: synchronous machines, power transformers, transmission lines, transmission substations, distribution lines, distribution substations, and different types of loads. These are located more or less far from the power consumption area (Fig. 5.1). Here, the electrical power is generated in large capacity power plants and then transferred long-distance via high voltage (HV) overhead transmission lines, to feed power substations for conversion to medium voltage (MV). MV overhead lines then dispatch the power in smaller regions, and final low voltage (LV) conversion is made to feed the "last mile" of the public power distribution network. This traditional "top-down structure" is now being challenged by the emergence of distributed energy resources (DERs).

Environmental problems related to greenhouse gases and the integration of RESs have greatly contributed to the restructuring of the power system. This is being drastically reconfigured, with the introduction of energy management (EM) functions at the distribution system level and of DERs, for which the distribution segment of the electrical grid is now becoming active. In turn, MV and LV networks are becoming active in all electricity mechanisms, with the direct involvement of users, progressively destined to participate as *prosumers* (producers and consumers) in the electricity market (Fig. 5.2).

In the modern structure of the power system (Fig. 5.3), the electric power generation is increasingly happening at the consumption point. The houses

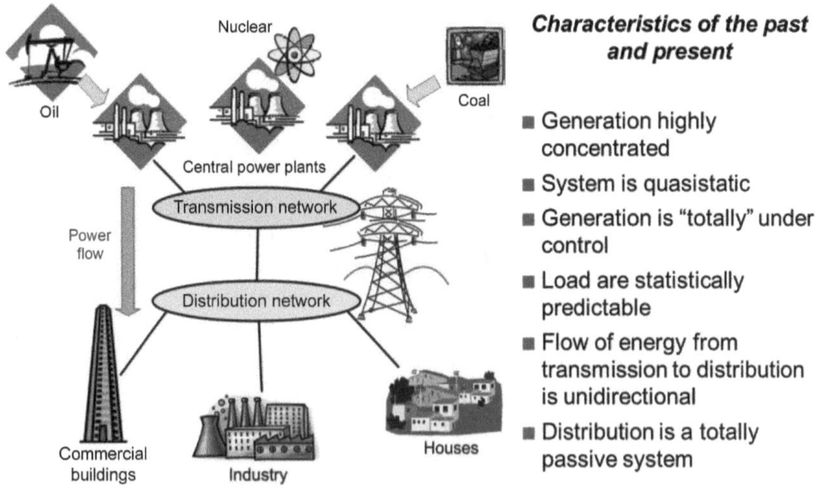

FIG. 5.1 Traditional "top-down" structure of the power system.

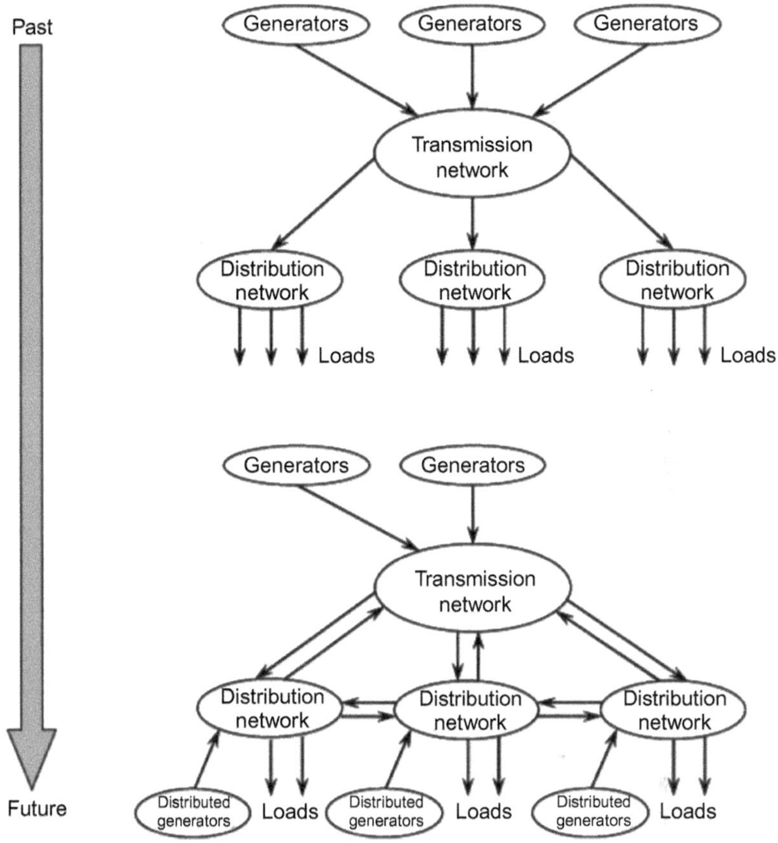

FIG. 5.2 Power system structure evolution, from the past to the future.

and buildings that were connected to the public power distribution network can generate power via solar panels, small wind turbines, or micro-hydro. With DC power generation, storage, and consumption, they can run as autonomous systems. The connection to the public distribution network can be used to trade power in and out depending on periods of over- or underpower production.

The term "smart grid" (SG) refers to the use and operation of a computer, communication, sensing, and control technology in an electric power grid, to enhance the reliability of power delivery at the minimum cost for consumers, and of facilitating the interconnection of new generating sources to the grid. The SG is a modern form of the traditional power grid and can provide a more secure and dependable electrical service thanks to the two-way communication between the utility and the electricity consumers. The SG is capable to monitor activities of the grid-connected system, as well as consumers´ preferences and needs in terms of electricity usage, and to provide real-time information about all events. Typically, an SG includes smart appliances, smart substations, smart

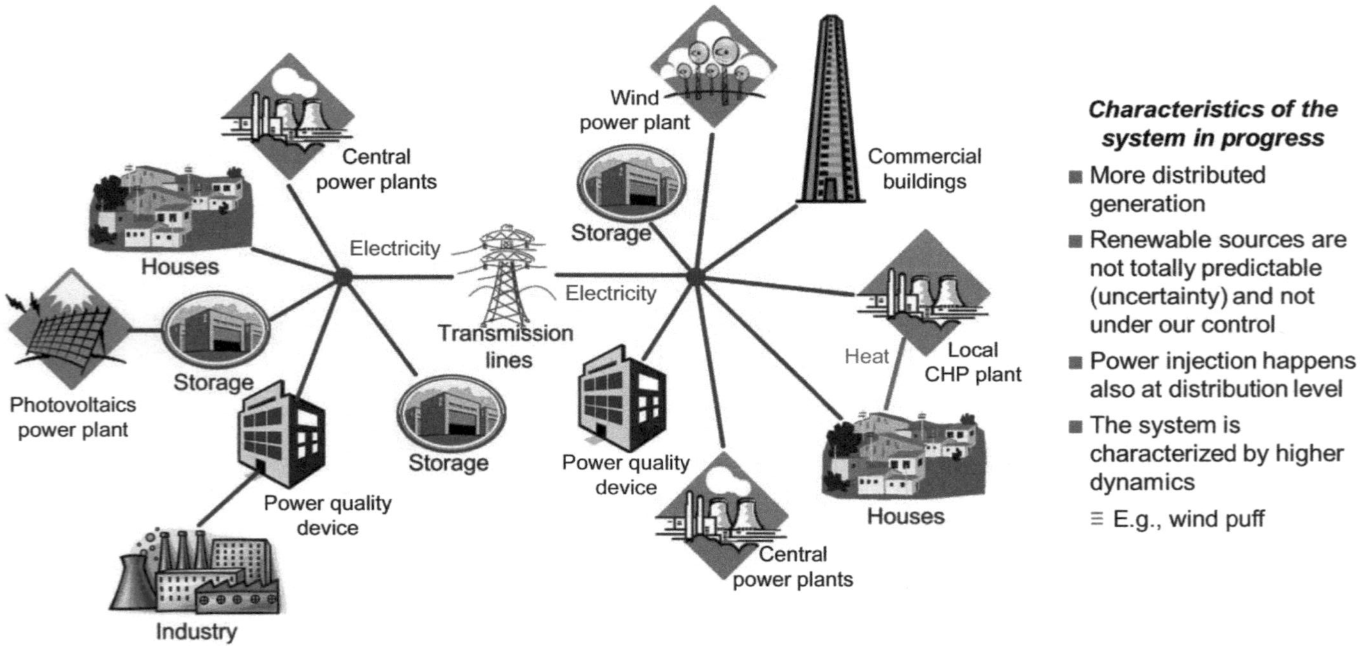

FIG. 5.3 The modern structure of the power system.

meters, and advanced monitoring and measurement technologies (e.g., *Wide Area Monitoring Systems—WAMS*—to support the development and control of transmission and distribution networks; *phasor measurement unit—PMU*—to estimate the magnitude and phase angle of electrical phasor quantities in the power grid) [2].

5.2.2 Impact of distributed energy resources on power system dynamics

The share of RESs in meeting global energy demand is expected to grow in the next years to reach 12.4% in 2023. RESs will have the fastest growth in the electricity sector, providing almost 30% of power demand in 2023, up from 24% in 2017. During this period, it is forecasted that RESs will meet over 70% of global electricity generation growth, led by solar PV and followed by wind, hydropower, and bioenergy. Hydropower remains the largest renewable source, meeting 16% of global electricity demand by 2023, followed by wind (6%), solar (4%), and bioenergy (3%).

The European Union had a binding target of 20% renewable energy use by 2020. The most recent statistics from Eurostat show that 11 member states were already above their national targets in 2017. As of 2017, the European Union countries have reached 17.5% of the target set for 2020.

RESs produce clean, inexhaustible, and increasingly competitive energy. They differ from fossil fuels as an energy source in that:

- They produce neither greenhouse gases—which cause climate change—nor polluting emissions. They are diverse, abundant, and potentially usable anywhere on the planet.
- Their costs are also falling and at a sustainable rate (whereas the general cost trend for fossil fuels is in the opposite direction) despite their volatility.

Nevertheless, the impact of RESs penetration in the structure and the operation of the electricity grid is significant, bringing benefits but also challenges in terms of stability and dynamics.

RESs can be integrated into the power systems at both the transmission and the distribution levels, depending on their generation capacity. Large-scale wind and solar power plants (hundreds to thousands of MW) are integrated through the transmission system, whereas small-scale wind turbines and solar panels plants (15–10 MW) are integrated through distribution systems. RESs with high MVA generation capacity is three-phase and connected to the MV network, whereas smaller capacity RESs are generally single-phase and mostly connected to the LV network [3]. Renewable distributed generation (RDG) can have a positive and negative impact on the system robustness, in relation not only to the reliability of the generating source (wind, solar, etc.)—which changes the actual generation capacity stressing the system—but also to the fraction of RDG that is used over the total generation capacity.

The volatile and variable nature of RESs, alongside the structural changes that they impose on the system, have increased the concerns about dynamics, control, and automation of the power systems.

5.2.3 Impact of renewable distributed generation on stability and dynamics of transmission systems

At the transmission level, wind and solar power plants are typically installed in scales of hundreds of MWs and are often located significantly distant from densely populated regions, or—in the case of offshore wind power plants—at a distance from the shore.

The intermittent and stochastic nature of wind and solar power represents a first main concern based on their integration into the transmission system, as it can have a high impact on the steady-state stability of the system itself (that is to say, the behavior of the system when it operates at a given equilibrium point). This can affect the voltage stability (voltage regulation, i.e., the ability of the system to maintain a constant voltage level) as well as the thermal stability (active/reactive power transfer capacity and loading of the transmission system). Typically, the voltage stability margin is much smaller than the thermal stability margin and, therefore, the voltage stability margin determines the overall stability margin for the system.

In terms of dynamics, when wind or solar power plants are present in the transmission system, dynamic changes in the level of power they produce can be seen as disturbances to the system. While changes in wind power are usually smooth (except for wind gusts), for solar power the changes could be more sudden, especially as a consequence of cloud formation and movement. For this reason, system operators (SOs) should ensure that sufficient levels of fast-responsive spinning reserve are available to damp out the oscillatory modes brought by the RESs.

Furthermore, in terms of inertia (which strongly influences dynamic stability), wind and solar power have different dynamic characteristics and inertia response compared to conventional rotating machine-based generation. Wind turbines are generally equipped with back-to-back converters, in the form of Doubly Fed Induction Generators or Full Converter Synchronous Generators, which electrically decouple the generator from the grid. Therefore, no inertial response is delivered during a frequency event, although a certain amount of kinetic energy is stored into the blades and the generator. Solar panels are virtually inertia-less.

This means that contributions from wind and solar power generators to the grid system's inertia are not direct and mechanical, which can represent a concern for the overall stability of the system. If wind and solar plants were used to replace conventional synchronous generation power plants in the transmission system, this would cause loss of rotating inertia and damping torque, so that the overall transient stability of the system could be weakened. In this case, proper counteractions should be applied via control methods based on the PE devices.

5.2.4 Impact of renewable distributed generation on stability and dynamics of distribution systems

Integrating DG in a distribution network (voltage level: 35 kV or lower) represents a fundamental change from the traditional power system mainly because the resultant power flows are bidirectional (Figs. 5.2 and 5.3). DG is mainly based on wind and solar power units, but it can rely also on other sources such as micro-turbines, gas turbines, fuel cells, biogas, and combined heat and power.

Distribution networks in the presence of DG are normally referred to as *active distribution networks (ADNs)*.

DG can bring several advantages to the network compared to conventional power generators, such as:

- *Increased reliability of power supply*: with a network of smaller, more evenly distributed generating units, the electricity providers can offer the best service possible to the customers.
- *Reduction of transmission losses*: DG units are placed closer to the end-users points. Thus, there's less electricity waste.
- *Higher efficiency*: in the old power system structure, a loss in service at any point in the grid could have a negative impact at any point in the grid. In the new structure, due to the distributed placement of the generating unit, that is less likely to happen.
- *An emergency supply of power*: DG can serve as a backup to the grid, acting as an emergency source in case of contingencies
- *Reduction of peak power requirements*: producing energy locally, DG units can reduce demand at peak times in specific areas and alleviate congestion on the main grid
- *Environmental benefit*: since DG mainly comes from RESs, it is good for the environment and translates into a reduction of emissions.

DG brings, however, also challenges the distribution network in terms of:

- Power quality
- System stability
- Inertia
- Reverse power flow phenomenon
- Increased complexity

We discuss these challenges in the following section.

5.2.4.1 Power quality

Firstly, also in normal operation, the intermittent behavior of RESs can cause voltage variation not only at the point of interconnection with the main grid but also in the buses located nearby. Furthermore, power flows can increase or reverse their direction. Accordingly, the ADN could experience an operational state for which it is not designed, with resulting violations in thermal and voltage constraints.

In case of a contingency (state of the system with one or several simultaneous short circuits), the presence of DG units can alter the fault level (in fact, the fault current is determined by the combined short-circuit contributions of the HV/MV substation and the present DG). This can cause disconnection from the main grid (i.e., *islanding*), as well as major effects on protections. In particular:

- *Unintentional islanding* can occur if the DG continues to supply the system even though the fault is present and the main breaker is open. In most cases, this is an undesirable condition because the distribution system operator (DSO) cannot ensure safe operation and power quality in the islanded area.
- *Failure of protection* can occur when the protection device (circuit breaker) is placed between the DG unit and the fault. In this case, the breaker sees the combined short-circuit current of the HV/MV substation and the RDG current, which can be higher than the current for which it is designed, causing failure.
- *Overcurrent blinding of the protections* can also occur when a relatively large DG unit is connected between the breaker and the fault. In this case, the contribution of DG to the fault reduces the fault current seen by the protection device, which causes the protection device to fail to react if the current falls below its pick-up value.
- *False tripping* occurs when a generator which is installed on a feeder, contributes to the fault in an adjacent feeder connected to the same substation. The DG contribution to the fault current can exceed the pick-up level of the overcurrent protection, causing a trip of the healthy feeder before the actual fault is cleared.

5.2.4.2 System stability

The interconnected structure makes an ADN a complex system where different factors, including uncertainties in power generation and demand, can lead to instability. The natural intermittency of RESs production increases the chance of several transient disturbances—such as a loss of power generation or system components, as well as faults—and makes them more complex to address than in a traditional (passive) distribution network. In particular, the uncertainty related to the change of distributed RES generation makes sudden and unpredictable the practical effects of load variation or disconnections and increases the risk of overcurrent and short circuit.

5.2.4.3 Low inertia

In a conventional generator, the inertia (a form of kinetic energy stored in a rotating mass of a synchronous generator) can support the network's frequency and voltage regulation during disturbances that lead to a generation-demand imbalance. The inertia can minimize the impact of disturbances within a time of up to 10 s, whereas other controllers react after this initial step. The rising penetration of converter-based RESs that have nonexistent inertia, or lower inertia characteristics than synchronous generators, reduces the inertia of an entire network, leading to voltage or frequency fluctuations during a contingent event. If control measures are not correctly applied to minimize these fluctuations within specified limits, the network may experience instability [4].

5.2.4.4 Reverse power flow (back-feeding)

Back-feeding or "reverse power flow" is the flow of electrical energy in the reverse direction from its normal flow. For example, back-feeding may occur when electrical power is injected into the local power grid from a source that is not owned by the SO, which can often happen in ADNs with high penetration of DG. As most of the renewable power generation comes from intermittent sources such as solar and wind, it is particularly hard to balance the intermittent production with a variable consumption in the system [5].

Reverse power flow upstream of DG in a distribution grid could occur during times of light load and high renewable power production. Reverse flow can cause problems for the protection system, as noted earlier, and for the voltage regulators if they are not bidirectional and not designed to accommodate the reverse flow. This problem has been extensively addressed in a report by IEEE PES Task Force on Contribution to Bulk System Control and Stability by DERs connected at Distribution Network [6].

5.2.4.5 Technological and managerial complexity

The rising penetration of dispersed RESs into the distribution network makes the entire power grid more complex and in need of control. Besides, the structural complexity related to the increase of new components and the change in generation patterns, there is another functional complexity brought by the roll-out of new models of use, reliability, and maintainability. This calls for a review and enhancement of the traditional methodologies for modeling, identification, status evaluation, operation, regulation and control of the system, and all its components. These methodologies must rely on advanced tools and technologies as those envisaged in the context of a smart grid (SG), such as Information and Communication Technology (ICT), computers, and PE devices.

5.2.5 Impact of renewable distributed generation at the end-user point: LV microgrids

An MG is a localized group of electrical power sources and loads that normally operates connected to, and synchronous with, the traditional wide area synchronous grid (*macrogrid*), but can also disconnect to "island mode," operating autonomously, for technical or economic requirements. An MG can integrate various sources DG, especially RESs, and can supply power locally to the loads as well as emergency power to the grid, changing between islanded and connected modes [7]. Typically, MG structures can be found at subregional and province levels (e.g., suburban and residential areas; industrial areas; logistics infrastructures).

The core concept behind an MG is integrating a limited number of DG units that can be optimally controlled without creating a complex network. The integration of DG units in the distribution network to export power cannot be considered an MG unless it has control over the network within a defined region. For this, the U.S. Department of Energy Microgrid Exchange Group defines the MG an "a group of interconnected loads and DERs within clearly defined electrical boundaries that acts as a single controllable entity for the grid [8]. It can connect to and disconnect from the grid to enable it to operate in both grid-connected and islanded modes." CIGRE C6.22 Working Group in Microgrid Evolution Roadmap [9] describes MGs as "electricity distribution systems containing loads and DERs (such as distributed generators, storage devices and controllable loads) that can be operated in a controlled, coordinated way either when connected to the main power network or islanded."

MGs can be classified into three categories based on the characteristics of their power supply:

- *AC MGs* supplying AC power in a distribution network. They can be connected to an existing grid utility without special requirements such as converters
- *DC MGs*, emerging due to the extensive application of modern PE equipment and the availability of environmentally friendly DC sources (solar and fuel cells). Typical commercial applications are for telecommunication systems, shipboard power systems, and electric vehicles.
- *Hybrid MGs* consisting of both AC and DC distribution networks, enable both AC and DC power to be supplied to a distribution network so that electricity customers can use electricity according to their needs.

5.2.5.1 Control of dynamics in microgrids

A major difference between an MG and a traditional distribution system is that the former generally must also include a control strategy to maintain, on an instantaneous basis, real and reactive power balance when the system is islanded and, over a longer time, to determine how to dispatch the resources.

The control system must also identify when and how to connect/disconnect from the grid. This control strategy can be realized through a centralized, decentralized, or hierarchical control scheme.

In a centralized scheme, an MG central controller (MGCC)—a physical computer system consisting of a software platform in which various modules for generation and load forecasts, human-machine interfaces, and supervisory control and data acquisition (SCADA) are used—executes the various processes for the effective operation of a power system, namely, forecasting power generation, power demand, and electricity market prices; data monitoring, analysis, and optimization aimed at ensuring power supply at the minimum cost. Fully centralized control relies on a large amount of information exchange between involving units, while the decision-making is carried out at a single point. This is not ideal when interconnected power systems usually cover extended geographic locations and involve an enormous number of units.

In a fully decentralized control, each MGCC is controlled by its local controller without knowing the situation of others [10].

A compromise between these two extreme control schemes can be achieved through a hierarchical control scheme consisting of three control levels: primary, secondary, and tertiary:

- Primary control serves for (a) stabilizing voltage and frequency, (b) offering plug-play capability for DERs and properly sharing active and reactive power among them; (c) mitigating circulating currents that can cause overcurrents in PE devices. It generally applies droop control concepts.
- Secondary control restores the MG voltage and frequency and compensates for the deviations caused by variations of loads of RESs. It is realized through a multiagent system (MAS), with local controllers making autonomous decisions about their own DG units, and communicating with neighboring ones. It can also be designed to support power quality requirements, such as voltage balancing at critical buses
- Tertiary control executes optimization functions for the economic operation of the MG (sampling time is from minutes to hours) and manages power flows between MG and the main grid. This level often includes the tasks of weather, grid tariff, and loads forecasting in the next hours or day. In emergencies (e.g., blackouts), this control could be also applied to manage a group of interconnected MGs (so-called "MG clustering") that could act as a virtual power plant and keep supplying at least the critical loads.

5.2.5.2 Microgrid stability

As addressed in detail in other chapters of this book, in traditional power systems, stability problems regard rotor angle (i.e., the balance between electromagnetic and mechanical torque of synchronous generator required to maintain synchronism), as well as voltage and frequency stability (i.e., the balance between the power supply and demand). In an MG with PE-interfaced

DGs, rotor angle stability is no longer a relevant issue because of the low inertia of DGs (as discussed in Section 5.2.3 of this chapter).

Furthermore, an MG also suffers from the buffered energy of an interconnected transmission system where energy is stored in the inductance and capacitance of line parameters. Although voltage and frequency stabilities in an MG are monitored and controlled in islanding, only the former is controlled in grid-connected operation due to the small effect of the frequency on the grid (in fact, the voltage represents a local variable whereas the frequency acts as a global one). Both stabilities are determined through either small-disturbance or large-disturbance analysis methods, with the stability of the MG dependent on its operational modes, control topology, types of DGs and loads, and network parameters.

5.2.6 The role played by power electronics in modern power systems

Nowadays, most of the electrical energy is still produced by fossil fuels and nuclear power plants. Fossil fuels cause environmental pollution associated with global warming and climate change problems. Nuclear energy is clean in that respect but brings issues in terms of safety and radioactive waste disposal. For these reasons, the whole world is now turning towards renewable energy, where PE is an indispensable ingredient.

PE is an extremely important element in modern SGs. PE plays a relevant role in modern industrial automation and high-efficiency energy systems that include RESs, energy storage, electric and hybrid vehicles, and energy-efficiency improvement of electrical equipment. In modern electric power grids, PE is indispensable in HVDC systems, static VAR compensators (SVCs), FACTS-based active and reactive power flow control, uninterruptible power systems (UPS), fuel-cell-based energy systems, etc. After a dynamic technology evolution for nearly five decades, PE has now grown possibly as the most important technology of the 21st century.

The application of PE in power systems can increase transmission network capacity and flexibility, and, specifically, can enhance system reliability and controllability with a limited environmental impact.

The next section will present how PE can be relevant in modern power systems, with the application at all levels of the electrical energy flow (transmission, distribution, end-user).

5.3 Controllability of transmission systems via power electronics: HVDC and FACTS

This section presents the two most relevant and promising PE-based technological solutions in modern transmissions systems, namely, HVDC and FACTS. The section aims at investigating the role that FACTS and HVDC can play towards the development of the future smart transmission grid, as they may provide European Transmission System Operators (TSOs) with effective solutions to several criticalities experienced, nowadays, in grid planning processes.

5.3.1 High voltage direct current systems

When large amounts of power have to be transferred over long distances, AC transmission can become nonviable, due to voltage variation limits, stability constraints, and the economic impact of additions necessary to correct these limits and constraints. In these cases, direct current (DC) transmission can be a most appropriate solution, since there is no stability limit related to the amount of power or the transmission distance.

An HVDC electric power transmission system uses DC for the bulk transmission of electrical power over long distances (several hundred to a few thousand kilometers), in contrast with the more common AC systems. It consists of a DC circuit (a cable or line in a full HVDC scheme, or simply a capacitor in a back-to-back HVDC scheme), combined with two PE converters, each at one link terminal, for AC/DC and DC/AC conversion. HVDC converters capable of converting up to 2GW [11] and with voltage ratings of up to 1100kV [12] have been realized, and higher ratings are technically viable. A converter station could comprise several converters in series and/or parallel.

5.3.1.1 HVDC conversion systems

HVDC converters have evolved over the years, taking different forms. The first HVDC systems, built in the 1930s, were rotary converters, that is to say, electrical machines acting as a mechanical rectifier. These used electromechanical conversion with motor-generator sets connected in series on the DC side and parallel on the AC side. Since the 1940s, HVDC has then implemented PE technologies: first line-commutated converters (LCC) and then, from the late 1990s, voltage-source converters (VSC).

Line commutated converter-HVDC

These converters are called *line-commutated* because the conversion necessary to carry out the commutation from a switching device to its neighbor, depends on the line voltage of the AC system to which the converter is connected to [13]. LCCs use switching devices that are either uncontrolled (diodes) or can be only turned on (not off) by control action (such as thyristors). In such converters, the DC flows through a large inductance and is unidirectional and almost constant. On the AC side, the converter behaves approximately as a current source, injecting both grid-frequency and harmonic currents into the AC network. For this reason, an LCC is also a "current-source converter." The current being unidirectional, a reversal of the power flow—where required—can only be achieved by reversing the polarity of DC voltage at both the AC and the DC stations.

Early LCC systems, realized until the late 1970s, used mercury-arc valves, with ratings of up to 150kV, 1800A [14]. The last and most powerful system of this type is the Nelson River DC Transmission System HVDC in Canada, completed in 1977 [15]. The last operating system, the HVDC interisland link between Northern and Southern islands of New Zealand, was closed in 2012.

Thyristor technology appeared in LCCs in 1972, on the Eel River Converter Station in Canada [16]. For applications at HV levels (order of several kV), multiple thyristors were used connected in series (each sustaining a breakdown voltage of a few kV), and passive components such as grading capacitors and resistors were connected in parallel with each thyristor, to ensure uniform voltage distribution across the valve. The thyristor plus its grading circuits and other auxiliary equipment are defined as a *thyristor level*. Every thyristor valve typically contains tens or hundreds of thyristor levels, each operating at a different (high) potential concerning the earth.

As of today, thyristors have been used on more than a hundred HVDC schemes, with more still under development or planning. The world's largest capacity thyristor-based HVDC system is the 800kV, 6400MW Xiangjiaba-Shanghai link in China, realized by ABB in 2010. The most recent system is the 800kV, 6000MW Raigarh-Pugalur link in India, expected for completion by ABB at the end of 2019.

The main limitation of LCC is that thyristors can only be turned on by control action, and it comes to the external AC system to effect the turn-off process. This means that the AC system in which the HVDC is connected to must embed synchronous machines to provide the commutating voltage.

LCC HVDC systems can assume different structures based on their application and placement, the most used being:

- Back-to-back, used to interconnect two possibly asynchronous AC systems. In this case, both converters are in the same area, and the length of the connecting DC line is kept as short as possible (Fig. 5.4).
- Monopole, mainly used in congested networks, undersea transmission, or areas with high earth resistivity. In this configuration, one of the terminals of the rectifier is connected to earth ground. The other terminal, which is at a relatively HV compared to ground, is connected to a transmission line.

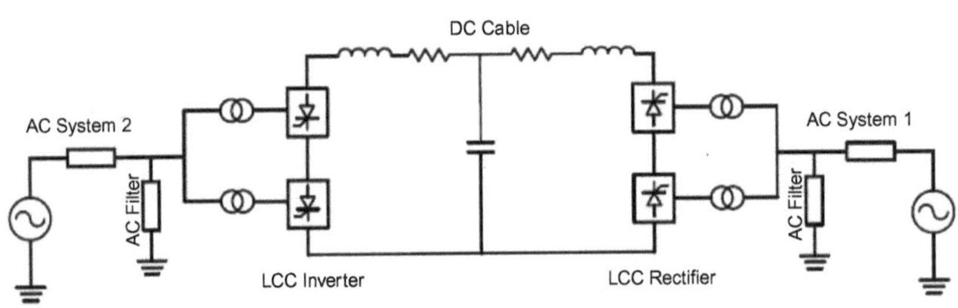

FIG. 5.4 Typical back-to-back structure of line-commutated converters.

The earthed terminal can be connected to the corresponding connection at the inverting station using a second conductor.

- Bipolar, most common for overhead HVDC transmissions, which uses a pair of conductors at a high potential concerning ground, and in opposite polarity.

These structures can easily be generalized to multiterminal HVDC systems.

VSC-HVDC

An HVDC converter using insulated-gate bipolar transistors (IGBTs) is usually referred to as a voltage-source converter (VSC) [17]. These converters are self-commutated because, with IGBTs, both turn-on and turn-off processes can be controlled, and synchronous machines in the AC system are not required for operation. In such converters, the polarity of DC voltage is usually fixed and the DC voltage, being smoothed by a large capacitance, can be considered constant. Pulse-width modulation (PWM) is usually used to improve the harmonic distortion of the converters.

VSC technology has been used for point-to-point HVDC transmission since the late 1990s, but unlike the LCC, which is mature and well-proven, today it is still a developing technology.

The possibility to switch IGBTs on and off many times per cycle ensures operation with improved harmonic performance. Furthermore, the possibility to reverse the direction of the current and the power, makes the VSCs capable of power supply to an AC network made only of passive loads, something which is impossible to achieve with LCC HVDC. VSCs are also considerably more compact than LCCs (since less harmonic filtering is needed) and are preferable in locations where space is limited, for example on offshore platforms.

VSC HVDC systems normally use the six-pulse connection, with the converter producing less harmonic distortion than a comparable LCC. However, these systems can assume different configurations and research is continuing to take place into new alternatives.

A typical two-terminal VSC-HVDC system is shown in Fig. 5.5, where two VSCs are interconnected through a DC transmission line (cable or overhead transmission line).

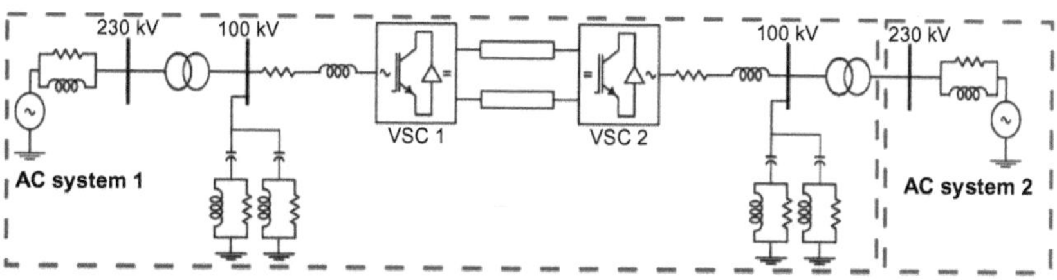

FIG. 5.5 Two-terminal VSC-HVDC system.

Multiterminal HVDC links (with more VSC terminals interconnected) are nowadays still rare, although possible. If realized, the configuration of multiple terminals can be series, parallel, or hybrid (a mixture of series and parallel). Parallel configuration tends to be used for large capacity stations, and series for lower capacity stations. An example is the 2000 MW, Hydro-Québec–New England (Canada) multiterminal HVDC system opened in 1992, which is the largest multiterminal HVDC system in the world.

While LCC-based multiterminal systems are difficult to realize due to the issues related to the reversals of power, which can be carried out only by reversing the polarity of DC voltage affecting all converters connected to the system, instead, with VSCs, parallel-connected multiterminal systems are much easier to control as power reversal is achieved simply by reversing the direction of the current. For this reason, multiterminal VSC systems are expected to become much more common shortly.

LCC scheme over VSC scheme

As of today, both LCC and VSC technologies are important, with the former being used mainly where very high capacity and efficiency is needed, and the latter being used for (i) interconnection of weak AC systems, (ii) connection of large-scale wind power to the grid or (iii) for HVDC interconnections that are likely to be further developed into multiterminal HVDC systems. The market for VSC HVDC is growing fast, especially about the rise in investments for offshore wind power, with one particular type of converter, the Modular Multilevel Converter emerging as a front-runner.

5.3.1.2 Advantages of HVDC

HVDC technology enhances power transfer capacity to the transmission system while ensuring control of power flow and the absence of reactive power. VSCs represent state-of-art technology for offshore wind power connections and multiterminal applications. Unlike LCC technology, it also can feed reactive power to a network for voltage support. Essential advantages of HVDC systems are:

- Transmission losses: these are quoted as 30%–40% less than with AC lines at the same voltage levels [18]. This is because DC transfers only active power which develops lower losses than AC.
- Absence of length limitation for transmission lines: compared to HVAC, HVDC transmission offers long submarine or underground cable transmission with a low level of losses and without the need for reactive power compensation.
- Increased transmission capacity: for a certain conductor cross-section, HVDC can transfer more current compared to traditional HVAC transmission [19].
- Quick and bidirectional control of power flow. Besides, the amount of active power can be set to a fixed value that can be kept in all operating conditions,

including faults in adjacent network sections. This ensures protection from overload. Given these features, the HVDC can support the stability of the surrounding AC system or even of the total network.

- The integration of HVDC lines into an existing system does not require an upgrade of the existing downstream equipment (circuit breakers, transformers, etc.) for short-circuit protection. In the case of short circuits at one of the HVDC terminals, the converters can be switched off within milliseconds preventing them from contributing to the short circuit currents.
- HVDC can be used to connect asynchronously operated power systems, providing cross-border trade of active power, and active power exchange in emergencies.
- HVDC brings environmental advantages in that: (i) it requires fewer overhead lines to deliver the same amount of power as HVAC systems; (ii) it enables power systems to use generating plants more efficiently, for example substituting thermal generation with available hydropower resources.

5.3.2 Flexible alternate current transmission system

FACTS devices are used to increase the usable transmission line capacity and to control power flow over designated transmission corridors while maintaining satisfactory steady-state and transient margins. FACTSs can achieve all this thanks to their speed and flexibility, through the simple control of one or more AC transmission parameters, i.e., series impedance, shunt impedance, voltage, phase angle, current, and damping of power system oscillations [20, 21].

5.3.2.1 Technical aspects

FACTS devices have different structures, based on how power systems components (e.g., transformers, reactors, switches, and capacitors) and PE components (transistors and thyristors) are combined. The advancement of FACTSs has followed the progress made by PE. Over the years, thyristors' rating has raised towards higher nominal values, making PE capable of high power applications up to thousands of MWs. Based on the type and rating of the device, and the voltage level and conditions of the network, FACTSs installation can today bring up to 40%–50% transmission capacity enhancement.

In general, FACTS devices are classified based on their connections, as

1. Shunt connected controllers: Static VAr compensator (SVC) and Static Synchronous Compensator (STATCOM)
2. Series connected controllers: Thyristor Controlled Series Capacitor (TCSC) and Static Synchronous Series Compensator (SSSC)
3. Combined controllers: Thyristor Controlled Phase Shifting Transformer (TCPST), Interline Power Flow Controller (IPFC), Dynamic Flow Controller (DFC), and Unified Power Flow Controller (UPFC)

FACTS devices can be also classified according to the PE technology used for the converters, as:

- Thyristor-based controllers (SVC, TCSC, TCPST and the DFC)
- Voltage source-based controllers (STATCOM, SSSC, IPFC, and UPFC).

Converter (VSC)-based FACTS controllers generally provide.

Based on more advanced technology, such as Gate Turn-Off (GTO) thyristors, Insulated Gate Commutated Thyristors (IGCT), and Insulated Gate Bipolar Transistors (IGBT), the voltage source-based FACTSs can provide better performance and uniform applicability for transmission voltage, effective line impedance, and angle control. If connected to a power source, they also can supply active AC power, in addition to providing reactive power compensation.

FACTS devices have been extensively discussed in the technical and scientific literature [20, 21]. Therefore, in the following, we will only review the basics of the technologies, with a closer eye to their relation to the dynamics of power systems.

STATCOM

This shunt FACTS is VSC-based and usually installed in electrical networks for voltage stability purposes, such as improving the power factor, regulating voltage, or reduce voltage fluctuations.

A STATCOM is composed of the following components:

1. Voltage-Source Converter (VSC), which converts the DC input voltage to an AC output voltage. Two of the most common VSC are (i) the "square-wave inverter" using Gate Turn-Off (GTO) thyristors and (ii) the PWM (Pulse-Width Modulation) Inverter using Insulated Gate Bipolar Transistors (IGBT)
2. DC Capacitor, which provides the DC voltage for the inverter.
3. Inductive Reactance (X), connecting the inverter output to the power system. This is usually the leakage inductance of a coupling transformer.
4. D.Harmonic Filters, mitigating harmonics, and other high-frequency components due to the inverters.

The schema of a typical GTO-based STATCOM is depicted in Fig. 5.6.

Operating mode The STATCOM can be operated in two different modes.

Voltage regulation The STATCOM provides voltage regulation at its connection point by controlling the reactive power absorbed from or injected into the power system through the VSC. In steady-state operation, the voltage V1 generated by the VSC through the DC capacitor is in phase with the system voltage V2 ($\delta = 0$), so that only reactive power (Q) is flowing ($P = 0$). When system voltage is high, the STATCOM will absorb reactive power (inductive behavior),

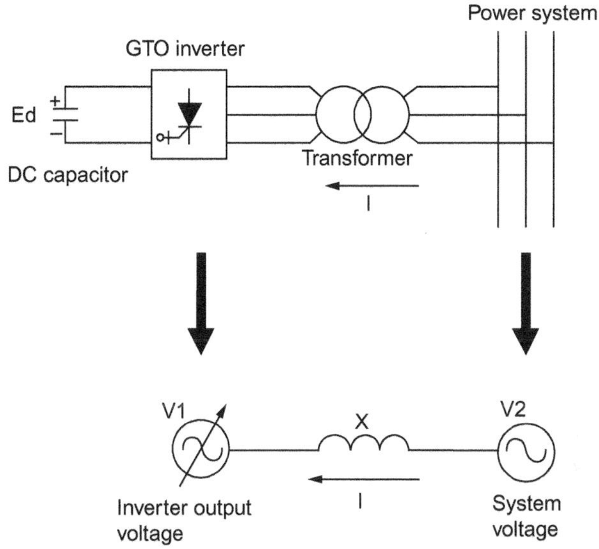

FIG. 5.6 Schema of a GTO-based STATCOM and its equivalent circuit. *(From N.G. Hingorani, L. Gyugyi. Understanding FACTS: Concepts and Technology of Flexible AC Transmission Systems, Wiley-IEEE Press, 1999.)*

whereas when the system voltage is low, the STATCOM will generate and inject reactive power into the system (capacitive).

The amount of reactive power flow put into play, is given by the equation: $Q = [V1(V1 - V2)]/X$.

Var control In this mode, the STATCOM reactive power output is kept constant independent of another system parameter.

Advantages

1. Small footprint, thanks to less primary equipment and the composition using compact electronic converters;
2. The modular architecture is made of separate components (modules) that can be connected, which allows replacement or addition of any module without affecting the rest of the system.
3. Low interaction with the network through a low generation of harmonics.

Static synchronous series compensator

The Static synchronous series compensator (SSSC) works similarly to the STATCOM, but it is serially connected instead of the shunt. It employs a VSC connected in series to a transmission line through a transformer. The injected voltage is not related to the line impedance and can be managed independently. This feature allows the SSSC to work satisfactorily with high loads as well as with lower loads.

The SSSC has three basic components:

(a) Voltage Source Converter (VSC)—main component.
(b) Transformer—couples the SSSC to the transmission line.
(c) Energy Source—provides voltage across the DC capacitor and compensate for device losses.

Operating mode The SSSC (Fig. 5.7) can inject an almost sinusoidal voltage in series with the line. This injected voltage can be considered as an inductive or capacitive reactance connected in series with the transmission line, allowing the provision of controllable voltage compensation [21].

The injected voltage V_S can be varied through the VSC (usually GTOs, IGBTs, or IGCTs) connected on the secondary side of a coupling transformer.

Advantages The SSSC is usually used to correct the voltage in case of faults in the power system. However, it also brings several advantages during normal conditions:

- Power factor regulation through continuous voltage injection combined with a properly structured controller.
- Load balancing in interconnected distribution networks.
- Support for balancing capacitive and reactive power demand.
- Power flow control.
- Reduction of harmonic distortion through active filtering.

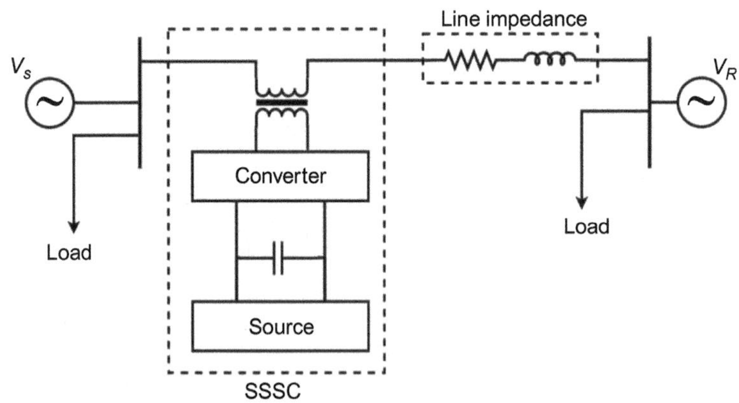

FIG. 5.7 The circuital equivalent of an SSSC. *(Adapted from A. El-Zonkoly, Optimal sizing of SSSC controllers to minimize transmission loss and a novel model of SSSC to study transient response, Electr. Power Syst. Res. 78 (11) (2008) 1856–1864.)*

Unified power flow controller

A UPFC comprises two back-to-back converters operated from a common DC link and connected to the AC system via shunt and series coupling transformers. It can control independently three power systems parameters within its operating limits, namely, line reactance, phase angle, and voltage.

Such a "new" FACTS device combines the features of STATCOM and SSSC. Figs. 5.8 and 5.9 show, respectively, the schematics and equivalent circuit of this FACTS device.

Operating mode The UPFC concept was described in 1995 by L. Gyugyi of Westinghouse [22].

The series inverter can be used to control both active and reactive power flows by injecting an appropriate voltage (V_{in}) with controllable magnitude and phase in series with the AC transmission line.

The shunt-connected converter is used mainly for voltage regulation and active power balance at the point of connection, injecting, and appropriate current I. The real component of this current is used to balance the active power

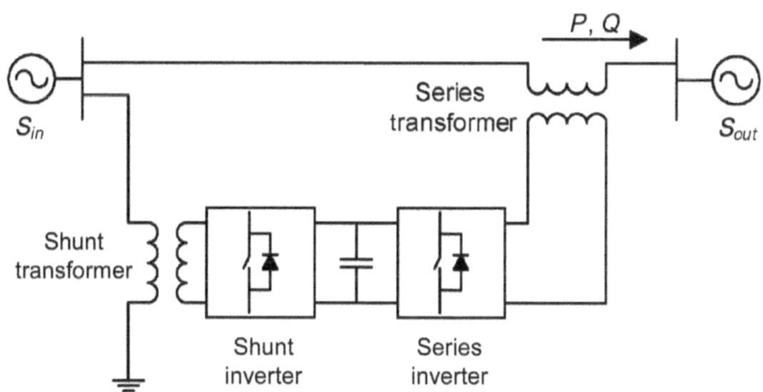

FIG. 5.8 Conventional schema of a unified power flow controller. *(Courtesy of H.-J. Lee, D.-S. Lee,Y.-D. Yoon, Unified power flow controller based on autotransformer structure, Electronics. 8 (2019) 1542L.)*

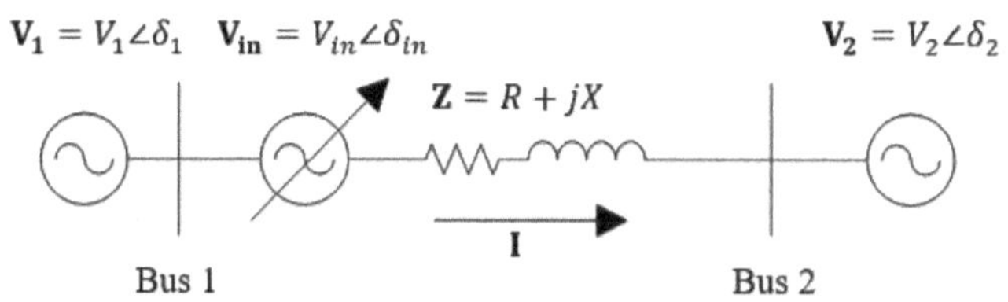

FIG. 5.9 Equivalent circuit of a unified power flow controller. *(Adapted from H.-J. Lee, D.-S. Lee, Y.-D. Yoon, Unified power flow controller based on autotransformer structure, Electronics 8 (2019) 1542. In IET Renewable Power Generation 1 (3) (2007) 160–166.)*

exchange between the series converter and the transmission line, whereas the reactive component is used to absorb/generate reactive power from/to the line.

Advantages

- Control of active and reactive power flows in the transmission line. This ensures the optimal utilization of transmission lines.
- Stability control to suppress power system oscillations. This translates into an improvement of the transient stability of the power system.

Interline power flow controller

The IPFC has been one of the latest developments of FACTS technology, specifically created to address the problem of compensating several transmission lines simultaneously at a given substation. Fig. 5.10 represents the conventional schematics of the device.

Operating principle This FACTS device consists of several DC/AC converters linked together at the DC terminal and providing series compensation, each for different lines. The converters are connected to the AC system through their series coupling transformers (ref. Fig. 5.10). Active power can be transferred between lines. Therefore, the IPFC is capable to level both active and reactive power flow between the lines, to reduce overloading of lines by active power transfer, and to increase the effectiveness of the overall compensating system for the dynamic disturbance.

Advantages and limitations

- Multiline compensation translates into more control capability, thus greatly improving the transmission capacity, stability, and reliability of the power system.

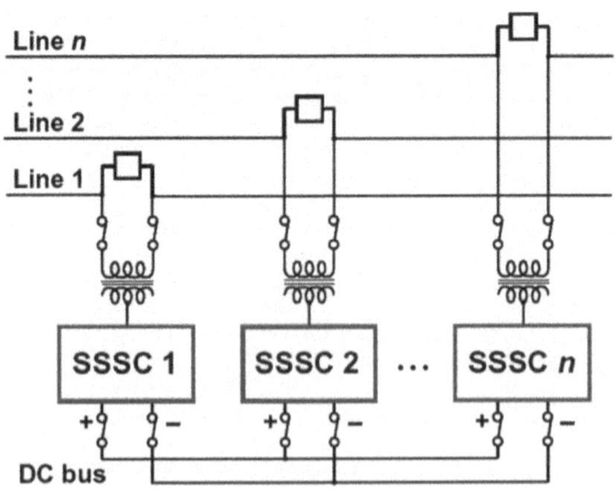

FIG. 5.10 General schema of a unified power flow controller. *(Adapted from L. Gyugyi, C.D. Schauder, S.L. Williams, T.R. Rietman, D.R. Torgerson, A. Edris, The unified power flow controller: a new approach to power transmission control, IEEE Trans. Power Del. 10 (2) (1085) 1995.)*

- Compared to the UPFC, the IPFC provides a relatively economical solution for multiple transmission line power flow control, since only one shunt converter is involved.
- The performance of the IPFC enhances when more series converter involves in the IPFC system.

However, the converters must be physically close to each other, to be linked through the common DC link, and this represents a location constraint that limits the commercial application of the IPFC in future networks. Technological solutions that could eradicate the common DC link and provide the active power exchange between converters will be interesting.

5.4 Controllability of distribution systems via power electronics: LVDC and custom power

DC power generation is constantly increasing in modern power systems with systems of smaller capacities that can be distributed in a large number of locations. The development of distributed DC generation can provide a unique opportunity to switch to smart distribution networks, as they enable easy and useful connection of energy storage in support of the local generating resources.

A key technological development that supports the progress towards DC power is the controllability of DC voltage via DC/DC voltage converters. These converters are suitable for application in fixed installations to control the voltage delivered, and in residential environments to adapt the voltage to the internal needs of the home appliances. They could also embed solid-state switching devices for functional safety purposes, and broadband communication for power management and home automation purposes.

5.4.1 Low voltage direct current systems

From a technological point of view, DC distribution is a new concept that represents a new area of business for PE manufacturers.

A group of PE converters and a DC link in-between forms an LVDC distribution system. The topology of an LVDC system can change based on locations and requirements: common to the different topologies is the AC/DC conversion that is always located near the MV line. The DC/AC conversion can instead be placed in different points, depending on which the LVDC system materializes into either an LVDC link (Fig. 5.11) or a wide LVDC distribution district at the customers' side. In an LVDC link (Fig. 5.11), 1 DC line interconnects 2 separate AC networks: the customers are connected to a common 3-phase LVAC network, to which the DC link must be connected via transformers to ensure compatibility with the existing AC system. In an LVDC district, the DC/AC conversion is placed at the customer's end, assuming the topology of an LVAC network topology with multiple branches.

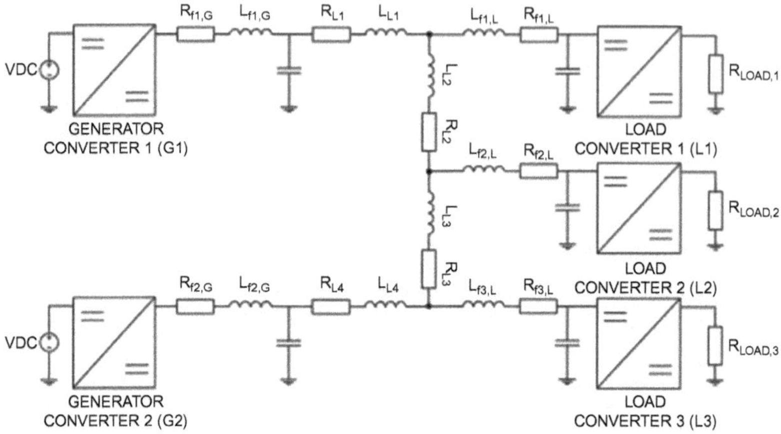

FIG. 5.11 Example of an LVDC link distribution system.

5.4.1.1 Advantages of LVDC systems

LVDC systems offer several advantages, primarily:

- Higher transmission capacity than the traditional 400 V AC system: this can be over 16 times at the voltage drop limit and over 4 times at thermal limit compared to the traditional 400 V AC system.
- Improved voltage quality, resulting from the capability of inverters to control the active power.
- Increased reliability: high transmission capacity enables LVDC systems to partly replace MV lines. Being the LVDC system equipped with its protection apparatus, a fault occurring in the DC district causes outage to the customers on that side of the system only. Therefore, the overall reliability increases while MV line length reduces [23].

5.4.1.2 Disadvantages of LVDC systems

The main concern related to LVDC systems is the electrical safety of the customers. Since LVDC systems at the customer end are exposed to the daily contact of untrained users, they should be properly configured to guarantee personal safety even under fault conditions. Therefore, guidelines should be put into place and followed for the selection of suitable voltage levels, as well as for the minimum requirements for protective devices from the perspective of personal safety.

5.4.1.3 Functional requirements

Similarly to an LVAC network, an LVDC system needs to satisfy several requirements. These requirements regard LV level standardization, as well as environmental, maintenance, and life issues.

5.4.2 Custom power devices

The concept of Flexible AC Transmission System (FACTS) was introduced in the late 1980s by Dr. Narain G. Hingorani, who proposed the application of PE technology in existing AC transmission systems as a solution to enhance power transfer capability, voltage regulation, and system security without adding new lines. With the concept of Custom Power (CP) devices proposed later in 1995, Dr. Hingorani extended the use of FACTSs to the distribution system. CP can be defined as the "application of PE at the distribution level for the benefit of a customer or group of customers." [24] In practice, CP is particularly applied to ensure a certain level of power quality (PQ) to "sensitive" customers such as commercial and industrial.

CP devices can be classified into two types (Fig. 5.12): *network reconfiguring* type and *compensating* type.

5.4.2.1 Network reconfiguring type CP devices

These CP devices are GTO- or thyristor-based and are generally applied as static keys in the system, playing the role of ordinary circuit breakers. The advantage of these switches is their high switching speed compared to conventional mechanical switches.

To the group belong: fault current limiter (FCL), static transfer switch (TS), static breaker, uninterruptible power supply.

Fault current limiter

An FCL (also called fault current controller, FCC) is a GTO-based protective device, used in distribution networks to reduce short circuit currents caused by faults occurring in the system.

Applying FLCs can be particularly convenient when it is aimed to integrate DG in a distribution system without large upgrades in the system itself. DG integration increases the amount of power that can be produced, and this in principle calls for an upgrade of all of the branch circuits with additional busbars, wiring,

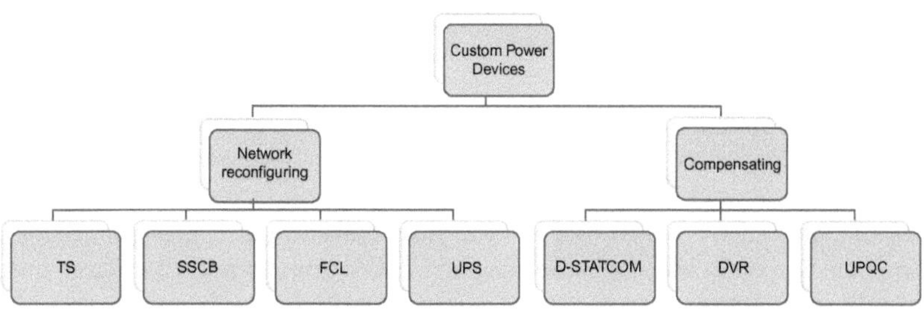

FIG. 5.12 Typologies of custom power devices.

and circuit breakers to handle the new higher fault current limit. Instead of adding new protective apparatus, a simple and noninvasive solution is to add an electric impedance to the circuit.

In this context, an FCL is just a nonlinear element having a low impedance at normal current levels and a higher impedance at fault current levels. This limits the increase rate of the fault current, as well as the level it can rise before the breaker is opened. After the faulting branch is disconnected, the FCL automatically returns to normal operation.

Transfer switch

A transfer switch (TS) is a critical component of the emergency power supply system for any facility. It selects a power source, either standard utility company-provided power or emergency generator power, and conducts that power to critical loads. TSs can be manually operated (static TS), or they can be automatic triggering when they sense one of the sources has lost or gained power.

An Automatic TS is often installed where a backup generator is located so that the generator can supply temporary electrical power if the utility source fails.

A static TS uses power semiconductors such as silicon-controlled rectifiers (SCRs) to transfer a load between two sources. Since there are no mechanical moving parts (as in the case of the automatic version), the transfer occurs rapidly, in less than 4ms (within 1/4 of an electrical cycle). Static TSs are used to transfer electric loads between two independent AC power sources without interruption.

Solid-state circuit breaker

The SSCB is based on GTO or thyristor switching technology. It is a high-speed switching device, applied to sense when the load terminal voltage exceeds a safe value, to protect from faults and short circuit currents in the distribution system.

Uninterruptible power supply

A UPS is a device that delivers emergency power to a load when the primary power source is lost. A UPS differs from an auxiliary generator, in that it contains a battery that "kicks in" when the device senses a loss of power from the primary source.

A UPS is typically used to protect hardware such as computers, data centers, telecommunication equipment, or other electrical equipment where an unexpected power disruption could cause injuries, fatalities, serious business disruption, or data loss. UPS units range in size from units designed to protect a single computer without a video monitor (around 200V-ampere rating) to large units powering entire data centers or buildings. Small UPS systems provide power for a few minutes (enough to power down a computer in an orderly manner), whereas larger systems have enough battery back for several hours. The world's largest UPS, the 46-MW Battery Electric Storage System (BESS), in Fairbanks, Alaska, powers the entire city and nearby rural communities during outages [25].

5.4.2.2 *Compensating type custom power devices*

Compensating CP devices are applied for active filtering, load balancing, power factor improvement, and voltage regulation [26]. To this category belong: distribution static shunt compensator (D-STATCOM), dynamic voltage restorer (DVR), and unified power quality controller (UPQC).

Distribution static compensator

The D-STATCOM is a STATCOM used at the distribution level, typically connected near the load, for active and reactive power compensation [27].

Dynamic voltage restorer

The dynamic voltage restorer (DVR) is a device that can be used in distribution networks to inject 3-phase voltage in series and synchronism with the distribution feeder voltages, to correct short voltage reductions (the so-called, voltage sags).

In practice, DVR systems can inject up to 50% of nominal voltage, but only for a short time (up to 0.1 s). This is not an issue since most voltage sags are much less than 50%.

DVRs can represent a good solution for the end-users subject to unwanted short power quality disturbances. On the contrary, they are not appropriate for systems that are subject to prolonged reactive power deficiencies (resulting in LV conditions) and in systems that are vulnerable to voltage collapse. In fact, in these systems DVRs make collapses more difficult to prevent and can even lead to cascading interruptions since they maintain appropriate supply voltage [28].

Unified power quality controller

The UPQC is employed in distribution systems to perform altogether shunt and series compensation, allowing simultaneous suppression of current in shunt and voltage in series.

Since a power distribution system may contain DC components, distortions, and unbalance in both voltages and currents while performing shunt and series compensation, the UPQC can compensate for voltage-related power quality issues (e.g., sags, swells, unbalance, flicker, harmonics) as well as for load current-related power quality problems (e.g., harmonics, unbalance, reactive current, and neutral currents) [29].

5.5 The smart transformer and its role in the electrical power grid

So far in this chapter, it has been discussed how the integration of generation mixes and ICT infrastructure has changed distribution systems to accommodate bidirectional power flows, hybrid distribution systems (ac and dc), voltage/frequency regulation, and the integration of small and medium DG. It has also been discussed how these changes have motivated the development and deployment of PE interfaces in distribution systems, for the sake of efficient and reliable operation of the system. Alongside the PE solutions already presented in the previous sections, another solution can be considered, which could allow implementing new functionalities in the distribution system, in the form of ancillary services. This is a distribution transformer based on PE, presented as a "concept" for the first time in the project HEART (The Highly Efficient And Reliable smart Transformer), with the name of Smart Transformer (ST) [30].

5.5.1 From the solid state transformer to the smart transformer

The concept of Solid-State Transformer (SST) was first proposed by McMurray in [31], as a device based on solid-state switches with high-frequency isolation, which behaved like a traditional transformer. A real application of the SST occurred only in the 1990s, in railway traction systems to replace magnetic-core power transformers. The operation of SSTs was found able to offer the typical functionalities of core-type transformers (namely, galvanic isolation of primary and secondary sides, and voltage matching between both sides), but with a highly desirable reduction in weight and volume. Unfortunately, the SST development didn't lead to an industrial product since its use could be justified only by the gain in volume and weight reduction, but the possible additional control functionalities played a rather limited role.

Further research and development stages over the last decades have shown that the potential of using SSTs at the distribution level for SG functionalities is much higher: in fact, the SST can replace the standard low-frequency transformer to connect the MV grid to the LV grid, and to offer services to LV and MV grids (such as power flow control, integration of DG, frequency and voltage regulation, power quality enhancement, fault currents reduction) [31].

These briefly introduced functionalities, combined with the control and communication features enabled by the application of ICT, can make the SST an intelligent device, the so-called "Smart Transformer," (ST) in the following briefly presented from the conceptual point of view.

5.5.2 The concept of smart transformer

Smart transformer (ST) has been proposed in [30] as a solution for increasing the hosting capacity of RESs in "smart" distribution systems, that is to say, in distribution systems integrated with ICT apparatus. The ST can control the

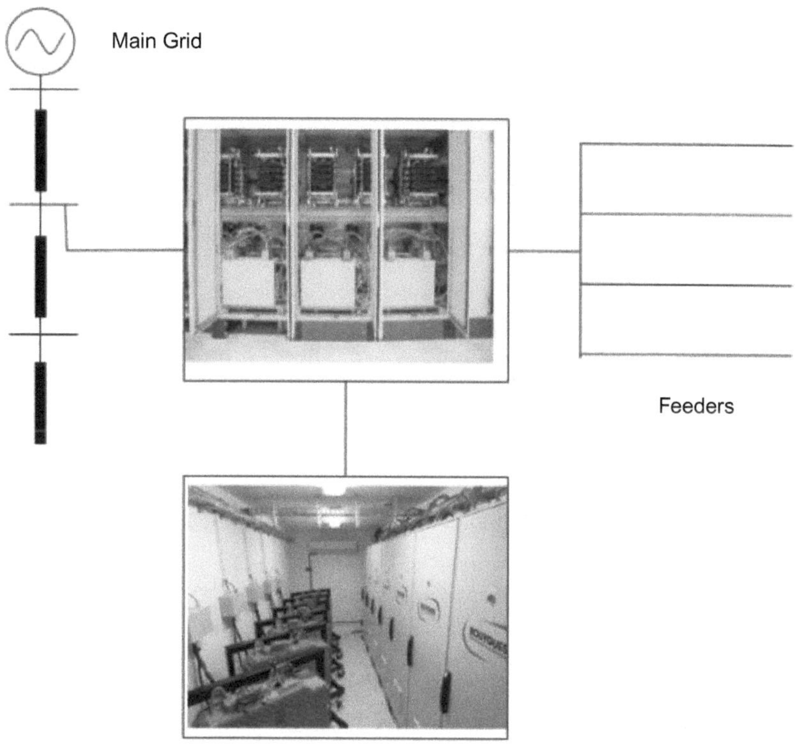

FIG. 5.13 Smart transformer-based distribution grid.

downstream-connected LV grid in islanded mode, given its two-fold capability of connecting and disconnecting from the main grid, and of delivering power from the LV grid to the MV grid (Fig. 5.13).

On the MV side, an ST can control the current absorption from the main grid to ensure the power demand from the loads on the LV side is satisfied while accounting for the losses. It can also support voltage control in the MV grid by injecting reactive power. Furthermore, the ST can work as an active filter in case of nonlinear loads producing high harmonic content, so reducing the stress on the HV/MV transformer and supporting the general power quality of the grid.

The ST could also provide another ancillary service on the MV side, that is to say, the limitation of reverse power flow by control of the local power generation. When the power generated in the LV grid exceeds the demand, the exceeding power flows back to the MV/LV substation, causing a voltage increase in the MV grid. In this case, if supported by the communication infrastructure, the ST could interact with the local generating units of the LV to decrease the power production, hence limiting power back-feeding to the MV grid.

5.5.3 The challenges to the realization of the smart transformer solution

Despite the application in the distribution grid that could seem the most futuristic, also due to the traditional skepticism of the power system community, in a scenario where ICT is opening new frontiers, it is possible to envisage than the solid-state transformer could impose itself becoming the "Smart Transformer."

However, any research and development initiative aimed at realizing the ST proposal should consider and resolve different challenges, such as efficiency, reliability, and cost, to be successful.

The main challenge to the realization and rollout of the ST concept is represented by the possible hardware and software implementations involved. In this respect, R and D should address four main areas:

1. Services: identify which SG services could better benefit from the implementation of the ST.
2. Control: define the suitable control which could guarantee the safe and efficient operation of the ST about the services it has to offer.
3. ICT: identify what grid conditions challenge most the ST components and how communication, sensing, and control can provide virtually the same robustness that traditional transformer had.
4. Physical structure: define for each stage of the ST, the optimal power level, topology, and size related to the offered services and to the disturbances to which it can be subjected
5. Composition: identify and select all the needed technologies starting from modules and passive components but not neglecting electronics, sensors, and protections.

Moreover, costs, efficiency, and reliability represent three major challenges for the ST, compared to a traditional transformer. A modular architecture could be beneficial regarding efficiency and reliability. The operation of modern distribution grids is typically characterized by highly dynamic power profiles and frequent contingencies, which could cause a high thermal excursion in the power semiconductors inside the ST, and, in turn, mechanical fatigue of the ST packaging, leading eventually to failures of the component. The choice of a modular architecture to implement the ST would represent an advantage in terms of tolerance to faults [32].

5.6 Summary

Modern power systems are operating in a constantly changing environment. Understanding, assessing, and controlling such dynamic systems is important and challenging because of the heterogeneous nature of the electrical components that are integrated.

In the preceding sections, we have presented and discussed how PE can play an important role in the operation and control of modern power systems. Focusing the three levels of the electrical power grid—namely, transmission, distribution, and end-user/consumption point—the different issues affecting the electrical grid from the dynamic point of view have been introduced, and the PE-technological solutions available nowadays to resolve them have been discussed. The categorization presented in this chapter helps focus and understand the needs of modern power systems from the operation and control point of view and helps identify and select the most suitable technological solutions.

Given the continuous and long-term evolution of the power systems towards the Smart Grid realization, it will be anyway important to follow the future advancements in technological innovation, standardization, and market environment, to address and resolve the new challenges and issues that the power system of the future will possibly face with the help of PE or other technological solutions.

References

[1] R.H. Lasseter, A.A. Akhil, C. Marnay, J. Stephens, J.E. Dagle, R.T. Guttromson, A.S. Meliopoulous, R.J. Yinger, J.H. Eto, Integration of Distributed Energy Resources: The CERTS MicroGrid Concept, Report (2003).

[2] US Department of Energy (DOE), Smart Grid System Report, Tech. Report (2009) Washington, DC.

[3] A. Sajadi, L. Strezoski, V. Strezoski, M. Prica, K. Loparo, Integration of Renewable Energy Systems and Challenges for Dynamics, Control, and Automation of Electrical Power Systems, Wiley Interdisciplinary Reviews: Energy and Environment, 2018 August.

[4] A. Ulbig, Impact of Low Rotational Inertia on Power System Stability and Operation, (2014).

[5] L.M. Cipcigan, P.C. Taylor, Investigation of the reverse power flow requirements of high penetrations of small-scale embedded generation, IET Renew. Power Gener. 1 (3) (2007) 160–166.

[6] N. Hatziargyriou, T. Cutsem, J. Milanović, et al., Contribution to Bulk System Control and Stability by Distributed Energy Resources connected at Distribution Network, IEEE Power Energy Society, 2017.

[7] N. Hatziargyriou, Microgrids Architectures and Control, John Wiley and Sons Ltd, 2014.

[8] DOE, Microgrid Workshop Report, (2011) August 30-31, San Diego, California.

[9] C. Marnay, et al., Microgrid evolution roadmap, in: 2015 International Symposium on Smart Electric Distribution Systems and Technologies (EDST), Vienna, 2015, pp. 139–144.

[10] A. Kaur, J. Kaushal, B. Prasenjit, A review on microgrid central controller, Renew. Sust. Energ. Rev. 55 (2016) 338–345.

[11] C.C. Davidson, R.M. Preedy, J. Cao, C. Zhou, J. Fu, Ultra-high-power thyristor valves for HVDC in developing countries, in: IET 9th International Conference on AC/DC Power Transmission, London, 2010.

[12] The World's First HVDC Transformer Passes the Test for ±1,100 kV Level, Siemens, 2017.

[13] O. Peake, The history of high voltage direct current transmission, in: 3rd Australasian Engineering Heritage Conference, 2009.

[14] H. Rissik, Mercury-Arc Current Convertors, Pitman & Sons Ltd., 1947

[15] A.G. Siemens, Ultra HVDC Transmission System, Siemens AG, 2010.
[16] J.E. Skog, H. van Asten, T. Worzyk, T. Andersrød, T. Norned, World's Longest Power Cable, CIGRÉ Session, Paris, (2010) paper reference B1–106.
[17] K.R. Padiyar, HVDC Power Transmission Systems: Technology and System Interactions, New Age International, 1990.
[18] M. Guarnieri, The alternating evolution of DC power transmission, IEEE Ind. Electron. Mag. 7 (3) (2013) 60–63.
[19] L. Reed, M. Granger Morgan, P. Vaishnav, D.E. Armanios, Converting existing transmission corridors to HVDC is an overlooked option for increasing transmission capacity, Proc. Natl. Acad. Sci. U. S. A. 116 (28) (2019) 13879–13884.
[20] N.G. Hingorani, L. Gyugyi, Understanding FACTS: Concepts and Technology of Flexible AC Transmission Systems, Wiley-IEEE Press, 1999.
[21] X.P. Zhang, C. Rehtanz, B. Pal, Flexible AC Transmission Systems: Modelling and Control, second ed., Springer, 2012.
[22] L. Gyugyi, C.D. Schauder, S.L. Williams, T.R. Rietman, D.R. Torgerson, A. Edris, The unified power flow controller: a new approach to power transmission control, IEEE Trans. Power Deliv. 10 (2) (1995) 1085.
[23] T. Kaipia, P. Peltoniemi, J. Lassila, P. Salonen, J. Partanen, Impact of Low Voltage DC System on Reliability of Electricity Distribution, CIRED, 2009.
[24] N.G. Hingorani, Introducing custom power, IEEE Spectrum 32 (6) (1995) 41–48.
[25] T. DeVries, J. McDowall, N. Umbricht, G. Linhofe, Battery Energy Storage System for Golden Valley Electric Association, ABB Review, 2004.
[26] Y. Pal, A. Swarup, B. Singh, A review of compensating type custom power devices for power quality improvement, in: 2008 Joint International Conference on Power System Technology and IEEE Power India Conference, New Delhi, 2008, pp. 1–8.
[27] W. Freitas, A. Morelato, W. Xu, F. Sato, Impacts of AC generators and DSTATCOM devices on the dynamic performance of distribution systems, IEEE Trans. Power Deliv. 20 (2) (2005) 1493–1501.
[28] M.H.J. Bollen, Solving Power Quality Problems: Voltage Sags and Interruptions, IEEE Press, New York, 2009, 139.
[29] V. Khadkikar, Enhancing electric power quality using UPQC: a comprehensive overview, IEEE Trans. Power Electron. 27 (5) (2012) 2284–2297.
[30] ERC Consolidator Grant Project HEART, The Highly Efficient and Reliable smart Transformer, http://www.heart.tf.uni-kiel.de, 2014–2018.
[31] S.A. Saleh, et al., Solid-state transformers for distribution systems: technology, performance, and challenges, in: 2019 IEEE/IAS 55th Industrial and Commercial Power Systems Technical Conference (I&CPS), Calgary, AB, Canada, 2019, pp. 1–15.
[32] M. Liserre, G. Buticchi, M. Andresen, G. De Carne, L.F. Costa, Z.-X. Zou, The smart transformer: impact on the electric grid and technology challenges, IEEE Ind. Electron. Mag. 10 (2016) 46–58.

Chapter 6

Frequency definition and estimation in modern power systems

Álvaro Ortega Manjavacas[a] and Federico Milano[b]
[a]Instituto de Investigación Tecnológica, Escuela Técnica Superior de Ingeniería ICAI, Universidad Pontificia Comillas, Madrid, Spain, [b]School of Electrical and Electronic Engineering, University College Dublin, Dublin, Ireland

6.1 Introduction

Any basic module on power systems taught in engineering programs sooner or later states that "the frequency in a power system is the same everywhere." This information is an oversimplification and is incorrect. A more accurate definition of *frequency* is provided in [1], as follows: *frequency means the electric frequency of the system expressed in hertz that can be measured in all parts of the synchronous area under the assumption of a consistent value for the system in the time frame of seconds, with only minor differences between different measurement locations. Its nominal value is 50 Hz.* An important observation from this definition is that the frequency slightly fluctuates from bus to bus due to local load variations. More importantly, after a contingency, such fluctuations can become large and, in some cases, lead to the loss of synchronism of some generator and, even, to the collapse of the whole system.

The study of the loss of synchronism (transient stability analysis) is one of the most important dynamic analyses carried out hundreds of times per minute, every minute, by system operators all around the world. Hence, the statement above is an oversimplification of a complex topic. The fundamental concept, however, is that the frequency of an electrical energy system is a common "property" of the system itself, at least in an ideal steady-state condition.

First, let us recall the simplified electromechanical equations in per unit of the ith synchronous generator connected to the grid [2, 3]:

$$\dot{\delta}_i(t) = \omega_b(\omega_i(t) - \omega_o), \tag{6.1}$$

Converter-Based Dynamics and Control of Modern Power Systems
https://doi.org/10.1016/B978-0-12-818491-2.00006-7

$$M_i\dot{\omega}(t) = p_{m,i}(t) - p_{e,i}(t) - D_i(\omega_i(t) - \omega_o), \tag{6.2}$$

$$p_{e,i}(t) = \frac{e'_i v_i}{x'_{\mathrm{d},i}} \sin\delta_i(t), \tag{6.3}$$

where δ_i is the rotor angle position, ω_b is the base angular speed in rad/s, ω_i is the rotor speed deviation, ω_o is the reference speed, e'_i is the internal machine electromotive force (EMF), v_i is the voltage at the terminal bus of the machine, $x'_{\mathrm{d},i}$ is the transient reactance, M_i is the inertia constant, D_i is the damping, $p_{m,i}$ is the mechanical power, and $p_{e,i}$ is the electrical power, which is a nonlinear function of the rotor angle position of the machine.

When more than one machine is connected to the grid, the expression of the electrical power $p_{e,i}$ of the ith machine depends on the physical connections among the machines and their relative rotor angle positions:

$$p_{e,i}(t) = \sum_{j=1}^{m} b_{i,j} \sin\left(\delta_i(t) - \delta_j(t)\right), \tag{6.4}$$

where m is the total number of machines, and coefficients $b_{i,\,j}$ depend on the topology of the grid and machine and network parameters.

It is easy to show that the only condition for which all machines are in steady state is when they are synchronous, that is, when their rotor speed deviations are null:

$$\omega_1 = \omega_2 = \cdots = \omega_i = \cdots = \omega_m, \quad \forall i = \{1,2,\ldots,m\}. \tag{6.5}$$

Without digging into the maths (see, for example, [2] and [3] for more details), it can be useful to say that the principle with which synchronous machines synchronize is the same that leads to the synchronization of pendulum clocks in the famous empirical observation described by Huygens in 1665 (see, for example, Oliveira and Melo [4]).

The following two remarks are relevant:

- The condition for which the machines are synchronous does not imply that the frequency is equal to the nominal one, that is, 50 Hz (or 314.16 rad/s). Surprisingly enough, the machines are most of the time rotating at speeds different than the nominal one as shown in Fig. 6.1. This means that for the vast majority of the time, the system is actually in transient conditions and, hence, the frequency is not the same everywhere. It is important to note, however, that large frequency deviations are not allowed. This is illustrated in Table 6.1, which shows frequency quality parameters for the synchronous areas that compose the ENTSO-E, namely, continental Europe area (CE), Great Britain (GB), the all-island Irish system (IRE), and the synchronous inter-Nordic system (NE).[a]

a. ENTSO-E network codes are available at: http://annualreport2016.entsoe.eu/network-codes/.

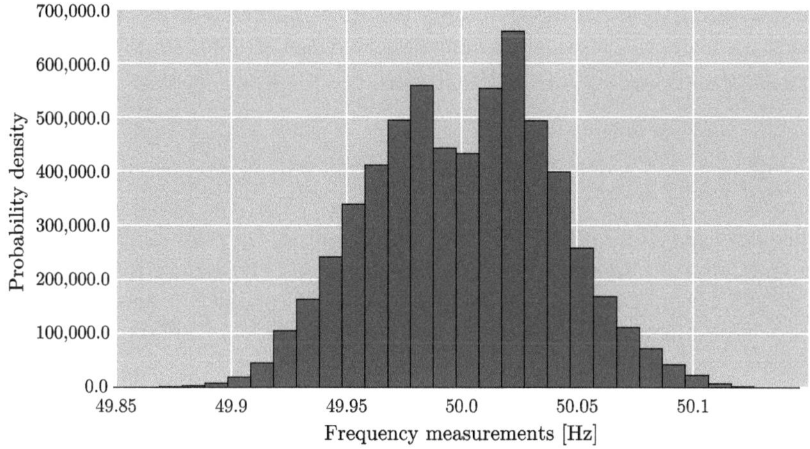

FIG. 6.1 Probability density of the frequency of the Irish system measured with a frequency disturbance recorder (FDR) [5] from December 16–22, 2016 [6].

TABLE 6.1 ENTSO-E: frequency quality parameters per synchronous area [1].

Quantity	CE	GB	IRE	NE
Standard frequency range	±50 mHz	±200 mHz	±200 mHz	±100 mHz
Max. instantaneous freq. deviation	800 mHz	800 mHz	1000 mHz	1000 mHz
Max. steady-state freq. deviation	200 mHz	500 mHz	500 mHz	500 mHz
Time to recover frequency	*Not used*	1 min	2 min	*Not used*
Time to restore frequency	15 min	10 min	20 min	15 min
Alert state trigger time	5 min	10 min	10 min	5 min

From Commission Regulation (EU) 2017/1485 of 2 August 2017 establishing a guideline on electricity transmission system operation (https://eur-lex.europa.eu/legal-content/EN/TXT/PDF/?uri=CELEX:32017R1485&from=EN), p. 116.

- Satisfying such a condition is an intrinsic property of synchronous machines. In other words, the machines tend to synchronize even without the help of controllers.

The latter remark does not mean that synchronous machines do not need control. In actual power systems, primary machine controllers, namely, turbine governors and automatic voltage regulators, are mandatory and fundamental for the operation of the whole system. However, the behavior of synchronous machines is intrinsically stable even without such controllers. Instability arises due to mainly three, often combined, effects:

- high loading conditions;
- large disturbances, for example, three-phase faults; and
- poor tuning of machine controllers.

The remainder of this chapter discusses the importance of estimating accurately and reliably the frequency signal to obtain optimal performance from fast and primary frequency control devices other than synchronous machines. The control schemes of a variety of nonconventional frequency regulation devices are described in Chapter 7.

6.1.1 Need for frequency estimation in power systems

The frequency of an electrical signal cannot be measured directly. When the penetration of power electronics converters and distributed energy resources (DERs) was not massive, the estimation of the frequency was limited to rotating machines through the measure of the mechanical angular speed of the shaft. This was the case of the primary frequency control of synchronous machines and speed/position control of motors in industrial applications.

The large penetration of power electronics-based devices, such as all voltage-sourced converter (VSC)-interfaced and current-sourced converter (CSC)-interfaced DERs and the development of phasor measurement units (PMUs), have led to the need to define the phase angle of voltage and/or current phasors. This is typically obtained through phase-locked loops (PLLs) [7, 8]. A by-product of PLLs is that an intermediate variable of the loop is a good estimation of the frequency deviation (see Section 6.3.2 below for further details).

Other devices, such as PMUs, estimate the frequency using some version of sliding-window discrete Fourier transform (see Chapter 3 of [9]) and also include a GPS signal to synchronize angle measurements. In recent years, there has been a proliferation of PMU-like devices able to estimate the frequency also at distribution, for example, micro-PMUs (μPMU) [10] and low voltage levels, for example, FDRs [11, 12]. However, since PMUs are not currently utilized for online control, in this chapter, we will focus exclusively on PLLs to estimate the local frequency at network buses.

The previous section states why the frequency of a power system is close to, but most of the time not exactly equal to the nominal frequency. However, for the correct operation of the power system, the frequency cannot deviate too much from the nominal value. Large variations, cause, in the short-term (seconds), the loss of synchronism of synchronous machines and, in the long-term (minutes), the malfunctioning of devices and protections, which, ultimately, can trigger system-wide blackouts. An example of this process is the blackout occurred in Italy on October 28, 2003 [13].

Based on the limits reported in Table 6.1, ENTSO-E defines relevant balancing principles and rules in the Network Code on Load-Frequency Control and Reserves. This network code defines three levels of controls, as follows.

- Frequency containment (FC): FC shall aim at containing the system frequency deviation after an incident within a predefined range which is similar to the traditional primary load-frequency control.
- Frequency restoration (FR): FR shall aim at restoring the system frequency to its nominal frequency of 50 Hz which is similar to the traditional secondary load-frequency control.
- Replacement reserves (RR): RR replaces the activated reserves to restore the available reserves in the system or for economic optimization which is similar to the tertiary load-frequency control.

There are two relevant time scales for frequency deviations, namely, the one due to the inertial response of synchronous machines and the one related to the primary frequency response of synchronous machines, as depicted in Fig. 6.2.

For the time scale of the inertial response and fast primary frequency control, that is, from a few hundred milliseconds to a few seconds following a contingency, the frequency of each machine cannot be assumed to be the same everywhere and hence, the location and the accuracy of the frequency estimation is important for the system.

Fig. 6.3 shows the schematic map of the Romanian transmission grid with several PMUs installed in the system. On October 23, 2017, at 12:48 hours, the 400 kV overhead line that connects Babaeski in Turkey with Nea Santa in Greece was suddenly disconnected due to overload. The event led to frequency oscillations visible in the Romanian system, as shown in Fig. 6.4.

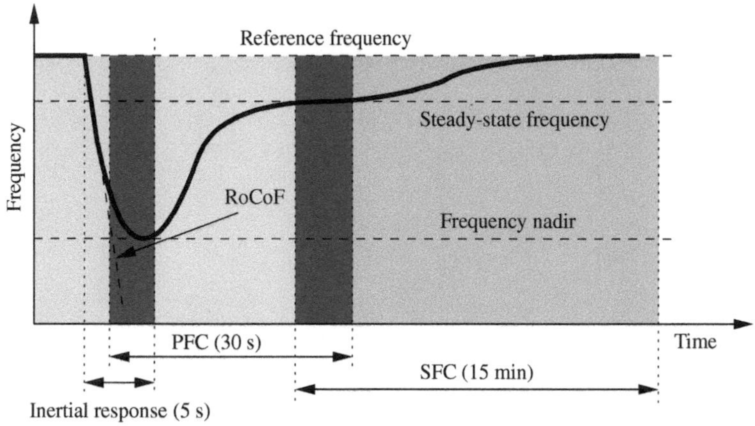

FIG. 6.2 Time scales of synchronous machine inertial response and frequency regulation. The figure qualitatively represents the transient response of the frequency of the system following a loss of generation.

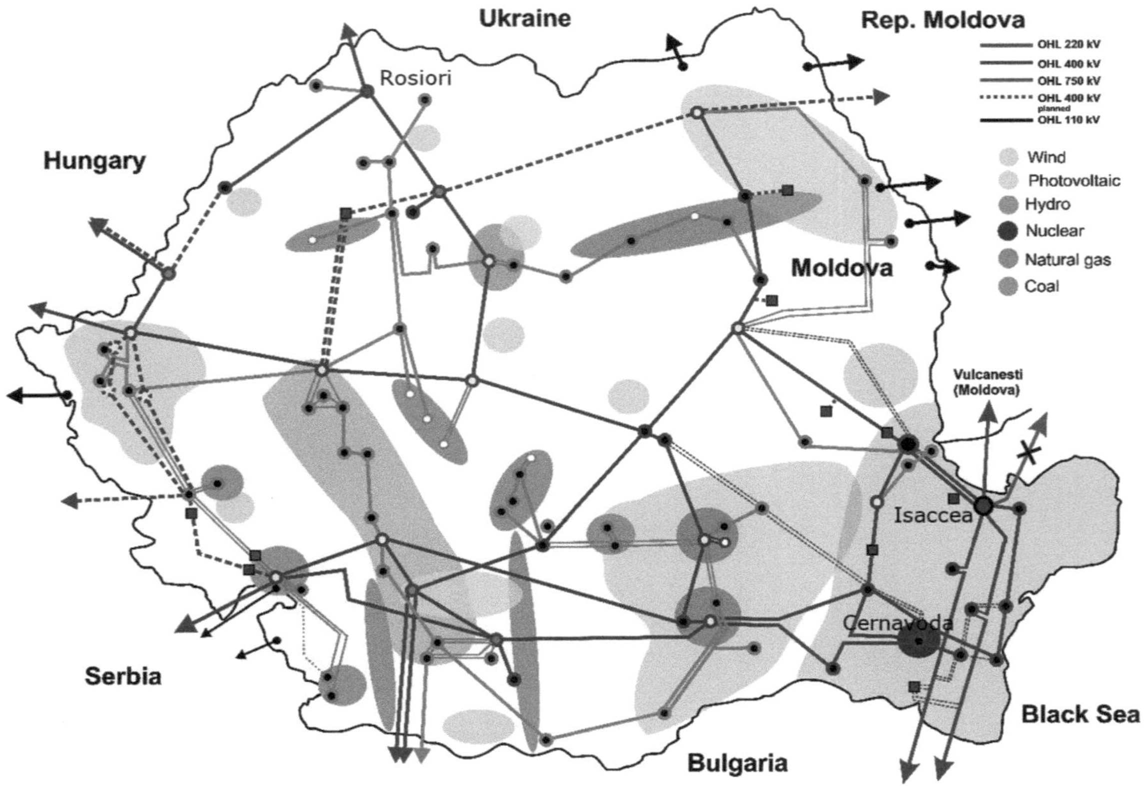

FIG. 6.3 Schematic map of the Romanian transmission grid. *(From Á. Ortega, F. Milano, L. Toma, D. Nouti, A. Musa, C. Chimirel (Transelectrica), A. Antemir, Drafting of Ancillary Services and Network Codes Definitions V2, No. 727481 RESERVE D2.7 v1.0, http://re-serve.eu/files/reserve/Content/Deliverables/727481_RESERVE_D2.7.pdf.)*

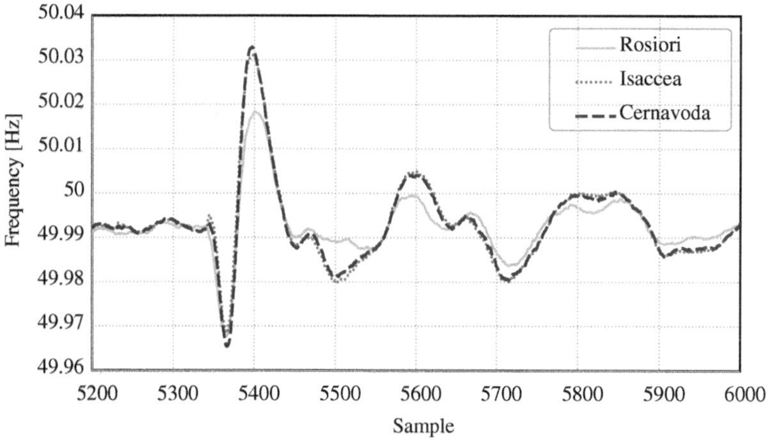

FIG. 6.4 PMU measurements of geographically separated nodes in Romania following the disconnection of the overhead line connecting Babaeski in Turkey with Nea Santa in Greece. *((Data provided by Transelectrica.))*

Frequency deviations can be consistently different depending on the location of the fault and the machines. It is thus impossible to predict a priori the behavior of each machine. Moreover, the oscillations of each machine rotor speed and hence, of the local frequency, depend on the *modes* of the system and the *clusters* that the synchronous machines form. Each coherent cluster oscillates differently than the others. In particular, at a given time, some clusters locally accelerate and other locally decelerate.

This behavior can be easily illustrated using an example where the machines form only two clusters. Fig. 6.5 shows the transient response of the rotor speed

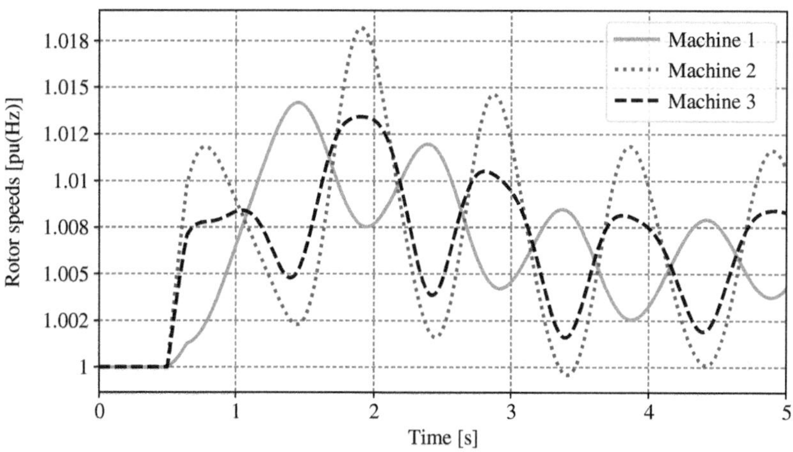

FIG. 6.5 Rotor speed dynamics following three-phase faults for a simple three-machine test system.

for the three-machine WSCC test system, which is largely used in the literature for transient stability analysis [14, 15]. We refer the reader to the case study discussed in Section 6.4 for an in-depth description of the WSCC system. The transient is originated by a three-phase fault. This leads the machines to oscillate and, as can be seen in Fig. 6.5, the rotor speeds of machines 2 and 3 are coherent (cluster 1) and oscillate in counter-phase concerning machine 1 (cluster 2). In more complex systems with several machines, there will be more clusters and more oscillating modes. For each mode, there will be one or more clusters oscillating *against* the others.

Therefore, from observing Figs. 6.4 and 6.5, the definition by European Network of Transmission System Operators for Electricity (ENTSO-E) [1] of *frequency* provided at the beginning of this chapter is not applicable during the first instants after a large disturbance. Moreover, it is to be expected that, as the penetration of nonsynchronous generation from renewable energy sources (RESs) increases, so will the local frequency deviations. Therefore, a new definition of *frequency* is required for high dynamic conditions that will characterize the power systems with up to 100% RES penetration.

From observing Fig. 6.5, one can also understand why it is important to measure the right (preferably, *local*) frequency signal. Whether the frequency is increasing or decreasing, is not a global property of the system, but depends on the cluster where the measure is taken. Hence, a controller aimed at regulating the frequency has to be fed with the proper frequency signal. In the past, when only the synchronous machines of large power plants were providing frequency control, the local frequency was always available and there was no doubt about which signal was to be used, that is, rotor speed deviations. However, it is not so clear which is the most appropriate frequency signal to use for DERs or loads equipped with frequency control. It is possible that the local frequency measured through a PMU or a PLL is not the best signal and that a remote measure or a signal obtained as the combination of more than one measure leads to better performances. The remainder of this chapter describes in detail several techniques aimed at estimating the frequency both locally and globally. Theoretical techniques, namely, the center of inertia (CoI) and the frequency divider formula (FDF), are discussed in Section 6.2, whereas the most common estimation techniques used in practice are described in Section 6.3. We refer the reader to Chapter 8 for an in-depth study of the effect of utilizing different strategies to measure and control the frequency.

6.2 Theoretical techniques to estimate the frequency

6.2.1 Center of inertia

The frequency of the CoI, ω_{CoI}, is a weighted arithmetic average of the rotor speeds of synchronous machines that are connected to a transmission system.

Assuming a set $\mathcal{G}$ of synchronous generators, the expression to compute the CoI is:

$$\omega_{\mathrm{CoI}}(t) = \frac{\sum_{j\in\mathcal{G}} M_j\omega_j(t)}{\sum_{j\in\mathcal{G}} M_j}, \tag{6.6}$$

where ω_j are rotor speeds and M_j are inertia constants.

The inertia-weighted nature of the CoI makes this quantity particularly suited to study interarea oscillations among machine clusters. However, local variations of the machines, especially those characterized by a small inertia, are lost. One can thus expect that the CoI is not fully adequate to simulate local frequency controllers. This fact is duly discussed in the case studies presented in Chapter 7, and in [16]. Moreover, from the modeling point of view, it is unrealistic to assume that distributed generators, microgrids, and consumers will receive the instantaneous signal of the CoI frequency from the system operator. The frequency is actually very likely measured locally, using well-assessed techniques based on the sampling of AC quantities (see, e.g., [17]). Thus, it is important to capture local variations of the frequency to properly model the response of such devices.

6.2.2 Frequency divider formula

The CoI is a well-accepted technique to estimate the frequency of AC transmission systems. However, as discussed earlier, the CoI shows relevant technical and theoretical drawbacks. An alternative approach is proposed in this section, namely the FDF. The FDF is based on the augmented admittance matrix of the system and on the assumption that the frequency along the impedances of transmission lines varies as in a *continuum matter* where synchronous machine rotor speeds define boundary conditions [18].

Let consider a network with n buses and m synchronous machines. The starting point is the augmented admittance matrix, which is discussed in many books, for example, [3]. Apart from the transmission system connections, this matrix also includes the m nodes of the EMFs behind the internal reactances of the synchronous machines. The resulting matrix has the following shape:

$$\overline{\mathbf{Y}} = \begin{bmatrix} \overline{\mathbf{Y}}_{\mathrm{GG}} & \overline{\mathbf{Y}}_{\mathrm{GB}} \\ \overline{\mathbf{Y}}_{\mathrm{BG}} & \overline{\mathbf{Y}}_{\mathrm{BB}} \end{bmatrix}, \tag{6.7}$$

where the subscripts G and B stand for generator buses, and for load and transition buses, respectively, and $\overline{\mathbf{Y}}_{\mathrm{GG}}\in\mathbb{C}^{m\times m}$, $\overline{\mathbf{Y}}_{\mathrm{BB}}\in\mathbb{C}^{n\times n}$; $\overline{\mathbf{Y}}_{\mathrm{GB}}\in\mathbb{C}^{m\times n}$; and $\overline{\mathbf{Y}}_{\mathrm{BG}}\in\mathbb{C}^{n\times m}$. The submatrix $\overline{\mathbf{Y}}_{\mathrm{BB}}$ is:

$$\overline{\mathbf{Y}}_{\mathrm{BB}} = \overline{\mathbf{Y}}_{\mathrm{bus}} + \overline{\mathbf{Y}}_{\mathrm{G}}, \tag{6.8}$$

where $\overline{\mathbf{Y}}_{\mathrm{bus}}$ is the well-known network admittance matrix, and $\overline{\mathbf{Y}}_{\mathrm{G}}$ is a diagonal matrix whose ith diagonal element is 0 if no machine is connected to bus i, and the inverse of a machine internal transient reactance if such a machine is connected to bus i. As discussed in [19], $\overline{\mathbf{Y}}_{\mathrm{BB}}$ is full rank.

Based on Eq. (6.7), the FDF is defined as:

$$\mathbf{0}_{n,1} = \mathbf{B}_{\mathrm{BB}}(\boldsymbol{\omega}_{\mathrm{B}}(t) - \omega_o \mathbf{1}_{n,1}) + \mathbf{B}_{\mathrm{BG}}(\boldsymbol{\omega}_{\mathrm{G}}(t) - \omega_o \mathbf{1}_{m,1}), \tag{6.9}$$

where $\boldsymbol{\omega}_{\mathrm{G}} \in \mathbb{R}^m$ is the vector of machine rotor speeds; $\boldsymbol{\omega}_{\mathrm{B}} \in \mathbb{R}^n$ are the frequencies at the system buses; ω_o is the nominal system frequency; and

$$\mathbf{B}_{\mathrm{BB}} = \mathrm{Im}\{\overline{\mathbf{Y}}_{\mathrm{BB}}\}, \quad \mathbf{B}_{\mathrm{BG}} = \mathrm{Im}\{\overline{\mathbf{Y}}_{\mathrm{BG}}\},$$

where $\mathbf{B}_{\mathrm{BB}}$ has same rank and symmetry properties as $\overline{\mathbf{Y}}_{\mathrm{BB}}$.

To simplify the notation, we rewrite Eq. (6.9) in terms of frequency deviations with respect to the nominal frequency ω_o, as follows:

$$\mathbf{0}_{n,1} = \mathbf{B}_{\mathrm{BB}} \Delta\boldsymbol{\omega}_{\mathrm{B}}(t) + \mathbf{B}_{\mathrm{BG}} \Delta\boldsymbol{\omega}_{\mathrm{G}}(t), \tag{6.10}$$

where $\Delta\boldsymbol{\omega}_{\mathrm{B}} = \boldsymbol{\omega}_{\mathrm{B}} - \omega_o \mathbf{1}_{n,1}$ and $\Delta\boldsymbol{\omega}_{\mathrm{G}} = \boldsymbol{\omega}_{\mathrm{G}} - \omega_o \mathbf{1}_{m,1}$.

Eq. (6.10) expresses the dependency of bus frequencies on the rotor speeds, which are weighted based on their electrical proximity, not their inertia. However, since the internal admittance of each machine is, in per unit, proportional to the capacity of the machine itself, bigger machines weight more than smaller ones in the system.

6.3 Practical techniques to estimate the frequency

6.3.1 Washout filters

Another common approach consists in defining the numerical derivative of the phase angle of bus voltage phasors through some sort of filtering, for example, a washout filter (WF). This approach was first discussed in [20] along with the CoI model, and is commonly used in proprietary software tools for power system simulation, for example, [21].

The frequency estimation obtained using a washout and a low-pass filter (LPF) is depicted in Fig. 6.6. As opposed to the CoI, this technique can properly capture local oscillation modes but is prone to numerical issues.

Differential equations of the WF are as follows:

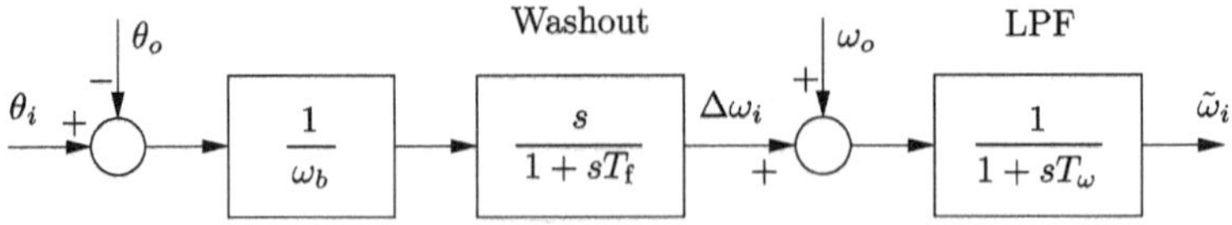

FIG. 6.6 Numerical derivative of the bus voltage phase angle composed of a washout and an LPF [18].

$$\dot{x}_\theta(t) = \frac{1}{T_\mathrm{f}}\left(\frac{1}{\omega_b}(\theta_i(t) - \theta_o) - x_\theta(t)\right) \tag{6.11}$$

$$\dot{\tilde{\omega}}_i(t) = \frac{1}{T_\omega}(\omega_o + \Delta\omega_i(t) - \tilde{\omega}_i(t)) \tag{6.12}$$

where θ_o is the initial bus voltage phase angle, for example, the phase angle as obtained with the power flow analysis; ω_b is the system nominal frequency in rad/s; ω_o is the synchronous frequency in pu (typically, $\omega_o = 1$ pu); ω_b is the nominal frequency of the system in rad/s; T_f and T_ω are the time constants of the WF and of the LPF, respectively; x_θ is the state variable of the washout filter; and the frequency deviation $\Delta\omega_i = \dot{x}_\theta$.

If using the dq-frame rectangular coordinates, to compute the frequency variation, $\Delta\omega_i$, the bus voltage phase angle θ_i has to be defined first. For example, using the convention defined in [14], the link between dq quantities (Park vector) and phasors of the AC voltage at bus i are:

$$v_{i,\mathrm{dq}}(t)\exp(j\theta_{i,\mathrm{dq}}(t) - j\pi/2) = v_i(t)\exp(j\theta_i(t)), \tag{6.13}$$

or, equivalently:

$$\begin{aligned} v_{i,\mathrm{d}}(t) &= -v_i(t)\sin(\theta_i(t)), \\ v_{i,\mathrm{q}}(t) &= v_i(t)\cos(\theta_i(t)). \end{aligned} \tag{6.14}$$

Most books on power electronic converters, such as [22], use a different convention for the components of the Park vector that appear in Eq. (6.14). The reader more familiar with power electronics than electrical machines is advised that, throughout the book, d- and q-axis quantities are swapped concerning usual references utilized for the control of power electronic converters.

Instead of computing directly θ_i, which might lead to numerical issues, one can define two fictitious state variables, namely $s_{\theta,i} = \sin\theta_i$ and $c_{\theta,i} = \cos\theta_i$, whose dynamics are defined as follows [21]:

$$\dot{c}_{\theta,i}(t) = \frac{1}{T_\mathrm{f}}\left(\frac{v_{i,\mathrm{q}}(t)}{v_i(t)} - c_{\theta,i}(t)\right), \tag{6.15}$$

$$\dot{s}_{\theta,i}(t) = \frac{1}{T_\mathrm{f}}\left(\frac{v_{i,\mathrm{d}}(t)}{v_i(t)} + s_{\theta,i}(t)\right), \tag{6.16}$$

where $v_i = \sqrt{v_{i,\mathrm{d}}^2 + v_{i,\mathrm{q}}^2}$. Then, $\Delta\omega_i$ is obtained as:

$$\Delta\omega_i(t) = \begin{cases} \dfrac{\dot{s}_{\theta,i}(t)}{\omega_b c_{\theta,i}(t)}, & \text{if } |c_{\theta,i}(t)| > |s_{\theta,i}(t)|, \\ -\dfrac{\dot{c}_{\theta,i}(t)}{\omega_b s_{\theta,i}(t)}, & \text{otherwise.} \end{cases} \tag{6.17}$$

The washout block is necessary as the input quantity, that is, the bus voltage phase angle θ_i is an algebraic variable and thus can *jump* as a consequence of discrete events, such as faults and line outages. The discontinuity of the derivative of θ_i is the main issue of the WF. The LPF mitigates numerical issues but also introduces a delay that can be detrimental for the performance of local frequency controllers. A commonly accepted trade-off between accuracy and numerical efficiency is obtained with $T_f = 3/\omega_b$ s and $T_\omega = 0.05$ s.

The issues of the numerical differentiation of voltage angles are well-known. The literature on this subject has mainly focused on the definition of analytical expressions, for example, [23], or more accurate numerical methods, for example, [24], to define the derivative of the bus voltage angles.

6.3.2 Phase-locked loop

The main purpose of PLLs is the synchronization of power electronic converters to a three-phase AC grid. Since, in turn, a PLL is a closed-loop controller with the inclusion of a filter, its implementation is not unique. In the literature, the main focus so far has been to propose and test PLL designs that properly filter harmonics, compensate unbalanced conditions, and reduce the delay and the error with which the phase is tracked.

Recent publications have recognized the impact of PLLs in the regulation provided by nonsynchronous devices [25, 26], but also the potential instabilities that these devices can cause to electronic converters [27–29]. Ortega and Milano [16] show how the delays and fast-flux dynamics introduced by PLLs can affect the ability of nonsynchronous devices to properly regulate the frequency. On the other hand, the noise of frequency signals can be filtered and is, generally, less harmful [30].

A drawback of efficient noise filtering is the introduction of delays in the estimation of the phase and, consequently, of frequency deviations. With this regard, alternative solutions to PLLs for grid synchronization aimed at improving the compromise between speed and noise filtering have been proposed in the literature [31]. In this reference, the authors propose an algorithm in the discrete-domain based on the well-known Kalman filter. The main drawback of these solutions is their generally high computational burden, which justifies the current wide use of the *conventional* PLL. In this section, we compare a variety of continuous-domain PLL implementations to define the proper trade-off between accuracy and responsiveness [32]. With this aim, we need the *ideal* signal—that is, virtually free of noise and delay—provided by the FDF.

6.3.2.1 *Generalities*

The basic scheme of the fundamental-frequency model of a PLL is shown in Fig. 6.7. The main elements are as follows.

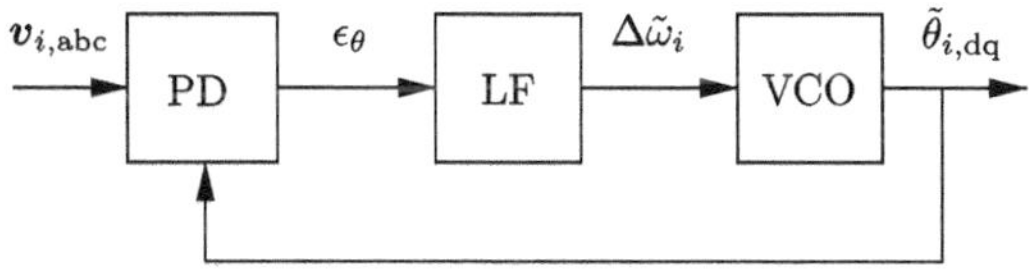

FIG. 6.7 Basic scheme of a PLL [32].

- The phase detector (PD) measures the three-phase voltage vector at the bus of connection i, $\mathbf{v}_{i,\mathrm{abc}}$, and calculates the phase angle difference, ϵ_θ, between the measured phase angle, θ_i, and the phase angle of the dq frame estimated by the PLL, $\tilde{\theta}_{i,\mathrm{dq}}$. The voltage is converted from "abc" into "$\alpha\beta\gamma$" first and then into "dqo" components [33].
- The loop filter (LF) filters the error ϵ_θ between the measured and estimated voltage phase angles. While there exist several configurations of the LF, they are generally based on a perfect tracking controller that, in steady state, imposes $\epsilon_\theta = 0$. Remarkably, the output of the LF is an estimation of the frequency deviation at the bus of connection, namely, $\Delta\tilde{\omega}_i$.
- The voltage controlled oscillator (VCO) takes the bus frequency deviation, $\Delta\tilde{\omega}_i$, and provides the estimation of the dq-frame phase angle $\tilde{\theta}_{i,\mathrm{dq}}$. The VCO typically consists of an integrator that, in steady state, leads to $\Delta\tilde{\omega}_i = 0$.

In transient stability models such as the ones presented in this book, $\mathbf{v}_{i,\mathrm{abc}}$ is not available, as the model consists in a balanced, fundamental frequency and quasisteady-state phasor representation of network branches and devices. The transformations operated within the PD are thus not explicitly implemented. Depending on the network model, however, either the components $v_{i,\mathrm{d}}$ and $v_{i,\mathrm{q}}$ or the polar representation in terms of voltage magnitude $v_i = |v_{i,\mathrm{d}} + jv_{i,\mathrm{q}}|$ and phase angle $\theta_i = \angle(v_{i,\mathrm{d}} + jv_{i,\mathrm{q}}) - \pi/2$ are available at the point of connection of the converter. In the following, polar coordinates for bus voltage phasors are used. But any component of the dq-frame representation of the grid voltage, namely $v_{i,\mathrm{d}}$ or $v_{i,\mathrm{q}}$, would work equally well.

A final remark is the following. For most PLL configurations, $\epsilon_\theta = 0$ in steady state, which, in turn, implies $v_{i,\mathrm{d}} = 0$. This allows decoupling, at least in stationary conditions, the control of the active and reactive powers of DERs, p_i and q_i, based on the quadrature and direct components of the AC-side current, $i_{i,\mathrm{q}}$ and $i_{i,\mathrm{d}}$, respectively [34].

6.3.2.2 PLL implementations

There are several PLL solutions specifically designed for power electronic converters. The synchronous reference frame PLL (SRF-PLL) is likely the simplest and the most commonly utilized scheme [8]. Other configurations are aimed at improving the SRF-PLL to reduce noise, distortions, and internal parameter uncertainties. Within this category, we cite the Lag PLL (Lag-PLL) [35]; the low-pass filter PLL (LPF-PLL) [36]; the enhanced PLL (E-PLL) [37]; and the second-order generalized integrator (SOGI) with frequency-locked loop (FLL) [38–40]. All these configurations are described below.

Synchronous reference frame PLL

Fig. 6.8 shows the commonly used SRF-PLL, where the PD includes a constant delay τ_θ (hence $\widetilde{\theta}_i(t) = \theta_i(t-\tau_\theta)$); and the LF consists of a proportional integral (PI) regulator.

The equations that represent the SRF-PLL shown in Fig. 6.8 are:

$$
\begin{aligned}
T_{2,\mathrm{LF}}\,\dot{x}_i(t) &= \epsilon_\theta(t) = \theta_i(t-\tau_\theta) - \widetilde{\theta}_{i,\mathrm{dq}}(t),\\
T_{1,\mathrm{LF}}\dot{x}_i(t) &= \Delta\widetilde{\omega}_i(t) - x_i(t),\\
\dot{\widetilde{\theta}}_{i,\mathrm{dq}}(t) &= K_{\mathrm{VCO}}\,\Delta\widetilde{\omega}_i(t),\\
0 &= \widetilde{\omega}_i(t) - (\omega^{\mathrm{ref}} + \Delta\widetilde{\omega}_i(t)),
\end{aligned}
\tag{6.18}
$$

where ω^{ref} can either be the synchronous frequency $\omega_o = 1$ pu, or the frequency of the CoI.

Eq. (6.13) can be thus rewritten taking into account the transient effect of the SRF-PLL, as follows:

$$
v_{i,\mathrm{d}}(t) + jv_{i,\mathrm{q}}(t) = v_i(t)\,(-\sin(\epsilon_\theta(t)) + j\cos(\epsilon_\theta(t))). \tag{6.19}
$$

Lag PLL

Fig. 6.9 shows a variation of the SRF-PLL presented earlier. In this configuration, referred to as Lag-PLL, an LPF is added in the LF prior to the PI regulator [35]. The aim of the LPF is to reduce the sensitivity of the PLL to noises and to prevent numerical issues when computing $\Delta\widetilde{\omega}_i$.

Low-pass filter PLL

Another possible implementation of the PLL is shown in Fig. 6.10, originally proposed in [36] and called LPF-PLL. Despite its name, the implementation of the LF proposed in [36] resembles more of a lead compensator. The purpose of

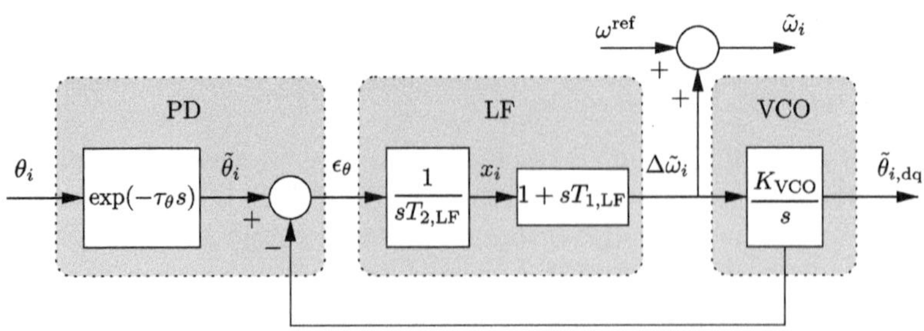

FIG. 6.8 Scheme of an SRF-PLL [32].

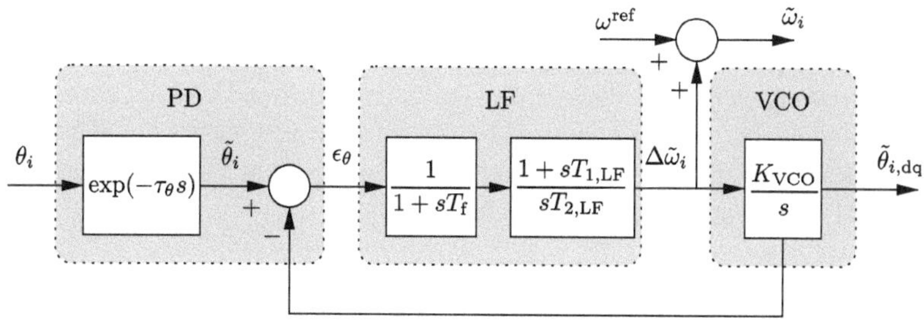

FIG. 6.9 Scheme of the Lag-PLL [32].

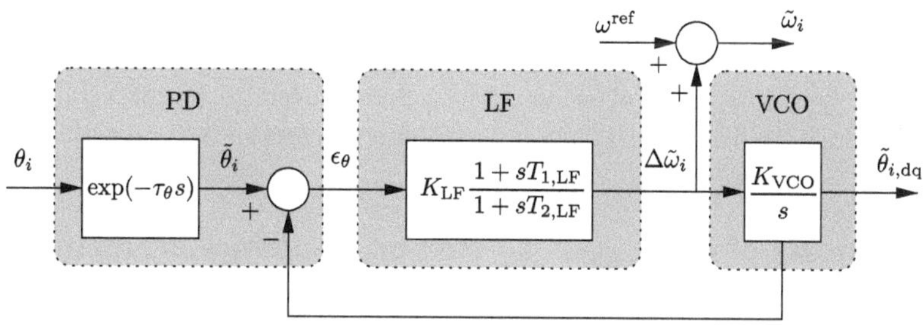

FIG. 6.10 Scheme of the LPF-PLL [32].

this block is to filter noises and possible numerical issues that may arise when computing $\Delta\widetilde{\omega}_i$, while preventing the addition of large delays in the process.

Enhanced PLL

Fig. 6.11 shows the single-phase implementation of the E-PLL that was originally proposed in [37]. The LF is composed of a PI regulator for the SRF-PLL.

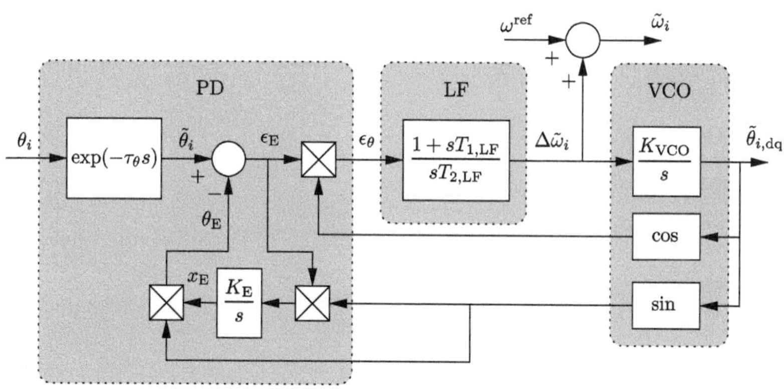

FIG. 6.11 Scheme of the E-PLL [32].

The main differences reside in the PD, as it is implemented as a combination of the pure delay τ_θ and an adaptive notch filter [41]. The E-PLL can be designed to be robust against noise, distortions, and uncertainties of internal parameters setting, and to be able to adaptively follow frequency variations. The main drawback of the E-PLL is its relatively slow response, since the estimation process takes more than one cycle.

Second-order generalized integrator FLL

The last PLL configuration discussed in this section is depicted in Fig. 6.12 [38]. This implementation consistently differs from the previous PLLs illustrated, as it includes an SOGI, that is, an adaptive filter structure whose input signal is the resonant frequency ω_S [39], and an FLL, which adapts ω_S [40]. In [38], the authors propose a cross-feedback, multiple SOGI-FLL tuned at different frequencies to estimate the sequence components of $\overline{v}_{i,\mathrm{dq}}$ under severe distortion conditions. The single SOGI-FLL configuration is considered in this book to estimate $\tilde{\omega}_i$.

6.4 Impact of noise and bad data on frequency estimation

This section compares the performance of the five PLL configurations discussed in Section 6.3.2, and bench-marked against the *ideal* frequency estimation of the FDF presented in Section 6.2.2. The well-known Western Systems Coordinating Council (WSCC) 9-bus system is considered for simulations. The system includes three synchronous machines, loads, and transformers, and six transmission lines. The machines include primary frequency and voltage regulation and an automatic generation control (AGC). Fig. 6.13 shows the scheme of the WSCC system. All static and dynamic data can be found in [14].

The case study considers two scenarios. In Section 6.4.1, a three-phase fault is simulated to study the accuracy of each PLL configuration following large

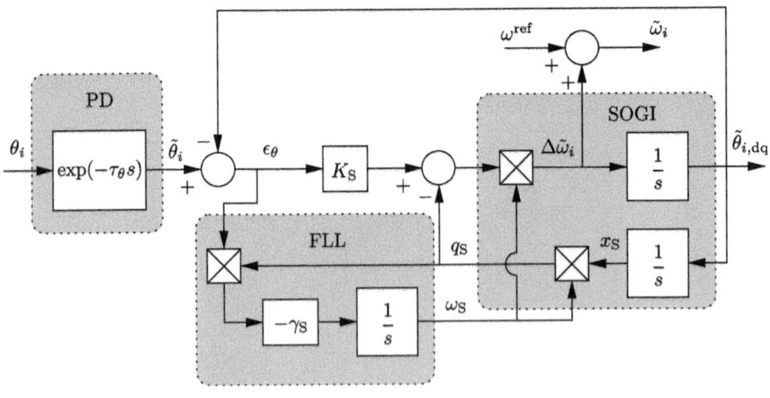

FIG. 6.12 Scheme of the SOGI-FLL [32].

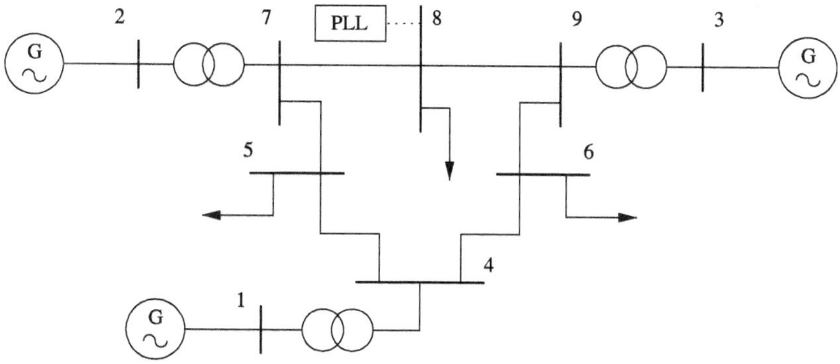

FIG. 6.13 WSCC 9-bus test system.

and fast frequency variations. Then, Section 6.4.2 discusses the sensitivity of the PLLs to noise in their input signal, that is, in the bus voltage angles.

All simulations and plots presented in this section were obtained using the software tool DOME [42] running on a 4 core 2.60 GHz Intel© Core i7 with 8 GB of RAM.

6.4.1 Three-phase fault

In this section, a three-phase fault is simulated on $t = 0.5$ s at bus 7. The fault is cleared 80 ms after its occurrence by opening the line connecting buses 7 and 5. The step of the time integration process is 1 ms, and every step is solved by using the dishonest Newton-Raphson method [3]. Finally, the integration method used is the implicit trapezoidal formula.

The parameters of the PLL schemes depicted in Figs. 6.8–6.12 are listed in Table 6.2. Such parameters have been tuned utilizing trial-and-error techniques to obtain the best performance for each configuration. The time delay τ_θ is zero for all configurations.

The frequency at the load bus 8 is estimated using each of the PLL configurations discussed in this chapter, and compared to the estimation provided by the FDF, and the trajectories are shown in Fig. 6.14. Besides, the absolute estimation errors ϵ_ω between each PLL estimation and the FDF signal are depicted in Fig. 6.15.

The most accurate frequency estimations are achieved with the LPF-PLL and the Lag-PLL, showing the latter the highest spikes of the two at the fault occurrence and the line opening. The SRF-PLL shows the lowest spikes during the two discontinuous events, but it also shows high ϵ_ω. Finally, the E-PLL and the SOGI-FLL show the worst performance overall, since their signals have the highest spikes (1.025 and 1.17 pu, respectively), and ϵ_ω during the first swings.

TABLE 6.2 Values of the parameters of the PLL controllers [32].

PLL	Parameter	Value	Unit
SRF-PLL	$T_{1,\,LF}$	2.0	s
	$T_{2,\,LF}$	20.0	s
	K_{VCO}	1.0	–
LPF	K_{LF}	1.0	–
	$T_{1,\,LF}$	0.01	s
	$T_{2,\,LF}$	0.5	s
	K_{VCO}	1.0	–
E-PLL	K_{E}	30.0	–
	$T_{1,\,LF}$	3.0	s
	$T_{2,\,LF}$	10.0	s
	K_{VCO}	1.0	–
Lag-PLL	T_{f}	0.01	s
	$T_{1,\,LF}$	5.0	s
	$T_{2,\,LF}$	10.0	s
	K_{VCO}	1.0	–
SOGI-FLL	K_{S}	2.0	–
	γ_{S}	0.5	–

6.4.2 Noise

PLLs are known to be sensitive to the noise of the measured signal, that is, the bus voltage angle. While the level of noise in transmission systems is generally small, this noise is of higher relevance in distribution systems due to the proximity of loads, unbalances, harmonics of power electronic converters, etc. This section considers a scenario where noise is applied to all bus voltage angles of the WSCC system. Such noise is modeled as an Ornstein-Uhlenbeck's process with Gaussian distribution [43].

The frequency at the load bus 8 is estimated utilizing each PLL configuration, and the trajectories are shown in Fig. 6.16, whereas the absolute frequency errors ϵ_{ω} are depicted in Fig. 6.17. Note that, for the short time simulated, the noise does not have an impact on the synchronous machine rotor speeds, and thus the frequency estimated by the FDF used to compute ϵ_{ω} is equal to 1 pu. The parameters of the different PLLs in Figs. 6.8–6.12 utilized to obtain the plots are the same as those listed in Table 6.2.

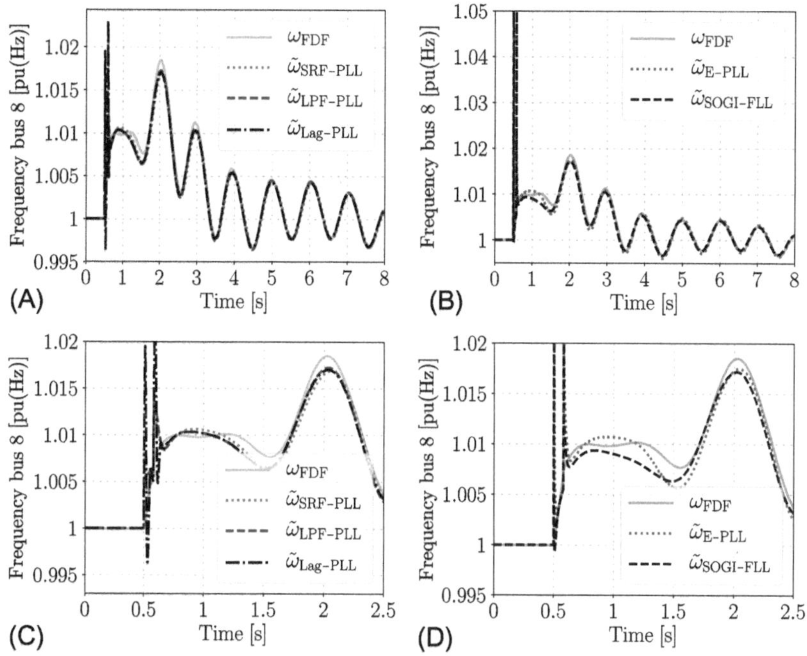

FIG. 6.14 Frequency of bus 8 measured with different PLLs configurations, and estimated with the FDF [32]. (A) SRF-PLL, LPF-PLL, and Lag-PLL; (B) E-PLL and SOGI-FLL; (C) SRF-PLL, LPF-PLL, and Lag-PLL: close-up; (D) E-PLL and SOGI-FLL: close-up.

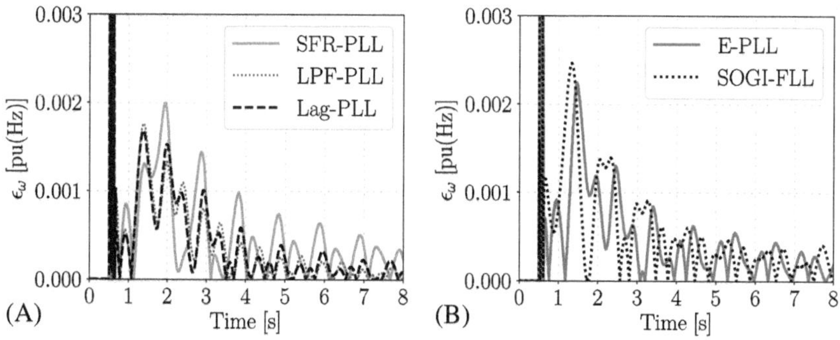

FIG. 6.15 Absolute frequency errors ϵ_ω between PLL measurements and the FDF estimation [32]. (A) SRF-PLL, LPF-PLL, and Lag-PLL; (B) E-PLL and SOGI-FLL.

The LPF-PLL is the least sensitive to the noise applied to the bus voltage angles, followed by the SRF-PLL. The Lag-PLL shows a smooth estimation, but the latency inserted by its LPF-PLL block leads to greater ϵ_ω. Finally, the E-PLL and the SOGI-FLL are the most sensitive to jitter, showing the latter the worst performance overall.

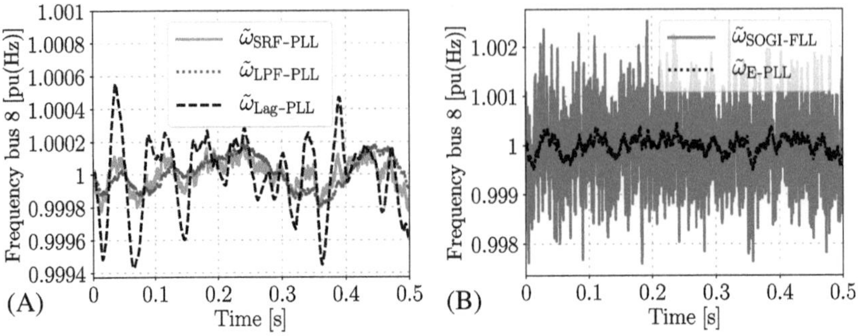

FIG. 6.16 Frequency estimated at bus 8 under the presence of noise in the PLLs input signal. (A) SRF-PLL, LPF-PLL, and Lag-PLL; (B) E-PLL and SOGI-FLL.

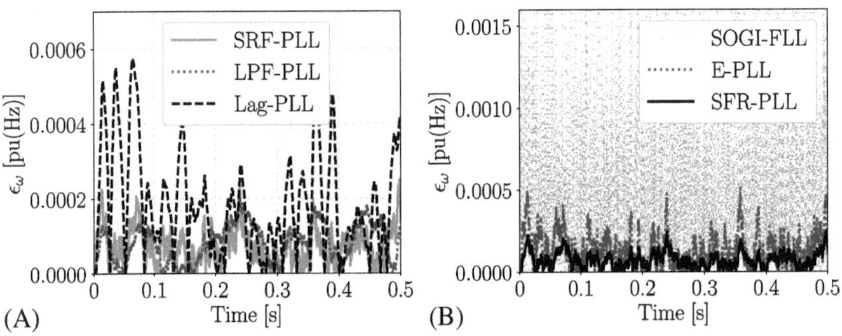

FIG. 6.17 Absolute error of the frequency estimated at bus 8 under the presence of noise in the PLLs input signal. (A) SRF-PLL, LPF-PLL, and Lag-PLL; (B) E-PLL, SOGI-FLL, and SRF-PLL.

6.4.3 Remarks

Simulation results allow concluding that the LPF-PLL has the best performance overall, as it provides the most accurate frequency estimation after fast and large frequency variations caused by contingencies such as faults and line outages, and it also shows the lowest sensitivity to noise present in the bus voltage angles. Good overall performance is also achieved by the commonly used SRF-PLL and the Lag-PLL. Finally, the worst accuracy, and the highest sensitivity to noise are observed from the E-PLL, and to a greater extent, from the SOGI-FLL.

References

[1] European Network of Transmission System Operators for Electricity (ENTSO-E), Commission regulation (EU) 2016/631 of 14 April 2016 establishing a network code on requirements for grid connection of generators, 2016.

[2] P. Kundur, Power System Stability and Control, McGraw-Hill, New York, 1994.

[3] F. Milano, Power System Modelling and Scripting, Springer, London, 2010.
[4] H.M. Oliveira, L.V. Melo, Huygens synchronization of two clocks, Sci. Rep. 5 (2015), Available at: www.nature.com/articles/srep11548.
[5] F.M. Mele, Á. Ortega, R. Zárate-Miñano, F. Milano, Impact of variability, uncertainty and frequency regulation on power system frequency distribution. in: 2016 Power Systems Computation Conference (PSCC), June, 2016, pp. 1–8, https://doi.org/10.1109/PSCC.2016.7540970.
[6] Á. Ortega, F. Milano, A. Musa, L. Toma, D. Preotescu, Definition of frequency under high dynamic conditions—deliverable 2.1, Tech. Rep., European Commission, 2017, www.re-serve.eu.
[7] F. Blaabjerg, R. Teodorescu, M. Liserre, A.V. Timbus, Overview of control and grid synchronization for distributed power generation systems. IEEE Trans. Ind. Electron. 53 (5) (2006) 1398–1409, https://doi.org/10.1109/TIE.2006.881997.
[8] A. Nicastri, A. Nagliero, Comparison and evaluation of the PLL techniques for the design of the grid-connected inverter systems. in: 2010 IEEE International Symposium on Industrial Electronics, July2010, pp. 3865–3870, https://doi.org/10.1109/ISIE.2010.5637778.
[9] F. Milano, Advances in Power System Modelling, Control and Stability Analysis, Energy Engineering, Institution of Engineering and Technology, Stevenage, UK, 2016.
[10] C. Roberts, E.M. Stewart, F. Milano, Validation of the Ornstein-Uhlenbeck process for load modeling based on μPMU measurements, in: 19th Power System Computation Conference (PSCC), Genoa, Italy, June, 2016, ISSN 2163-5137.
[11] L. Wang, J. Burgett, J. Zuo, C.C. Xu, B.J. Billian, R.W. Conners, Y. Liu, Frequency disturbance recorder design and developments, in: 2007 IEEE Power Engineering Society General Meeting, June, 2007, pp. 1–7.
[12] L. Zhan, Y. Liu, J. Culliss, J. Zhao, Y. Liu, Dynamic single-phase synchronized phase and frequency estimation at the distribution level, IEEE Trans. Smart Grid 6 (4) (2015) 2013–2022.
[13] A. Berizzi, The Italian 2003 blackout, in: IEEE Power Engineering Society General Meeting, 2004, June, vol. 2, 2004, pp. 1673–1679, https://doi.org/10.1109/PES.2004.1373159.
[14] P.W. Sauer, M.A. Pai, Power System Dynamics and Stability, Prentice Hall, Upper Saddle River, NJ, 1998.
[15] P.M. Anderson, A.A. Fouad, Power System Control and Stability, second ed., Wiley-IEEE Press, New York, NY, 2002.
[16] Á. Ortega, F. Milano, Comparison of bus frequency estimators for power system transient stability analysis, in: International Conference on Power System Technology (POWERCON), September, Wollongong, Australia, 2016.
[17] L. Wang, et al., Frequency disturbance recorder design and developments, in: IEEE PES General Meeting, 2007, pp. 1–7.
[18] F. Milano, Á. Ortega, Frequency divider, IEEE Trans. Power Syst. 32 (2) (2017) 1493–1501.
[19] A.M. Kettner, M. Paolone, On the properties of the power systems nodal admittance matrix, IEEE Trans. Power Syst. 33 (1) (2018) 1130–1131.
[20] IEEE Task Force on Load Representation for Dynamic Performance, Load representation for dynamic performance analysis [of power systems], IEEE Trans. Power Syst. 8 (2) (1993) 472–482.
[21] DIgSILENT, PowerFactory Technical Reference Ver. 15, Gomaringen, Germany, 2015.
[22] A. Yazdani, R. Iravani, Voltage-Sourced Converters in Power Systems. Modeling, Control and Applications, first ed., Wiley-IEEE Press, Hoboken, NJ, 2010.
[23] J. Nutaro, V. Protopopescu, Calculating frequency at loads in simulations of electromechanical transients, IEEE Trans. Smart Grid 3 (1) (2012) 233–240, https://doi.org/10.1109/TSG.2011.2173359.

[24] C.-S. Hsu, M.-S. Chen, W. Lee, Approach for bus frequency estimating in power system simulations, IEE Proc. Gener. Transm. Distrib. 145 (4) (1998) 431–435.

[25] S. Wang, J. Hu, X. Yuan, L. Sun, On inertial dynamics of virtual-synchronous-controlled DFIG-based wind turbines, IEEE Trans. Energy Convers. 30 (4) (2015) 1691–1702, https://doi.org/10.1109/TEC.2015.2460262.

[26] J. Hu, S. Wang, W. Tang, X. Xiong, Full-capacity wind turbine with inertial support by adjusting phase-locked loop response, IET Renew. Power Gener. 11 (1) (2017) 44–53, https://doi.org/10.1049/iet-rpg.2016.0155.

[27] Ö. Göksu, R. Teodorescu, C.L. Bak, F. Iov, P.C. Kjær, Instability of wind turbine converters during current injection to low voltage grid faults and PLL frequency based stability solution, IEEE Trans. Power Syst. 29 (4) (2014) 1683–1691, https://doi.org/10.1109/TPWRS.2013.2295261.

[28] P. Zhou, X. Yuan, J. Hu, Y. Huang, Stability of DC-link voltage as affected by phase locked loop in VSC when attached to weak grid, in: IEEE PES General Meeting, IEEE, 2014, pp. 1–5.

[29] F. Bizzarri, A. Brambilla, F. Milano, Analytic and numerical study of TCSC devices: unveiling the crucial role of phase-locked loops, IEEE Trans. Circuits Syst. I Regul. Pap. 65 (6) (2018) 1840–1849, https://doi.org/10.1109/TCSI.2017.2768220.

[30] K. Emami, T. Fernando, H.H.C. Iu, B.D. Nener, K.P. Wong, Application of unscented transform in frequency control of a complex power system using noisy PMU data, IEEE Trans. Ind. Inform. 12 (2) (2016) 853–863, https://doi.org/10.1109/TII.2015.2491222.

[31] X. Cai, C. Wang, R. Kennel, A fast and precise grid synchronization method based on fixed-gain filter, IEEE Trans. Ind. Electron. 65 (9) (2018) 7119–7128, https://doi.org/10.1109/TIE.2018.2798600.

[32] Á. Ortega, F. Milano, Comparison of different PLL implementations for frequency estimation and control, in: Proceedings of the 18th International Conference on Harmonics and Quality of Power (ICHQP), 2018, pp. 1–6.

[33] W.C. Duesterhoeft, M.W. Schulz, E. Clarke, Determination of instantaneous currents and voltages by means of alpha, beta, and zero components, Trans. Am. Inst. Electr. Eng. 70 (2) (1951) 1248–1255.

[34] F. Milano, Á. Ortega, Converter-Interfaced Energy Storage Systems—Context, Modelling and Dynamic Analysis, Cambridge University Press, Cambridge, UK, 2019.

[35] A. Cataliotti, V. Cosentino, S. Nuccio, A phase-locked loop for the synchronization of power quality instruments in the presence of stationary and transient disturbance, IEEE Trans. Instrum. Meas. 56 (6) (2007) 2232–2239, https://doi.org/10.1109/TIM.2007.908350.

[36] G.-C. Hsieh, J.C. Hung, Phase-locked loop techniques—a survey, IEEE Trans. Ind. Electron. 43 (6) (1996) 609–615, https://doi.org/10.1109/41.544547.

[37] M. Karimi-Ghartemani, M.R. Iravani, Robust and frequency-adaptive measurement of peak value, IEEE Trans. Power Deliv. 19 (2) (2004) 481–489, https://doi.org/10.1109/TPWRD.2004.824764.

[38] P. Rodríguez, A. Luna, I. Candela, R. Mujal, R. Teodorescu, F. Blaabjerg, Multiresonant frequency-locked loop for grid synchronization of power converters under distorted grid conditions, IEEE Trans. Ind. Electron. 58 (1) (2011) 127–138, https://doi.org/10.1109/TIE.2010.2042420.

[39] M. Ciobotaru, R. Teodorescu, F. Blaabjerg, A new single-phase PLL structure based on second order generalized integrator. in: Proceedings of the 37th IEEE Power Electronics Specialists Conference, June, Jeju, South Korea, 2006, pp. 1–6, https://doi.org/10.1109/pesc.2006.1711988.

[40] P. Rodríguez, A. Luna, M. Ciobotaru, R. Teodorescu, F. Blaabjerg, Advanced grid synchronization system for power converters under unbalanced and distorted operating conditions, in: Proceedings of the 32nd Annual Conference on IEEE Industrial Electronics (IECON), November, Paris, France, 2006, pp. 5173–5178, https://doi.org/10.1109/IECON.2006.347807.
[41] M. Mojiri, A.R. Bakhshai, An adaptive notch filter for frequency estimation of a periodic signal, IEEE Trans. Autom. Control 49 (2) (2004) 314–318, https://doi.org/10.1109/TAC.2003.821414.
[42] F. Milano, A Python-based software tool for power system analysis, in: Proceedings of the IEEE PES General Meeting, Vancouver, BC, 2013.
[43] F. Milano, R. Zárate-Miñano, A systematic method to model power systems as stochastic differential algebraic equations, IEEE Trans. Power Syst. 28 (4) (2013) 4537–4544.

Chapter 7

Architectures for frequency control in modern power systems

Álvaro Ortega Manjavacas[a], Mohammed Ahsan Adib Murad[b], Junru Chen[c], Muyang Liu[c], Terence O'Donnell[b], and Federico Milano[b]

[a]Instituto de Investigación Tecnológica, Escuela Técnica Superior de Ingeniería ICAI, Universidad Pontificia Comillas, Madrid, Spain, [b]School of Electrical and Electronic Engineering, University College Dublin, Dublin, Ireland, [c]School of Electrical Engineering, Xinjiang University, Urumqi, China

7.1 Introduction

Traditionally, synchronous machines were the main devices apt to provide primary frequency control (PFC) in AC transmission grids. This situation is rapidly changing due to the increasing penetration of distributed, nonsynchronous converter-interfaced generation (CIG) based on renewable energy sources (RESs), such as wind and photo-voltaic power plants, as well as other emerging devices, such as converter-interfaced energy storage systems (ESSs). These devices, which generally are connected to the grid through voltage-sourced converters (VSCs), reduce the overall system inertia and increase the risk of frequency and voltage instabilities. This fact has led, in recent years, to the development of a large variety of frequency regulation strategies for RESs [1–3].

RESs, if operated with an adequate power reserve, can respond to rate of change of frequency (RoCoF) and frequency variations [4, 5]. Virtual power plants are an example of combining devices together with different control purposes [6–8]. However, it is also recognized that, due to their intermittent nature, most RESs have a limited frequency control capability.

This is one of the main tasks of ESSs: to provide low-inertia power systems with the ability of maintaining the power balance during a transient in the interval of time between the end of the response of the transmission line dynamics (a few milliseconds) and before the beginning of the response of PFC (a few seconds). Depending on the response times and capacities of the storage

Converter-Based Dynamics and Control of Modern Power Systems
https://doi.org/10.1016/B978-0-12-818491-2.00007-9

technology, ESSs can provide either RoCoF control or frequency control, or both at the same time.

Fast frequency control of distributed energy resources (DERs) such as CIG and ESSs is conceptually similar to the PFC of synchronous machines, that is, droop control. However, while the turbine governors of the synchronous machines require several seconds to start regulating, converter-interfaced devices can respond to frequency deviations much faster. Depending on the technology, the response time of converter-interfaced droop control can span from a few tens of milliseconds to several hundred milliseconds. In this time scale, the frequency is not the best quantity to be regulated due to its relatively slow rate of change. In fact, to properly cope with frequency deviations shortly after a contingency, relatively high gains of the control are usually required, which may compromise the performance of the DER.

The frequency control techniques applied to the RESs and ESSs discussed above all assume the conventional grid-feeding converter configuration. In the grid-feeding configuration, the converters feed current to the grid to maximize their power output, under the assumption that the grid voltage is being established externally (by, for example, the synchronous generators [SGs]). With increasing RESs and decreasing SG, the RESs must move from being grid-feeding current sources to grid-forming voltage sources with the ability to independently establish the voltage and provide support serves such as emulated inertia, frequency, and voltage support. The virtual synchronous generator (VSG) control has been proposed to control RESs and ESSs grid interfacing converters to mimic the dynamics of the SG for the purpose of presenting similar stability performance and actively establishing the voltage in the system [9–12].

Less conventional is the frequency control through devices that regulate the voltage and are thus sources of reactive power rather than active one. The rationale behind the voltage-based frequency control (VFC) is that voltage-dependent loads can vary their active power consumption if the voltage at their point of connection is varied. While the voltage cannot vary too much, still the power variations that can be obtained with VFC are nonnegligible and can be part of the solution of the frequency control in systems with high penetration of renewables. The key point is that the VFC has to be fast and thus under-load tap changers are excluded. On the other hand, flexible AC transmission system (FACTS) devices, such as static VAR compensators (SVCs), STATCOMs, and, more recently, solid-state transformers, are ideal candidates for the provision of VFC.

The use of power-electronics-based smart transformers (STs) [13, 14] placed at the interface between the transmission and distribution systems is another approach that can be used to control the demand in the distribution side and thus to also support the frequency. The ST also has the advantage of being able to support the voltage in the transmission side via reactive power compensation. This can be achieved since the secondary-side voltage and primary-side reactive power regulation of STs can be fully independent.

In this chapter, the effect of the provision of RoCoF and PFC by nonsynchronous devices is studied. Wind energy conversion systems (WECSs) and solar photo-voltaic generation (SPVG) are discussed in Section 7.2. Section 7.3 discusses the fast frequency response of ESSs as well as of a particular kind of grid forming devices, namely the VSG that includes both an RES and an ESS. These sections also provide a comparison of the performance of the frequency controllers of WECSs, SPVG, ESSs, and VSGs when their input signals are generated by the frequency estimation techniques presented in Chapter 6. Finally, the provision of frequency control from devices whose primary purpose is the regulation of the voltage at their point of connection with the grid is discussed in Section 7.4. With this aim, the SVC and the ST are considered.

7.2 Frequency control through converter-interfaced generation

This section deals with the frequency control of CIG and consists of two parts: (i) Section 7.2.1 that focuses on wind power plants and (ii) Section 7.2.2 that focuses on solar photo-voltaic power plants.

7.2.1 Wind power plants

In recent years, the rapid growth of the penetration of wind power has led grid operators to face new challenges related to the control and the stability of power systems. Unlike power plants based on SGs, in fact, wind turbines are equipped with limited or null primary regulation capability. In particular, frequency regulation is often not available as wind turbines are operated to produce the maximum power according to wind speed conditions through the Maximum Power Point Tracking (MPPT) control. Moreover, the penetration of wind power plants reduces the overall inertia of the system as they are often based on non-synchronous machines and connected to the grid though power electronic devices. This situation is clearly prone to increase frequency and voltage variations following large disturbances and the risk of the occurrence of unstable conditions [15]. To cope with this emerging issue, new rules have been introduced in several grid codes aimed at defining primary and secondary frequency regulation for wind turbines [16].

Several proposals of frequency control strategies for WECSs can be found in the literature. In most cases, such controllers measure either the frequency deviation or the RoCoF and then accordingly modify the output of the MPPT control. The devices based on the frequency deviation are basically droop controllers commonly used on synchronous machines. The main difference is that they do not measure the speed of the rotor of the machine but the frequency of the grid at the point of connection of the wind power plant. Due to the limited inertia of wind turbines, such controllers can only provide a transient effect, whereas the long-term regulation is obtained through the synchronous machines [17]. The proposal of an additional power signal based on the RoCoF or the frequency deviation to simulate the inertial response of synchronous machines has been extensively explored [16, 18–21].

The droop and RoCoF controllers work similarly and are typically coupled to the MPPT control of the wind turbine. For this reason, they are described together later.

The droop controller, as the name suggests, is comparable to the PFC of synchronous machines and is also called proportional controller. As shown in Fig. 7.1, the droop controller consists of a constant gain $1/\mathcal{R}_w$. As discussed in [17], nonconventional generators cannot contribute permanently with extra real-power production, thus the need for a dead-band block, which inhibits steady-state frequency errors below a given threshold from the controller and helps preventing the machine from stalling.

Fig. 7.1 also shows the RoCoF controller. This is composed of a low-pass filter, with time constant $T_{2,\mathrm{w}}$, and the time derivative of frequency measurement, followed by an inertial factor, K_{w}. The low-pass filter is not only needed to reduce noise and numerical errors due to the derivative of a signal but also to avoid stress of other parts of the WECS, caused by an abrupt signal from the inertial controller.

The resulting frequency control signal is the sum of the output of the droop and inertial controllers, which is then added to the active power reference obtained from the MPPT. Each regulating signal is expected to play a different role for different time scales following a contingency. The RoCoF control is more relevant in the first instants after the occurrence of a contingency due to its sensitivity to the rate of change of the frequency, while the frequency deviation is more effective to mitigate the frequency nadir. Hence, one can expect that the effects of the two controllers are complementary.

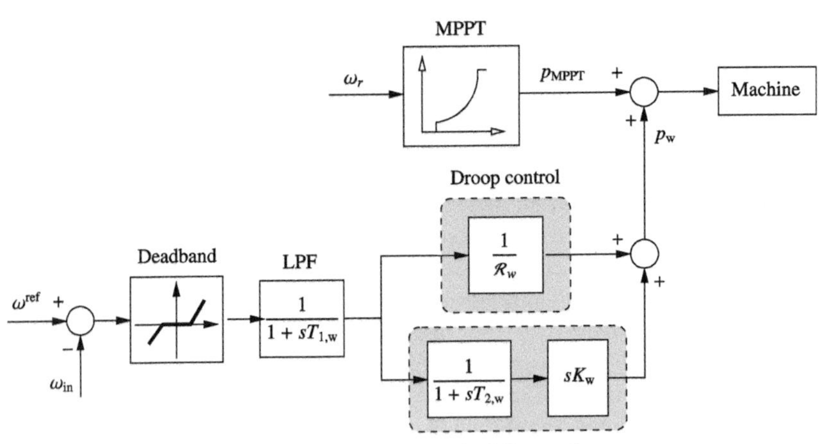

FIG. 7.1 Scheme of the droop and RoCoF controllers coupled to the MPPT [22].

7.2.1.1 Example

In this example, the performance of the PFC provided by a WECS when its input signal is provided by a phase-locked loop (PLL) or by the frequency estimation approaches based on the center of inertia (CoI) and the frequency divider formula (FDF) is compared (see Chapter 6 for details). The case study is based on the well-known WSCC 9-bus, 3-machine test system, described in [23] (see Fig. 7.13).

In this scenario, the synchronous machine at bus 3 has been replaced with a wind power plant of the same capacity, composed of 50 variable-speed wind turbines modeled with a fifth-order doubly-fed induction generator (DFIG) model [24]. The stochastic process applied to the wind follows a Weibull distribution [25]. The contingency is the outage of the line connecting buses 5 and 7 at $t = 30$ s. The values of the parameters of the WECS are based on [26], whereas the values of the parameters of the controller depicted in Fig. 7.1 are provided in Table 7.1.

Noise is modeled as an Ornstein-Uhlenbeck process with Gaussian distribution and is applied to the magnitudes and angles of all bus voltage phasors of the system. The interested reader can find in [27] a detailed description of the modeling and implementation of stochastic processes applied to bus voltage phasors.

The effect of the low-pass filter of the WECS frequency control in Fig. 7.1 is studied. With this aim, the response of the WECS is analyzed for the case when the filter time constant is $T_{1,\mathrm{w}} = 0.5$ s and for the case when such a filter is disabled, that is, $T_{1,\mathrm{w}} = 0$ s. The active power supplied by the WECS during the transient caused by the line outage and the rotor speed of the synchronous machine at bus 2, ω_{G2} are depicted in Figs. 7.2 and 7.3, respectively.

Comparing these figures, it appears that the inclusion of the filter highly reduces the oscillations of the active power output of the WECS, thus reducing the stress of the device, for both the FDF and the PLL signals, whose dynamic behavior is very similar, whereas the CoI shows little sensitivity to the value of

TABLE 7.1 Values of the parameters of the WECS controller [22].

Parameter	Value	Unit
K_{w}	80.0	–
$\mathcal{R}_{\mathrm{w}}$	0.05	–
$T_{1,\mathrm{w}}$	0.5	s
$T_{2,\mathrm{w}}$	4.0	s

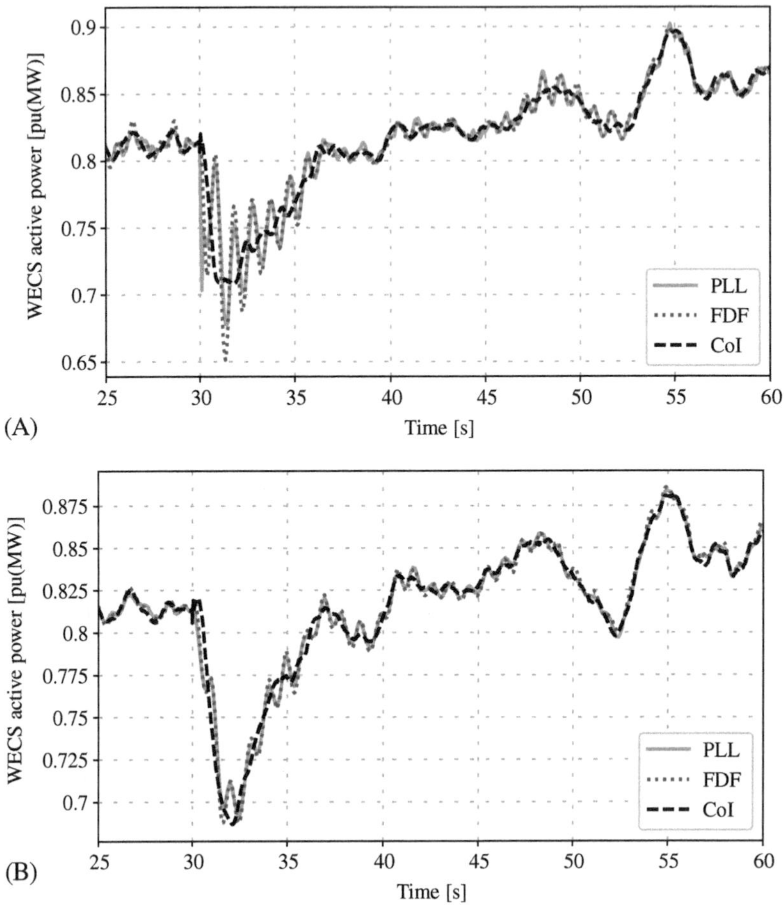

FIG. 7.2 Active power supplied by the wind power plant [28]. (A) $T_{1,\,w} = 0.5$ s; (B) $T_{1,\,w} = 0$ s.

$T_{1,w}$. However, while without the filter the oscillations of ω_{G2} are damped after about 10 s (FDF) and 25 s (PLL), these oscillations last over 30 s when the low-pass filter is included.

There is thus a trade-off between reducing the damping of the rotor speeds of the synchronous machines and the stress of the WECS. However, if fast dynamics such as the fluxes of the synchronous machines are included, the relevance of the low-pass filter becomes apparent, as shown in Fig. 7.4, where the synchronous machines are modeled using the fully fledged, eighth-order model. If the filter is disabled, the power drops abruptly to about 0.35 pu when the PLL signal is used. On the other hand, when the FDF is used, the power oscillations are similar to the previous case when sixth-order synchronous machine models were used (Fig. 7.2).

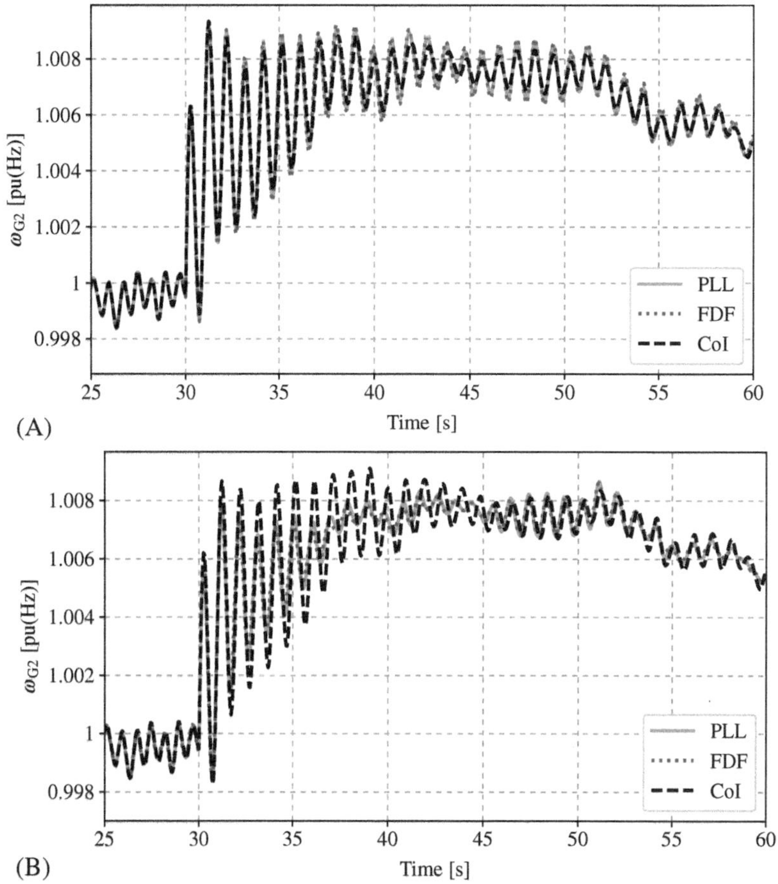

FIG. 7.3 Rotor speed of the synchronous machine at bus 2 [28]. (A) $T_{1,\,w} = 0.5$ s; (B) $T_{1,\,w} = 0$ s.

These results suggest that fast dynamics can negatively affect the response of controllers based on local frequency measures and that an appropriate filtering should be implemented if PLL devices are to be utilized. On the other hand, filtering should not introduce a delay in the frequency measure to prevent a possible deterioration in the dynamic response.

7.2.1.2 Remarks

Based on the simulation results, the following remarks are relevant.

- The FDF provides the *ideal* value of the local frequency at buses and can thus be utilized as a reference for testing the quality of the frequency estimated by the PLL and other estimation approaches.

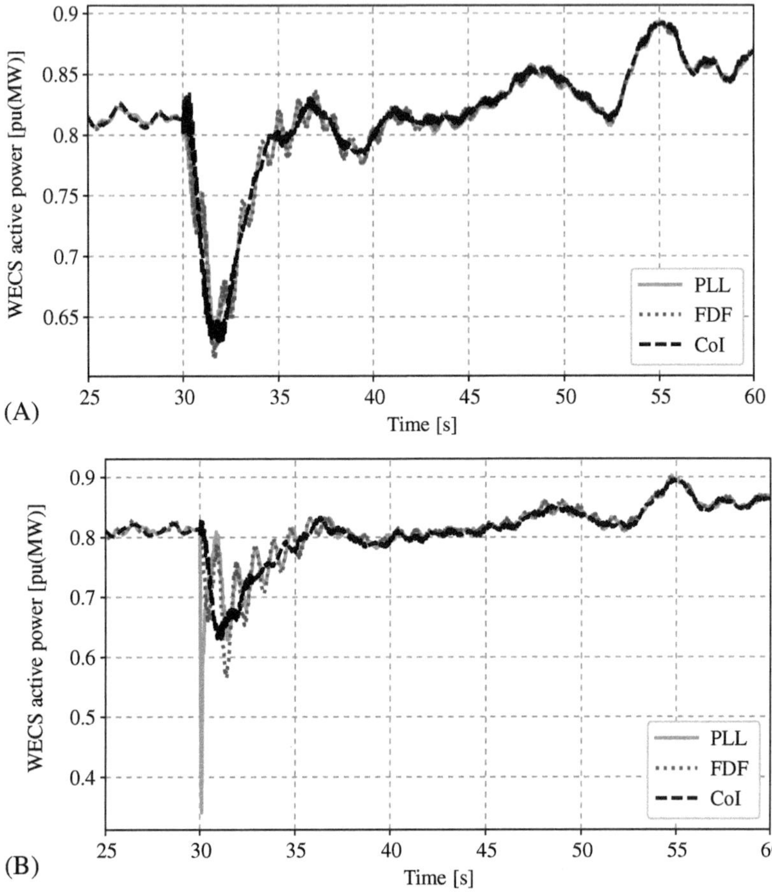

FIG. 7.4 Active power supplied by the wind power plant. Synchronous machines are modeled using the eighth-order model [28]. (A) $T_{1,w} = 0.5$ s; (B) $T_{1,w} = 0.0$ s.

- A standard synchronous reference frame model of the PLL works reasonably well compared to the FDF. Noise and numerical spikes do not deteriorate significantly the quality of the control, provided that WECSs include a proper low-pass filter within their primary frequency controllers. Fast dynamics of fluxes, however, can deteriorate the dynamic response of PLL-based frequency controllers.
- The CoI signal is inadequate to simulate the behavior of WECSs frequency controllers, although its average nature often leads to an overall smoother frequency response. This consideration could be further developed in the future considering coordinated area controllers sharing an average value of the frequency signal rather than utilizing a local one.

7.2.2 Solar photo-voltaic power plants

Concurrently with WECSs, the penetration of SPVG in power systems is experiencing a significant growth in recent years. SPVG shares with WECSs the fact that they are connected to the grid by means of power electronic converters, and thus, their integration into the power system implies a decrease of the overall inertia of the network, which leads to a poorer frequency response against contingencies. Therefore, to overcome potential system frequency and/or voltage stability issues due to large disturbances, it is important that SPVG provides both frequency and voltage regulation [29].

The control of the frequency and the voltage at the bus of connection of the SPVG with the grid is performed similarly to [29–31], and the scheme of the controllers and the converter is depicted in Fig. 7.5. Due to the resemblance between the connection of SPVG and WECSs with the rest of the grid, their controllers also show relevant similarities. With regard to the frequency regulation, a droop control is implemented. The droop control is composed of a droop gain and a low-pass filter and takes the deviation of the frequency at the bus of connection with respect to a given reference. The output signal is then added to the reference power provided by the MPPT, and processed by a PI regulator, which generates the reference current input signal of the SPVG converter.

7.2.2.1 Example

This example studies the capability of SPVG to provide primary frequency regulation through simulations. With this aim, a similar case study to that presented in Section 7.2.1.1 is conducted. The case study is thus based on the well-known WSCC 9-bus, three-machine test system described in [23] (see Fig. 7.13) and compares the SPVG PFC for different input signals, namely, PLL, CoI, and FDF.

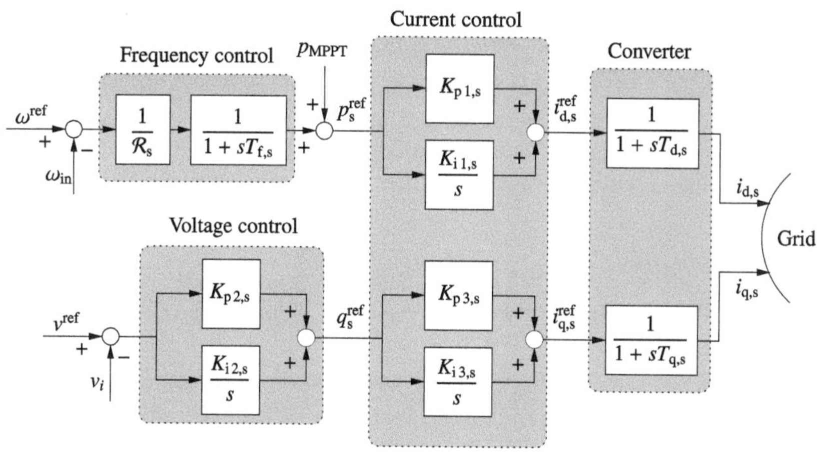

FIG. 7.5 Scheme of the frequency and voltage control of SPVG [28].

TABLE 7.2 Values of the parameters of the SPVG frequency controller [31].

Parameter	Value	Unit
$K_{i\ 1,s}$	5.0	–
$K_{p\ 1,s}$	10.0	–
R_s	0.05	–
$T_{d,\ s}$	0.015	s
$T_{f,\ s}$	1.0	s

Noise modeled as Ornstein-Uhlenbeck process is also applied to bus voltage magnitudes and angles in this case study [27]. The synchronous machine at bus 3 has been replaced in this case study with an SPVG of the same capacity. Given the short time scales of the simulations, the solar radiation, and consequently, the active power reference provided by the MPPT, p_{MPPT}, are assumed to be constant. The contingency is the outage of the line connecting buses 5 and 7 at $t = 30$ s. The values of the parameters of the frequency control loop depicted in Fig. 7.5 are provided in Table 7.2 [31].

Fig. 7.6A and B shows the input signal of the SPVG frequency controller and the SPVG active power output, respectively, for the three frequency estimations, namely, CoI, PLL, and FDF. The overall trends of the three signals are very similar, leading to small differences of the SPVG active power output. However, the FDF shows relatively high-amplitude fluctuations due to the local oscillations of the synchronous machines, which are averaged out from the CoI. On the other hand, the PLL shows high sensitivity to the bus voltage noises.

Fig. 7.6C compares the frequency of the CoI of the system without the frequency control of the SPVG and with the controller for the three frequency estimation signals. The controller reduces the frequency zenith by about 70% for the three cases, without significant differences between them. The high-frequency oscillations that are present in the PLL and FDF signals in Fig. 7.6A are filtered out by the low-pass filter of the SPVG frequency control.

7.2.2.2 Remarks

Remarks on frequency control provided by SPVG are listed in the following, based on the simulation results.

- The frequency control provided by SPVGs can considerably improve the response of the system in case of over frequencies, regardless of the frequency estimation technique used as input signal of the controller.
- The PLL shows a good accuracy and provides a good performance of the frequency controller, provided that the signal is passed through a low-pass

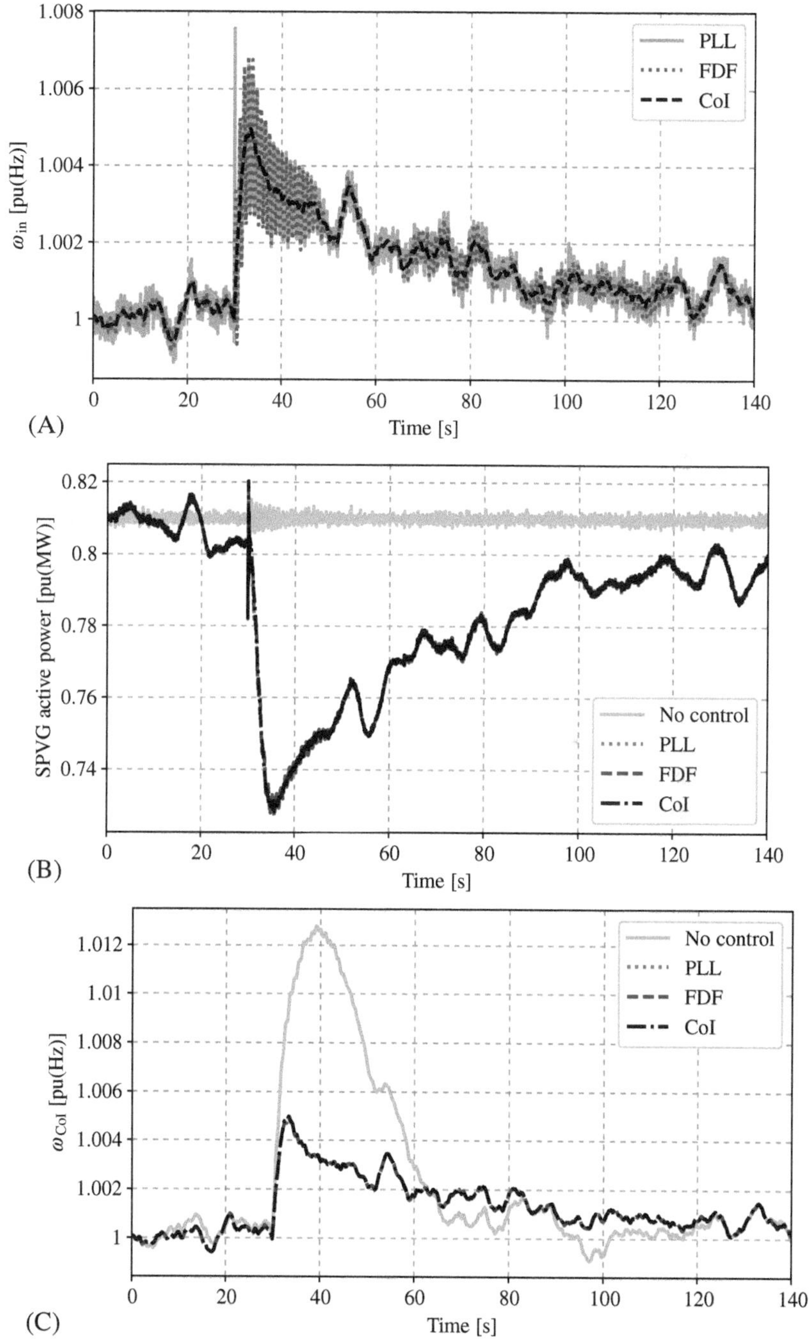

FIG. 7.6 WSCC system with SPVG coupled to a PFC system following a line outage [28]. (A) Input frequency signal of the SPVG control; (B) active power supplied by the SPVG; (C) frequency of the CoI.

filter to reduce the inherent sensitivity of the PLL to noises and numerical problems due to discontinuities.

- The performance of the SPVG frequency controller is not deteriorated when the CoI signal is used as the frequency estimation. This is due to the fact that the average nature of the CoI naturally imitates the effect of the low-pass filter in the FDF and the PLL signals, leading to similar output signals of the SPVG frequency control, and thus, to similar active power outputs of the SPVG.

7.3 Frequency control through energy storage systems

Among all currently existing alternatives to improve the performance, reliability, and resiliency of power systems with high RES penetration, converter-interfaced ESSs are one of the most promising [32–34]. ESSs have the potential to provide a large variety of ancillary services to the system thanks to their capability to supply/absorb active and reactive powers. These services include flattening of the power provided by RESs, active power regulation in transmission lines, local and/or global frequency regulation, RoCoF mitigation, and local voltage regulation. These features of ESSs have led, in recent years, to a huge investment in the research, development, prototyping, and installation of a large variety of technologies. Among all ESS technologies that are currently most promising, there are battery energy storage (BES), compressed air energy storage (CAES), fuel cell energy storage (FCES), flywheel energy storage (FES), pumped hydro energy storage (PHES), super capacitor energy storage (SCES), and superconducting magnetic energy storage (SMES) [35].

7.3.1 Energy storage systems

A general scheme of a converter-interfaced ESS connected to a power system is shown in Fig. 7.7. The main objective of the depicted ESS is to regulate an either measured or estimated frequency, for example, the frequency at the point of common coupling (PCC) with the grid, or the frequency of the CoI. While the storage device is responsible of the active power support, the VSC provides reactive power support by regulating the AC voltage at the bus of connection, and it links the storage device with the grid.

The charge/discharge process of the storage device is regulated by the storage control (see Fig. 7.8). The input signal of the control is the error between the measured/estimated frequency, $\omega_{\rm in}$, and a reference value ($\omega^{\rm ref}$). If $\omega_{\rm in} = \omega^{\rm ref}$, the storage device is inactive and its stored energy is thus kept constant. For $\omega_{\rm in} \neq \omega^{\rm ref}$, the storage device injects active power into the AC bus through the VSC (discharge process) or absorbs power from the AC bus (charge process). The control scheme can also include an additional channel to simultaneously provide droop and RoCoF regulation, similarly to the WECS scheme discussed in Section 7.2.1 [36].

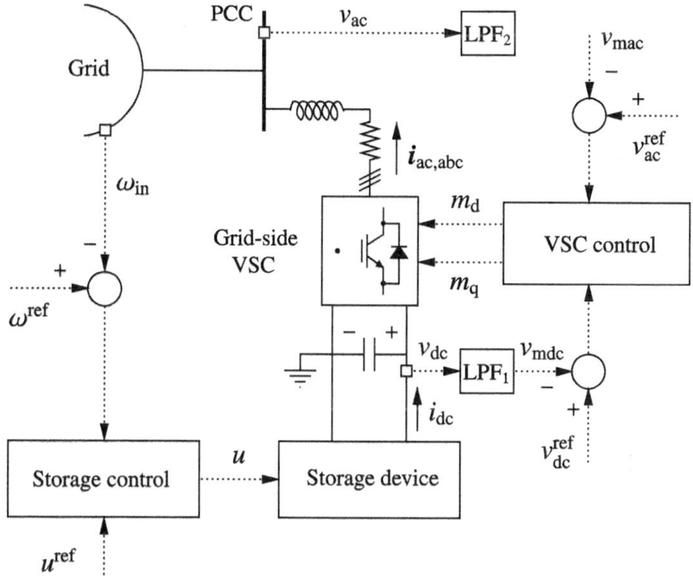

FIG. 7.7 Scheme of the ESS connected to a grid [28].

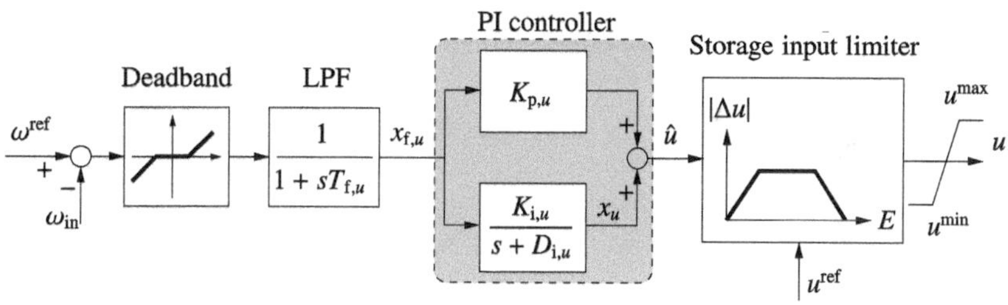

FIG. 7.8 Storage control scheme [28].

The typical configuration of the PI-based controller includes deadband and low-pass filter (LPF) blocks that are responsible for reducing the sensitivity of the storage control to small, high-frequency perturbations such as noises. The aim of these blocks is to reduce the number of charge/discharge operations, thus increasing the life of the ESS [35, 37]. The PI regulator is composed of a proportional gain, $K_{p,u}$, and an integrator with gain $K_{i,u}$ and integral deviation coefficient $D_{i,u}$. These parameters are commonly tuned by trial-and-error or pole-placement techniques. The simplicity of the implementation and design and the mass utilization of this controller in industrial applications are its main strengths. Note also that the structure of the PI control does not depend on the energy storage technology considered. The scheme shown in Fig. 7.8 also includes a block referred to as storage input limiter (SIL) [38]. The purpose of the SIL is to reduce the impact of energy saturation of the storage device on system transients. This block takes the actual value of the energy stored in the device, E, and regulates accordingly the input controlled variable of the storage device, u.

7.3.1.1 Examples

This section considers the IEEE 14-bus test system for the simulations (see Fig. 7.9). This benchmark network consists of 2 synchronous machines and 3 synchronous compensators, 2 two-winding and 1 three-winding transformers, and 15 transmission lines and 11 loads. The system also includes automatic voltage regulations (AVRs), turbine governors (TGs), and an automatic generation control (AGC). All dynamic data of the IEEE 14-bus system as well as a detailed discussion of its transient behavior can be found in [39].

Some modifications have been made to this network to study the interaction of the storage device with the rest of the system:

- The capacity of the SG placed in bus 1 is reduced by 5 times its original value.
- A 30 MW, 70 MVAr ESS is connected to bus 4.

Two scenarios have been considered to study the performance of the ESS: (i) a three-phase fault followed by a line outage and (ii) a wind power plant is included, and stochastic variations of the wind are considered.

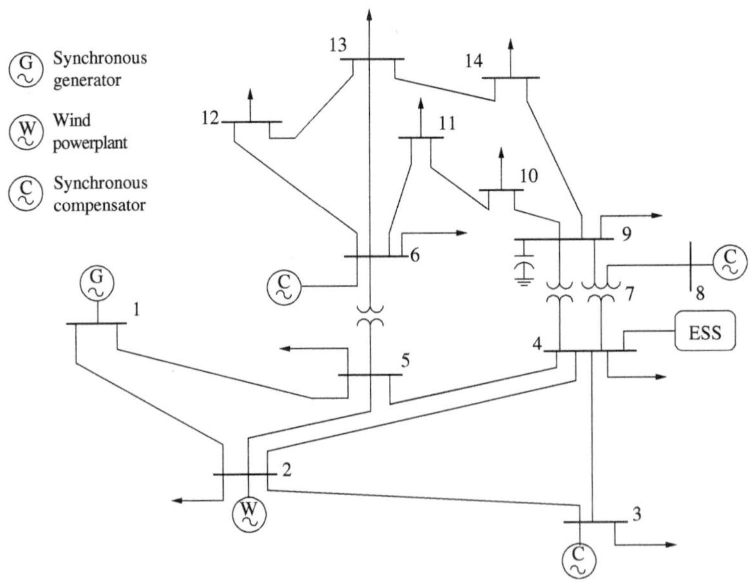

FIG. 7.9 IEEE 14-bus test system with an ESS device connected to bus 4.

Three-phase fault and line outage

In this section, the contingency is a three-phase fault occurring at bus 5 at $t = 1$ s and cleared after 70 ms by opening the line that connects buses 5 and 1. Fig. 7.10 shows the rotor speed of the synchronous machine at bus 2 for the cases without the ESS and with the ESS regulating the frequency of the CoI and of the bus of connection estimated by means of the FDF and a PLL.

The ESS can reduce the frequency deviation in the first swing by about 50%, and allows the system to reach steady state after about 5 s, as opposed to the required $\gtrsim$15 s without the ESS. On the other hand, a greater frequency nadir is present in the cases with the ESS due to current saturation of the storage device. The frequency nadir is about the half of the amplitude of the frequency variation of the first swing without the ESS.

The frequency at bus 4, that is, the bus of connection of the ESS with the rest of the system, is represented in Fig. 7.11.

From observing Figs. 7.10 and 7.11, it is worth noticing that the response of the ESS is virtually the same regardless of the frequency estimation technique used. However, the signals provided by each technique are considerably different, as shown in Fig. 7.12, where the input signal of the ESS control, ω_{in}, is represented.

As expected, the CoI filters local frequency oscillations, while the PLL includes spikes in the signal that span from 0.96 to 1.01 pu. The similarity of the ESS responses despite the differences in the control input signal is due to the presence of the low-pass filter with time constant $T_{f,\ u}$ (see Fig. 7.8). This filter removes to a large extent the amplitude of the PLL spikes, as well as the local frequency oscillations, leading to a similar output signal of the controller,

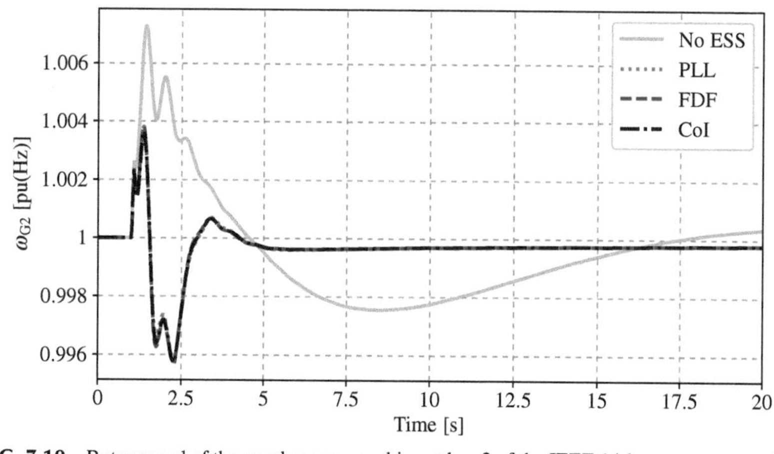

FIG. 7.10 Rotor speed of the synchronous machine at bus 2 of the IEEE 14-bus test system without and with an ESS regulating a system frequency [28].

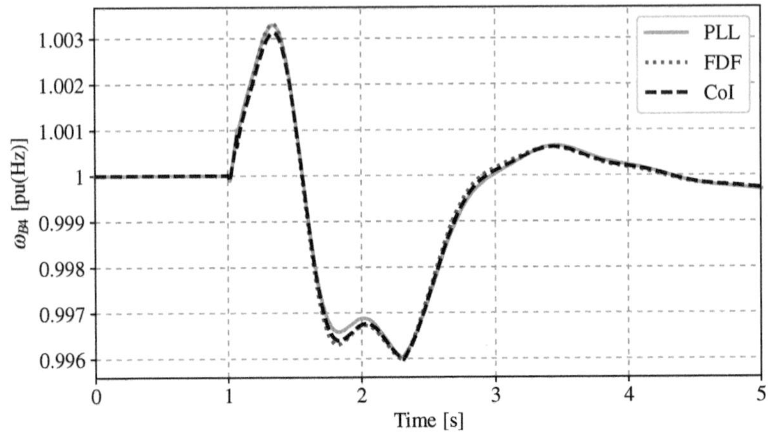

FIG. 7.11 Frequency at bus 4 estimated with the FDF when the ESS control input signal is generated by different frequency estimation techniques [28].

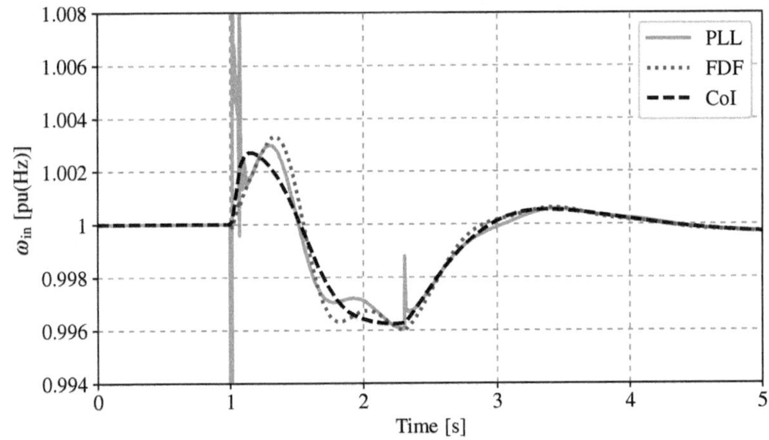

FIG. 7.12 Input signal of the ESS control generated by each frequency estimation technique [28].

u, and thus, of the active power supplied/absorbed by the ESS, as depicted in Figs. 7.13 and 7.14, respectively.

Stochastic variations of wind

In this case study, the SG placed in bus 2 is substituted with a 60-turbine wind power plant of the same power capacity. A stochastic process that follows a Weibull distribution is applied to the wind, and its profile is shown in Fig. 7.15 [25]. Values of the mean wind speed, scale, and shape factors are taken from [40] for the month of August at a height of 65 m.

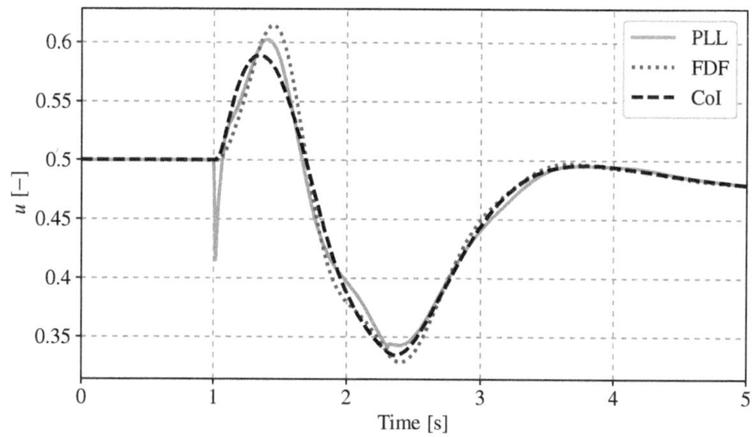

FIG. 7.13 Output signal of the ESS control for each frequency estimation technique [28].

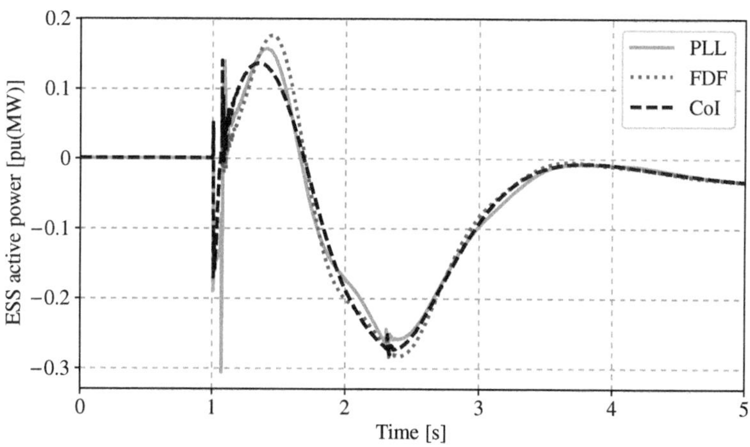

FIG. 7.14 Active power supplied/absorbed by the ESS for each frequency estimation technique [28].

The objective of the ESS installed at bus 4 is to regulate the frequency (in this section, the only frequency estimation technique considered is the FDF) and the voltage at the PCC, whose trajectories are depicted in Fig. 7.16A and B, respectively. The ESS is able to reduce considerably the frequency and voltage variations due to the wind variations. Moreover, despite the local nature of the ESS controllers, it nevertheless improves the overall system response, as shown in Fig. 7.16C and D, where the frequency and the voltage at the load bus 14 are depicted, respectively.

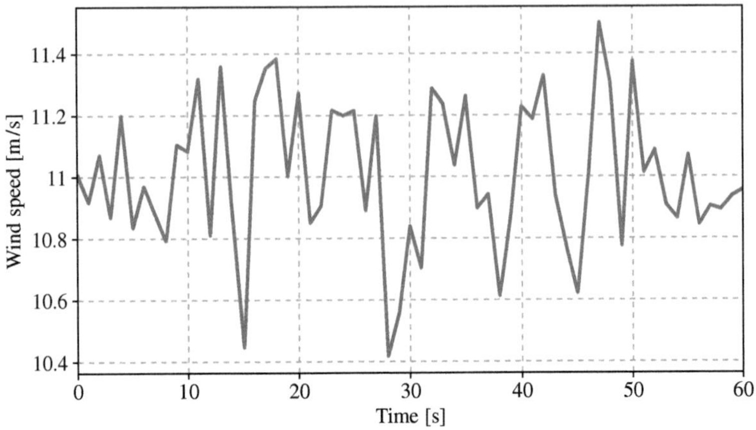

FIG. 7.15 Wind profile that follows a Weibull distribution [28].

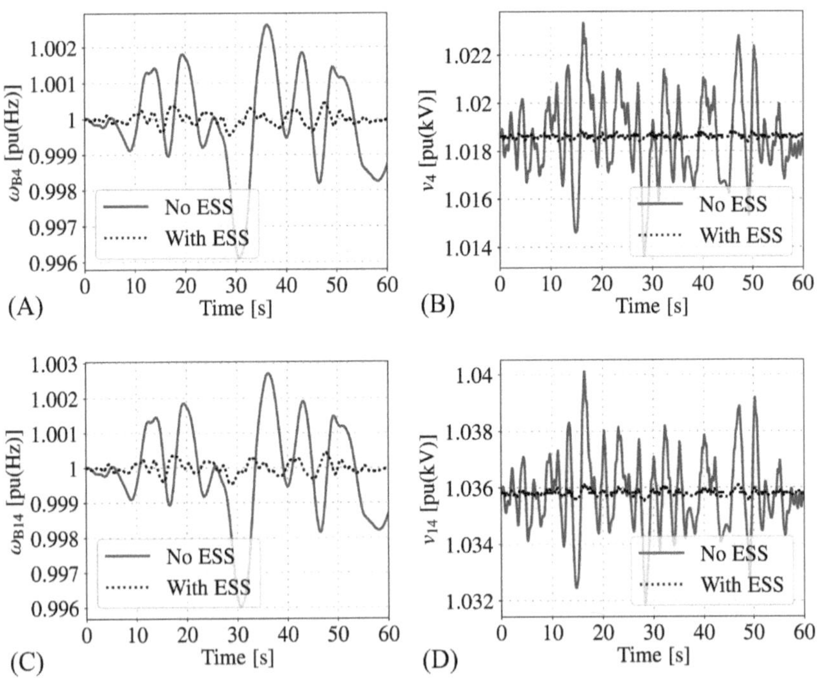

FIG. 7.16 Response of the IEEE 14-bus test system with stochastic wind perturbations [28]. (A) Frequency at bus 4, (B) voltage at bus 4, (C) frequency at bus 14, (D) voltage at bus 14.

7.3.1.2 Remarks

Based on the simulation results presented in this section, the following remarks are relevant.

- The inclusion of ESSs in the system allows reducing, to a large extent, frequency and voltage variations due to severe contingencies such as faults, and to the stochastic nature of RESs such as wind power plants, thanks to the capability of ESSs to inject/absorb both active and reactive power simultaneously.
- As for the case with WECSs, the low-pass filter included in the ESS frequency control loop allows reducing the impact of the noise and numerical issues present when the PLL is used to estimate the frequency.
- The response of ESSs can be significantly deteriorated if current saturation of the storage device is reached.
- Despite the fact that ESSs provide frequency and voltage regulation locally, their effect can be seen system-wide.

7.3.2 Virtual synchronous generator

The conventional grid-feeding converter applies outer power; inner current control (see Fig. 7.5), that is, its main goal is to deliver a set power to the grid with the assumption that the grid voltage is established. This control also decouples the active and reactive power output, while the converter voltage is only indirectly controlled to achieve the required power flows. A converter with this kind of control typically requires a PLL to detect and synchronize with the grid voltage and the design of this PLL is critical, especially as the frequency support provided relies on the accurate measurement of the frequency and its rate of change.

In contrast, VSG control works on a grid-forming converter concept [41], which applies outer-voltage, inner-current control and directly controls the voltage amplitude and phase. The power flow is indirectly controlled and determined by the voltage difference between the converter output and the grid voltages existing across any interconnecting impedance. Synchronization in this type of converter emulates the synchronization dynamics of the SG and can be considered to be based on a type of power feedback control.

The power difference between a reference (analogous to the SG mechanical input power) and its output (analogous to the SG electric power) is used to determine the converter output voltage phase [42]. The output voltage phase difference with the grid voltage in turn changes the converter output power until the power error is eliminated and balance between the input and output is achieved.

The relationship imposed by the VSG control between the VSG frequency and power emulates the conventional swing equation of synchronous machines, as follows:

$$M\Delta\dot{\omega}_{vsg}(t) = p^{ref} + p_d(t) + D\Delta\omega_{vsg}(t) - p(t), \quad (7.1)$$

$$\dot{\delta}_{vsg}(t) = \Delta\omega_{vsg}(t) + \omega^{ref}, \quad (7.2)$$

$$p_d(t) = K_d\left(\omega^{ref} - \omega_{pll}(t)\right), \quad (7.3)$$

where M is the virtual inertia, D is the damping, K_d is the droop gain, p^{ref} is a reference power, p_d is a droop power component, p is the output power from the converter, ω^{ref} is the nominal frequency, ω_{vsg} is the VSG internal frequency, ω_{pll} is the measured grid frequency via PLL, and δ_{vsg} is the converter output voltage phase angle. The block diagram of the VSG control scheme is shown in Fig. 7.17A. If the control parameters K_d and D are set to be the same, then the damping and droop can be combined and the need for a PLL can be removed. In general, the DC side of the converter may be fed from an RES, an ESS or a combination of both. If the DC side of the converter is an ESS only, then p^{ref} would typically be zero; if there is also an RES, then p^{ref} is the power generated by the RES.

The converter output voltage amplitude or electric potential e is determined by a voltage regulator as

$$e(t) = K_v\left(v^{ref} - v_g(t)\right) + e_o, \quad (7.4)$$

where K_v is the regulator gain, v^{ref} is the reference voltage, v_g is the grid voltage, and e_o is an initial voltage set-point which, for example, can be determined via an optimal power flow.

This control only emulates the electromechanical and power regulation capabilities of the SG, and it is not necessary to emulate the electromagnetic or subtransients of SGs. In addition to the controls discussed above, a virtual impedance control loop, including virtual resistance R_v and virtual inductance L_v, can be used to mimic the stator impedance of the SG. Essentially, this virtual impedance can be seen as connecting in series with the line impedance and can be used to modify the apparent R/X ratio of the grid, compensate for inaccurate power sharing due to different line impedances in multiconverter systems [43], and avoid synchronous resonance [44]. If a virtual impedance is implemented, the reference voltage to the VSC outer-voltage controller is derived from the electric potential e minus the voltage drop on the virtual impedance caused by the output current i_o as

$$v_o^{ref} = e(t) - \left(R_v + j\omega_{vsg}(t)L_v\right)i_o, \quad (7.5)$$

The VSG control scheme in a grid-forming converter is shown in Fig. 7.17B. The complete detailed dynamic model of the VSG controlled grid-forming converter is presented in [10]. Many aspects of the behavior of this type of VSG would be expected to be similar to the SG, that is, the converter synchronization mimics the SG electromechanical behavior including inertia, the frequency to active power droop emulates the TG, and the voltage regulation is similar to the traditional AVR of the SG.

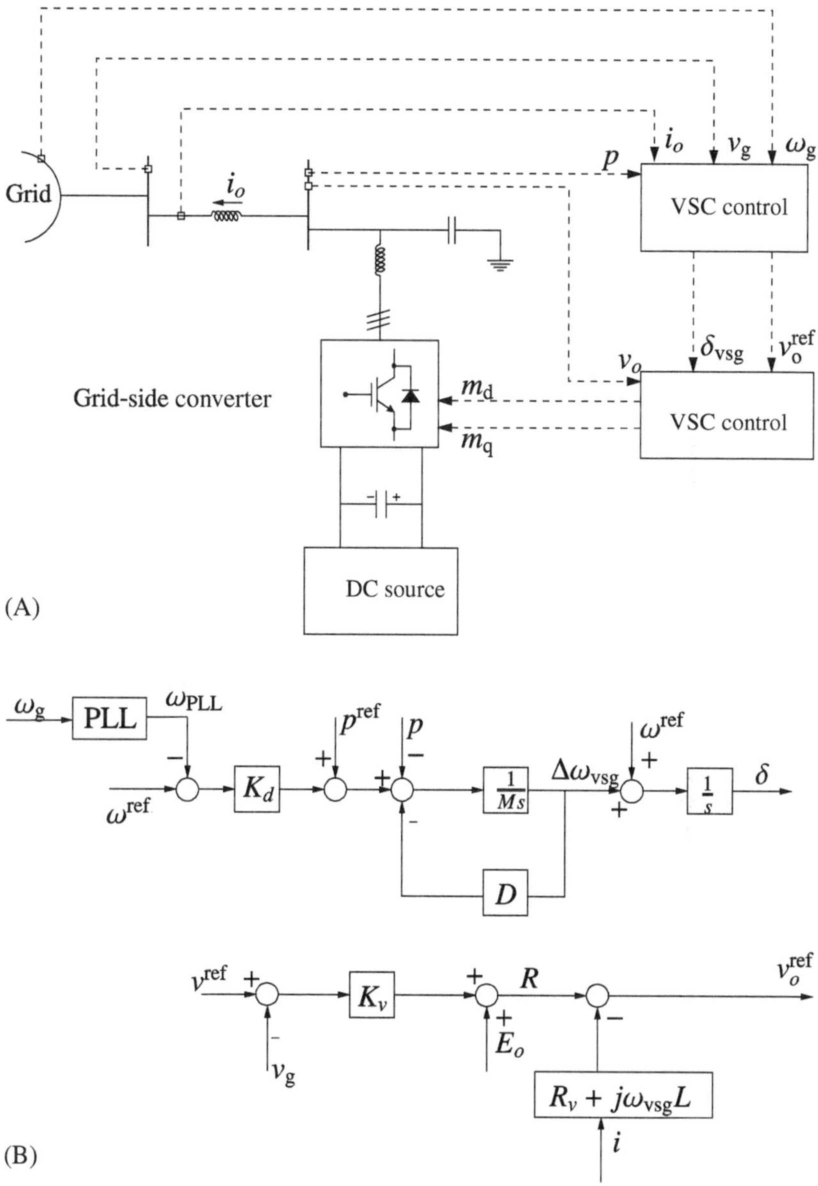

FIG. 7.17 VSG control scheme in a grid-forming converter: (A) high-level block diagram and (B) real power to angle controller and a voltage amplitude controller [10]. (A) Grid interface; (B) VSG control.

Reference [45] shows that if the parameters of the VSG are set to be identical to those of the SG then their responses as regards frequency dynamics from a system perspective will be identical. However, the response time of the AVR and TG are typically on the order of seconds, while the response times of the VSG can be on the order of milliseconds. Thus, the VSG has the potential to have improved performance compared to the SG with respect to fast frequency and voltage support. Even though it may be convenient to emulate the characteristics of a particular SG with a VSG, in general, the VSG parameters are simply controller settings that can be tuned to give any required response (within the limitations of the converter ratings).

For example, it can be shown that under certain conditions the VSG response can be approximated as a second-order system where appropriate choice of virtual inertia M and damping D can be used to tune the response from underdamped to overdamped [44]. In addition, these parameters can be time varying to give an adaptive response that is further discussed in the examples below. Another important consideration is that this VSG-controlled grid-forming converter is no longer following the grid frequency but actively participates in forming the grid voltage making it suitable for implementation in fully nonsynchronous systems.

7.3.2.1 Examples

This section illustrates the characteristics of the VSG in providing frequency support using the well-known IEEE 39-bus test system (see Fig. 7.18) as case study. This benchmark network comprises 10 SGs with primary controls (AVR and TG in each SG) and 18 loads. The frequency support provided after a contingency, namely the outage of generator Gen1 at 1 s, will be considered.

VSG vs. grid feeding with frequency support

The purpose of this section is to compare the VSG control with the grid-feeding control providing frequency support through a frequency to power droop, where both control approaches are applied to an ESS. Here, for brevity, the latter control approach will be referred to as a droop support control. A more detailed treatment of this comparison has been discussed in [46, 47]. It is assumed that the ESS is installed with a WECS with the topology shown in Fig. 7.19, where the ESS and the wind generator are connected to the same bus. This topology is used to replace the SG at bus 39 (Gen10) with the same initial power generation. WECS is modeled as a set of variable-speed wind turbines with doubly-fed induction machines working at a constant wind speed of 13 m/s. The system initially operates at the frequency of 1 pu (50 Hz) so that the initial active power output of the ESS is null.

To have a fair comparison, the following assumptions and parameter selections for the controls are employed:

- The DC voltage of the ESS is assumed to be constant in both cases.
- The VSCs and their filters are assumed to be the same. The inner-current loop PI control settings for both cases are identical.
- The droop gain, K_d for both controls are set to be identical.
- The VSG virtual inertia, M, and damping, D, are selected to give an overdamped response with no power overshoot so that the power limit of both controls are identical.

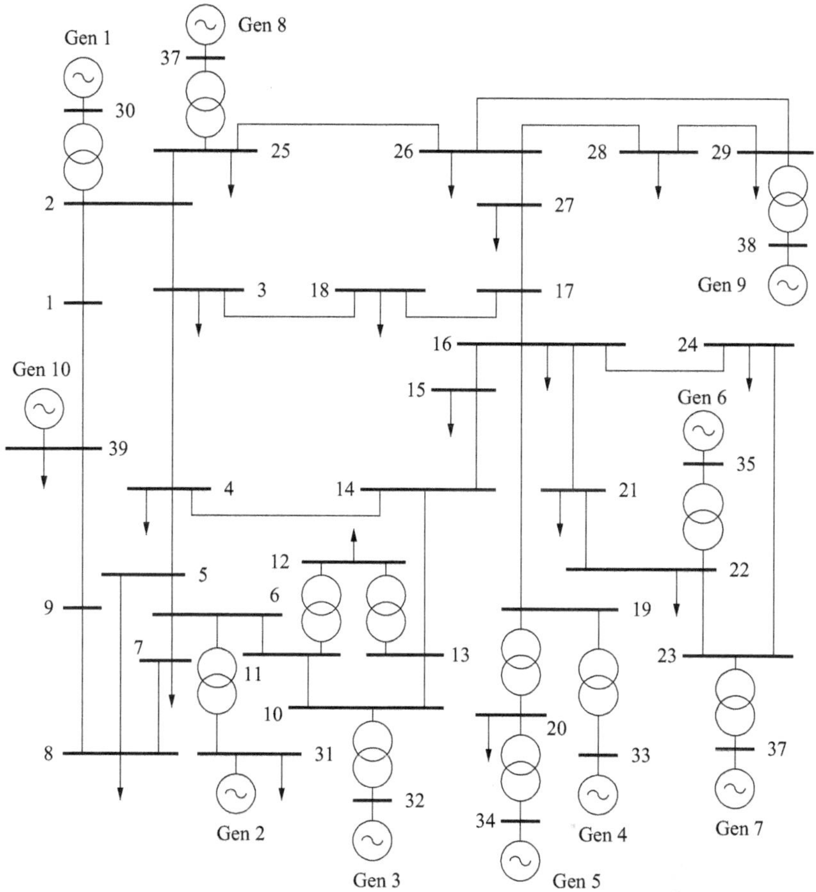

FIG. 7.18 IEEE 39-bus system.

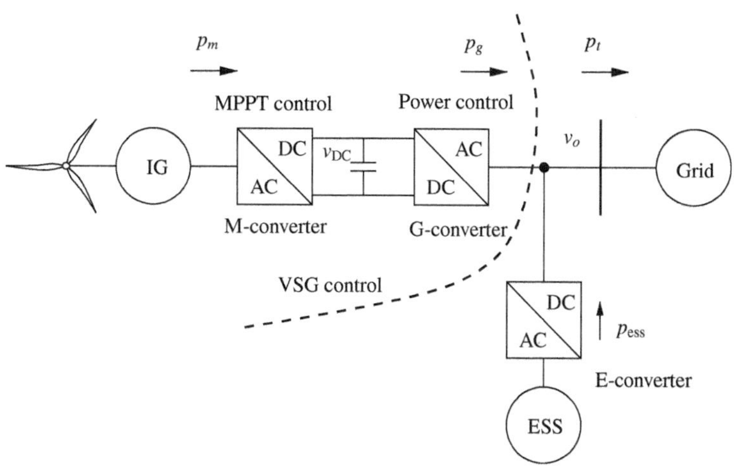

FIG. 7.19 ESS colocated WECS topology [48].

To illustrate the inherent differences between a simple grid-feeding frequency droop support approach and VSG control, the trajectories of the CoI frequency of the grid and the ESS active power output following the generator outage are presented in Fig. 7.20.

As it can be seen from Fig. 7.20A, the VSG control shows larger damping and fewer oscillations in the frequency response compared to the droop control. In steady state, both approaches stabilize at the same frequency due to the same K_d settings. The VSG control, due to the emulated inertia, gives a smaller RoCoF during transients. The emulated inertia can also be seen from Fig. 7.20B, which shows that the VSG-controlled ESS supplies more power after the contingency. The combination of damping and virtual inertia in the VSG control can be used to smooth its output, while the output from the droop-controlled ESS simply follows the grid frequency that has significant decaying

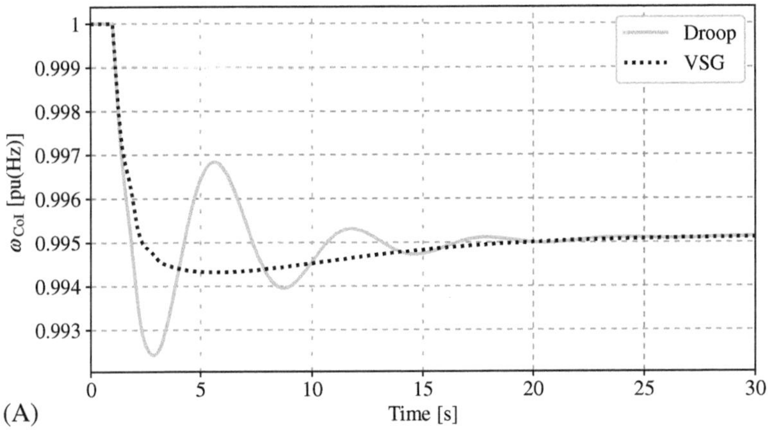

(A)

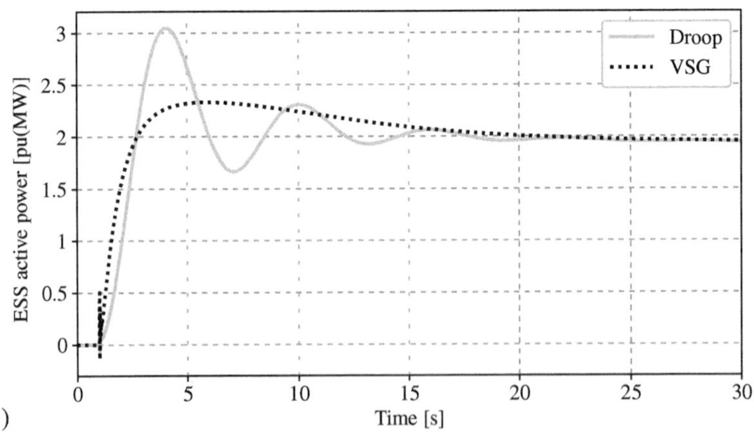

(B)

FIG. 7.20 Trajectories of the IEEE 39-bus system with VSG-controlled/droop-controlled ESS following the generator Gen1 outage [46]. (A) CoI frequency; (B) ESS output active power.

oscillations in this case. As a consequence, because of the lower system frequency nadir under droop control, the droop-controlled converter has a higher peak power output, thus requiring a higher converter power rating.

In general, the VSG control improves the frequency stability of the test system, that is, it results in improved frequency nadir, lower RoCoF, fewer oscillations, and faster stabilization.

Virtual synchronous generator vs. synchronous generator

The previous section highlighted the important differences between the VSG control and the grid-feeding droop control for frequency support. It is also of considerable interest to compare the frequency support from the VSG to that from a comparable SG. A more detailed discussion of the similarity between the VSG and SG can be found in [45, 49]. For this comparison, a slightly different structure of the VSG is considered, where the DC side of the converter is supplied by a renewable generation source and an ESS. The ESS is located on the DC side of the WECS, as shown in Fig. 7.21.

The VSG control is applied to the grid-side converter of the system, while the storage-side DC-DC converter is used to regulate the DC bus voltage and the turbine-side converter maintains its original MPPT function. In this configuration, the whole WECS can be seen as an SG, where all the power goes through the swing equation, while in the configuration of the previous section, only the power from ESS goes through the swing equation. The VSG replaces the generator located at bus 34, that is, Gen5. For the comparison, the control parameters are set as follows:

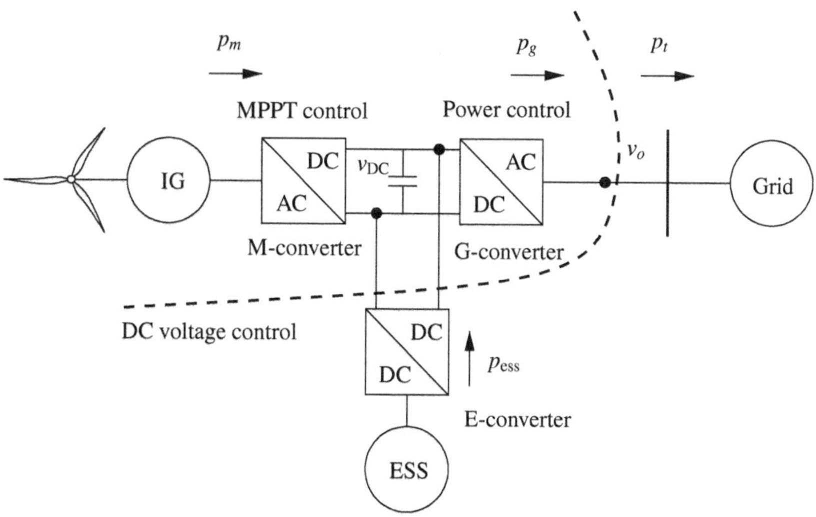

FIG. 7.21 ESS colocated with WECS topology [48].

- The VSG droop gain is set equal to the TG gain of the SG.
- The VSG voltage regulation gain is set equal to the AVR gain of the SG.
- Since the VSG only applies to primary control, auxiliary and secondary controls of the SG, including power system stabilizers and secondary frequency control, are removed.
- The inertia values of the VSG and SG are the same.
- The virtual impedance of the VSG is set equal to the SG stator impedance.

Three scenarios are considered. In the first scenario, all generators are conventional synchronous machines, labeled SG1 with TG time constant of 5 s. In the second scenario, the SG at Gen5 is instead replaced with an idealized SG, labeled SG2, with a very fast TG time constant of 0.1 s. In the third scenario, the SG at Gen5 is replaced with a VSG with all other generators unchanged.

The third scenario is provided to illustrate the similarity between the frequency response of the VSG and an SG with a very fast governor response. Fig. 7.22 shows the CoI frequency of the power system and the output active power of the devices following the outage of Gen1.

As it can be seen in Fig. 7.22A, the RoCoF and the initial power response from all three scenarios is similar, indicating that the inertial response is similar to all three devices. However, for some time after the contingency, the output power from SG1 is lower than that from the VSG and SG2, thus resulting in a lower frequency nadir in the SG1 scenario. Essentially, the slower TG response from SG1 leads to less active power output after the contingency, as shown in Fig. 7.22B. Fig. 7.22B also shows the same active power outputs from the VSG and SG2, indicating that the VSG acts like a fast SG, that is, the same inertial response but a much faster governor response. On the other hand, the VSG does not mimic the subtransient dynamics in the SG, which can be seen in the first few seconds, and thus its output is much smoother without the initial fast oscillations. This comparison illustrates that the VSG could potentially give improved frequency dynamics in the system due to its faster response.

Adaptive virtual synchronous generator

Unlike the SG where its parameters are linked to it physical construction, in the VSG, parameters such as inertia, damping, and to some extent impedance are controller settings that can be easily changed to tune the response time. For example, references [50–52] propose adaptive VSG parameter settings to improve the system stabilization. The basic concept is that, at the instant of the contingency, the system will benefit from a large inertia to reduce the RoCoF and nadir, while during the frequency recovery period of the grid, the inertia can be reduced and the damping increased to damp oscillations and reduce the settling time. The basic inertia and damping self-turning strategy are given in Table 7.3.

To illustrate the potential benefit of this adaptive VSG, this section compares the performances of the fixed VSG and adaptive VSG in terms of frequency support. The VSC controller parameters and power limits are the same in both cases.

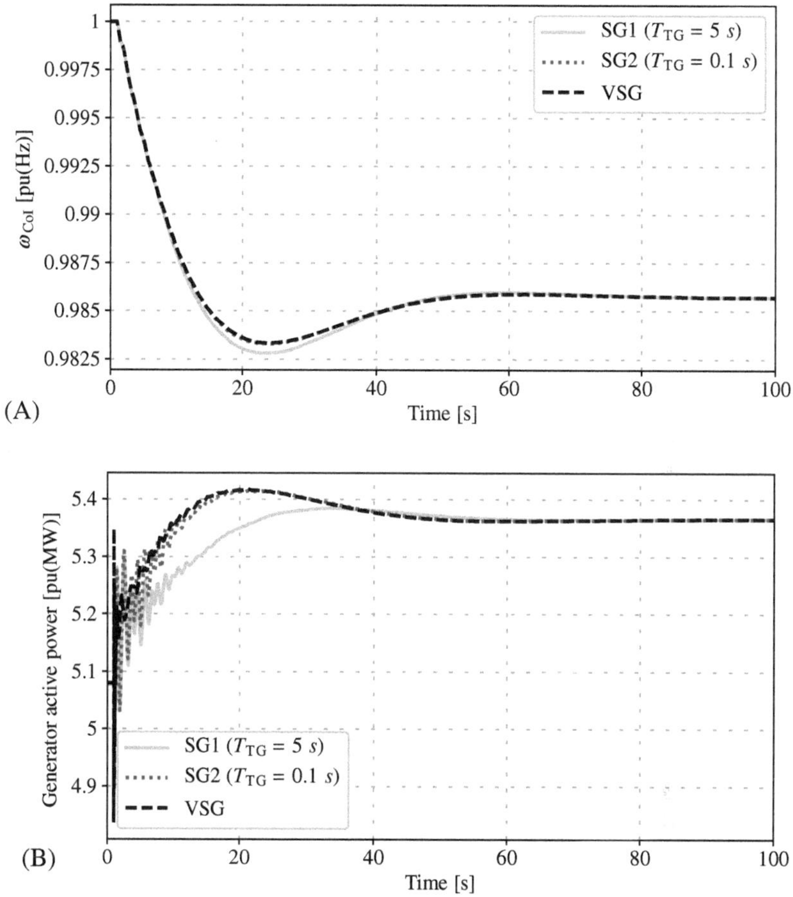

FIG. 7.22 Trajectories of the IEEE 39-bus system with different VSGs following the generator Gen1 outage [45]. (A) CoI frequency; (B) ESS output active power.

TABLE 7.3 Rules to adjust virtual inertia M and damping D during a transient.

$\Delta\omega$	$\Delta\omega/\Delta t$	Change of M	Change of D
$\Delta\omega > 0$	$\Delta\omega/\Delta t > 0$	Increase	Decrease
$\Delta\omega > 0$	$\Delta\omega/\Delta t < 0$	Decrease	Increase
$\Delta\omega < 0$	$\Delta\omega/\Delta t > 0$	Increase	Decrease
$\Delta\omega < 0$	$\Delta\omega/\Delta t < 0$	Decrease	Increase

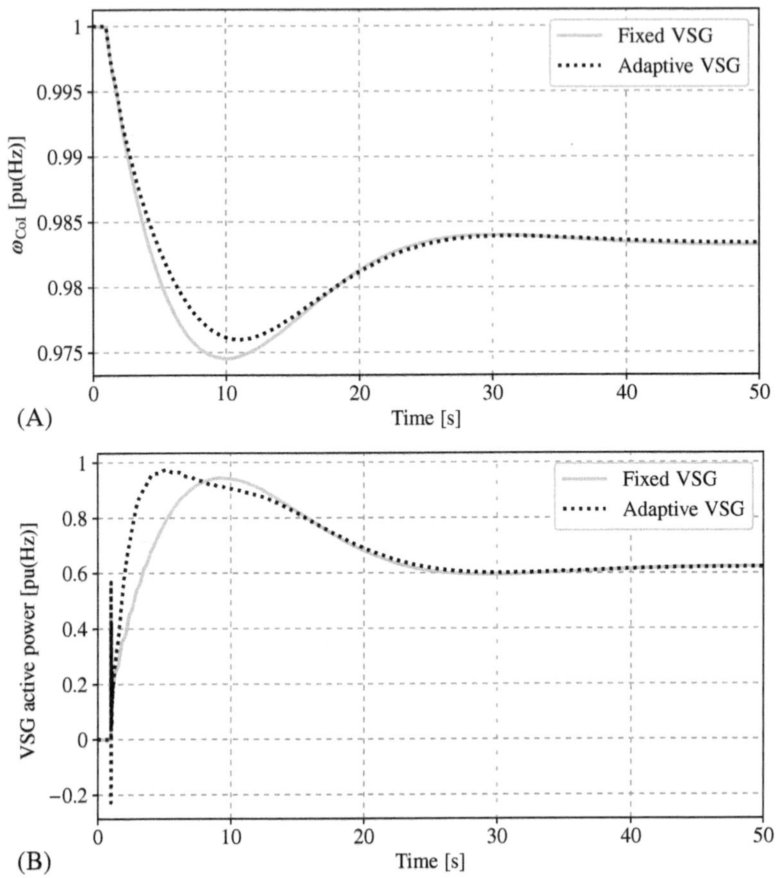

FIG. 7.23 Trajectories of the IEEE 39-bus system with different generators following the generator Gen1 outage [45]. (A) CoI frequency; (B) ESS output active power.

Fig. 7.23 shows the CoI frequency of the test system and the output active power of the different VSGs. Fig. 7.23 also shows that the adaptive VSG transiently outputs more power than the fixed parameter VSG after the contingency due to the large initial inertia setting, which significantly reduces the RoCoF and frequency nadir. After the first frequency swing, that is, just after the frequency nadir is reached, the adaptive VSG decreases the inertia setting and increases the damping, and thus, the output power variation becomes slower, and the frequency recovers smoothly.

7.3.2.2 Remarks

Based on the simulation results presented in this section, the following remarks are relevant.

- VSG control can endow power electronics-interfaced generation and storage with features of inertia emulation, primary frequency control, and voltage establishment or grid-forming capability. At least in terms of frequency support, and provided sufficient storage and converter power ratings, a VSG can have the same performance as a comparable SG that has identical settings.
- Since the VSG is based on power converters with fast response times, they have the potential to give improved frequency dynamics and stability compared to the SG dominated system.
- Compared to the SG, the VSG has more flexibility to better enhance the stability of power systems due to its adjustable parameters. Therefore, VSGs should have increasingly wide applications in low-inertia power systems.

7.4 Frequency control through FACTS devices

Frequency control is ultimately a regulation of active power and thus requires an active power reserve. It may thus seem inconsistent to regulate the frequency through reactive power sources such as FACTS devices. If one couples a FACTS device with a voltage-dependent load, however, one can vary the active power consumed by the load by regulating the voltage at the bus where the load is connected. This section elaborates on this idea and considers two devices: (i) a conventional and common SVC device and (ii) a last generation FACTS device, namely the smart transformer (ST).

Industrial and residential loads are generally modeled as aggregated power consumption in the dynamic analysis of power systems. These aggregated models can be static or dynamic [25]. A common static model expresses the active and reactive powers as functions of the bus voltage magnitude, as follows:

$$p_i(t) = p_{i,o}\left(\frac{v_i(t)}{v_{i,o}}\right)^{\alpha_p}, \tag{7.6}$$

$$q_i(t) = q_{i,o}\left(\frac{v_i(t)}{v_{i,o}}\right)^{\alpha_q}, \tag{7.7}$$

where p_i and q_i are the active and reactive power demand; $p_{i,\,o}$ and $q_{i,\,o}$ are the rated active and reactive power demand at the rated voltage ($v_{i,\,o}$) of the bus; α_p and α_q are the voltage exponents of active and reactive power, respectively, and v_i is the bus voltage magnitude. The exponents α_p and α_q vary depending on the load type [53]. Typical ranges for the exponents are $\alpha_p \in (0.9, 1.7)$ and $\alpha_q \in (1.9, 4)$ [54].

A change in the operation voltage, say Δv_i, results in the following change in power demand Δp_i:

$$\Delta p_i(t) = \frac{p_{i,o}}{v_{i,o}^{\alpha_p}}\left((v_i(t) + \Delta v_i(t))^{\alpha_p} - v_i(t)^{\alpha_p}\right). \tag{7.8}$$

For example, assume $v_i = v_{i,o} = 1$ pu and $\alpha_p = 1.5$. Then, a 5% voltage increase will lead to an increase of the active power demand of about 7.6%. The VFC scheme considered in this chapter exploits Eq. (7.8) to vary the active power consumption through the variation of the bus voltage magnitudes. In the remainder of the chapter, $\alpha_p = 1.5$ and $\alpha_q = 2$ are assumed.

7.4.1 Static VAR compensator

An SVC is a combination of a capacitor and a variable shunt reactor controlled through thyristor-based power electronic switches [55]. The SVC can generate and absorb reactive power and thus is conventionally utilized for voltage control at the bus to which it is connected. The control diagram of an SVC is shown in Fig. 7.24, where the controlled variable is the susceptance b_{SVC}.

The SVC model is defined by the following differential-algebraic equations [25]:

$$T_{\mathrm{r}}\dot{b}_{\mathrm{SVC}}(t) = -b_{\mathrm{SVC}}(t) + K_{\mathrm{r}}\left(v^{\mathrm{ref}} - v_i(t) - v_{\mathrm{g}}(t)\right), \tag{7.9}$$

$$q_{\mathrm{SVC}} = b_{\mathrm{SVC}}(t)v_i^2(t), \tag{7.10}$$

where v^{ref}, K_{r}, T_{r}, and q_{SVC} are the reference voltage, the regulator gain, the regulator time constant, and the output reactive power generated by the SVC, respectively. In conventional applications, the signal v_{g} is usually the output of a power oscillation damping (POD) control.

The POD control typically used to obtain v_{g} in Fig. 7.24, utilizes the rate of change of the input signal u_{in}, for example, active power flow or voltage measured either locally or remotely [57], and serves for damping electromechanical oscillations. Instead, we employ two different types of controllers to get v_{g}, which are designed to achieve VFC.

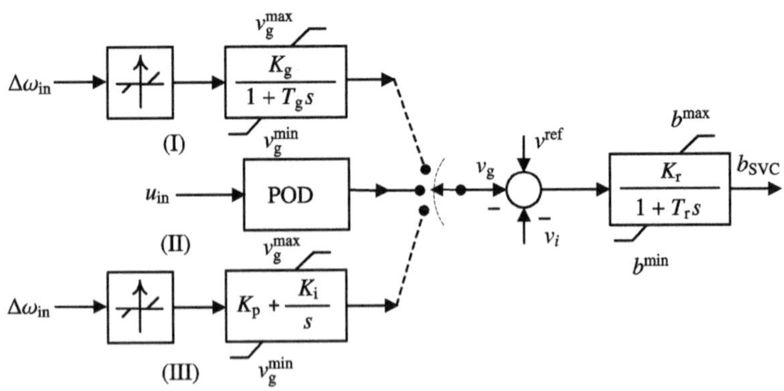

FIG. 7.24 Block diagram of conventional SVC model with additional control loops: (I) droop frequency control, (II) POD control, and (III) PI-based frequency control [56].

With this aim, two control methods are considered: (i) a droop (lag) controller and (ii) a PI controller (see (I) and (III) in Fig. 7.24, respectively). Both methods consider the frequency error $\Delta\omega_{\text{in}}$ as control input. The presence of the deadband, *db*, ensures that for a small variation of the frequency, the controller will not deteriorate the local voltage response.

In case that the droop controller is used, v_{g} is given by

$$T_{\text{g}}\dot{v}_{\text{g}}(t) = K_{\text{g}}\Delta\omega_{\text{in}}(t) - v_{\text{g}}(t), \tag{7.11}$$

where T_{g} and K_{g} are the time constant and gain of the lag controller, respectively; $\Delta\omega_{\text{in}}$ is the frequency error, that is, $\Delta\omega_{\text{in}} = \omega^{\text{ref}} - \omega_i$, where ω^{ref} is the reference frequency, and ω_i is the measured frequency at the SVC bus.

If the PI controller is used, v_{g} is as follows:

$$\begin{aligned} v_{\text{g}}(t) &= K_{\text{p}}\Delta\omega_{\text{in}}(t) + x_{\omega}(t), \\ \dot{x}_{\omega}(t) &= K_{\text{i}}\Delta\omega_{\text{in}}(t), \end{aligned} \tag{7.12}$$

where K_{p}, K_{i}, and x_{ω} are the proportional gain, the integral gain, and the state variable of the PI control, respectively.

To ensure that the bus voltage remains within its operational range, both frequency control types constrain the output signal to its respective limits ($v_{\text{g}}^{\max}$ and $v_{\text{g}}^{\min}$). Moreover, antiwindup-type limits are considered to get better overall transient response [58].

To obtain the bus frequency ω_i, a PLL is utilized. We make use of the Lag-PLL model that produces accurate bus frequency estimations (see Section 6.3.2.2).

7.4.1.1 *Examples*

WSCC 9-bus system

We consider an SVC connected at bus 8 of the WSCC 9-bus test system described in [23] (see Fig. 7.13). The parameters used for the SVC controllers are given in Table 7.4, where $v_{\text{g}}^{\text{lim}} = v_{\text{g}}^{\max} = -v_{\text{g}}^{\min}$.

TABLE 7.4 SVC and PLL parameters for the WSCC system [56].

Name	Values			
CSVC	$K_r = 20$	$T_r = 0.01$ s	$b^{\max} = 0.5$ pu	$b^{\min} = -0.5$ pu
LFC	$K_g = 25$	$T_g = 0.01$ s	$db = 0.015$ Hz	$v_g^{\text{lim}} = 0.05$ pu
PIFC	$K_p = 1.5$	$K_i = 25$	$db = 0.015$ Hz	$v_g^{\text{lim}} = 0.05$ pu
PLL	$K_p = 0.1$	$K_i = 0.5$		

Based on the discussion in Section 7.4, four scenarios are tested and compared by means of nonlinear time domain simulations: (a) without SVC (NSVC); (b) only conventional ($v_g = 0$) SVC (CSVC); (c) CSVC with lag frequency controller (LFC); and (d) CSVC with PI frequency controller (PIFC).

A three-phase fault at bus 6 at $t = 1$ s is simulated. The fault is cleared after 60 ms by tripping the line that connects buses 6 and 9. The trajectories of the frequency of the CoI and the voltage at bus 8 are depicted in Figs. 7.25 and 7.26, respectively.

Fig. 7.25 shows that the utilization of LFC and PIFC leads to a significant improvement of the initial frequency deviation. Concerning the voltage response (see Fig. 7.26), after the disturbance, the CSVC provides reactive power support and, therefore, improves the bus voltage. This also leads to increase the load consumption, as imposed by Eq. (7.8). In this case, the CSVC leads to a relatively good frequency response even without the frequency control loop, but this result

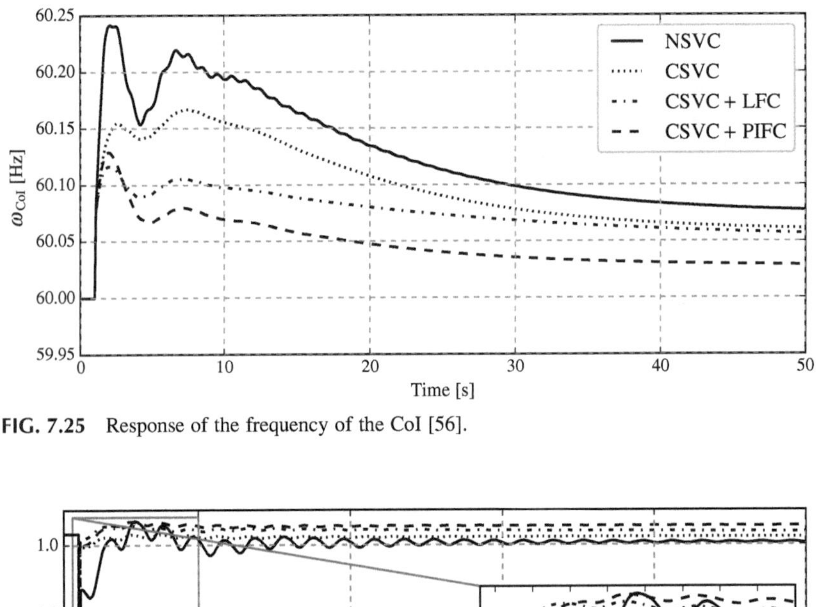

FIG. 7.25 Response of the frequency of the CoI [56].

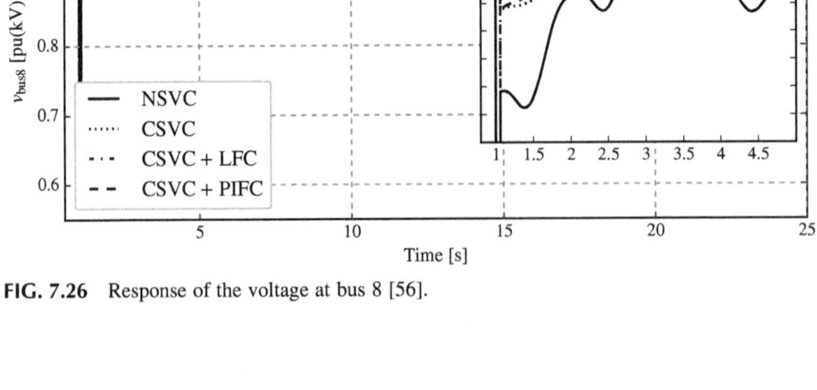

FIG. 7.26 Response of the voltage at bus 8 [56].

is not always guaranteed. How effective the CSVC for frequency control is depends on the disturbance and cannot be determined a priori.

No secondary frequency control through AGC is considered in this study. Hence, the trajectories of the frequency reach a postdisturbance equilibrium with a nonzero steady-state error. Comparing all cases, this error is smaller when the PIFC is employed. The PIFC also leads to a higher-steady-state voltage magnitude at the bus of the SVC. This is a consequence of the perfect tracking behavior of the PI control. Note that the PIFC integrator does not eliminate the steady-state frequency error, due to the deadband in the VFC input and the limits on the output.

Overall, the improvement of the frequency response obtained in Fig. 7.25 when either LFC or PIFC is included is significant (>0.1 Hz). It is clear that this improvement varies depending on several factors: size of the system, number and location of installed SVCs, etc.

All-island Irish transmission system

To better quantify the real impact of the VFC provided by SVCs, in the next section we study the effect of the inclusion of frequency control loops in the SVC that the Irish system operator has planned to install in 2019.

To carry out a realistic case study, we first validate the Irish test system by applying a real severe overfrequency event. On February 28, 2018, the VSC-HVDC link East-West Interconnector (EWIC) [59] that connects the Irish transmission system with the Great Britain (GB) transmission system was tripped. At that moment, Ireland was exporting 470 MW to GB. Due to the loss of the EWIC, the frequency in the Irish grid rose to 50.42 Hz. Overfrequency protections were triggered, and several wind farms were curtailed.

In the test system model, the 470 MW active power export is considered as a constant load. A comparison of simulated and actual frequency response following the outage of the EWIC is shown in Fig. 7.27. With a proper tuning of turbine governor parameters, a satisfactory match between the simulated

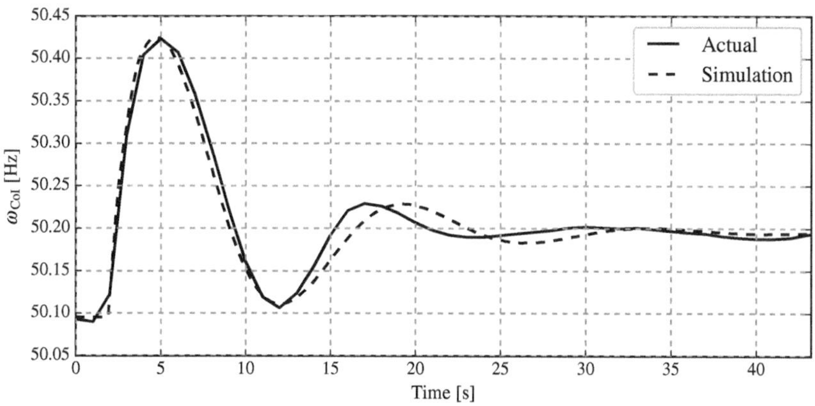

FIG. 7.27 Comparison of simulated and actual frequency of the CoI responses after the loss of the EWIC [56].

transient and the actual one was obtained. For the purpose of the model validation discussed previously, no VFC is considered.

Next, the frequency response of the Irish system model with inclusion of SVC-based VFC is examined. The same four scenarios discussed for the example mentioned previously based on the WSCC 9-bus system are considered.

The current Irish transmission system includes several shunt capacitors and shunt reactors but only two SVCs for reactive power compensation. The installed SVCs have a capability of +90 and −10 MVAr and are installed at the 110 kV voltage level (see Table B-7 in [60]). Three more SVCs are expected to be integrated at the same voltage level in 2019. The capacity of the new SVCs is ±470 MVAr (see Table B-11 in [60]).

In the simulations presented in this section, all five SVCs have been connected to the actual/expected buses and their limits have been imposed based on their nominal ratings. The parameters of the SVC controllers and the PLLs are given in Table 7.5. Except for reactive power limits, all parameters are the same for the five SVCs.

The test system is simulated by applying the disturbance discussed previously, namely the loss of the EWIC. The comparative trajectories of the frequency and the voltage at an SVC bus (Omagh Main) are shown in Figs. 7.28 and 7.29, respectively. The NSVC case is the same as the simulated response shown in Fig. 7.27. Compared to NSVC and CSVC, the utilization of CSVC with LFC and PIFC improves the frequency response. This improvement (for LFC $\approx$ 0.037 Hz and for PIFC $\approx$ 0.03 Hz) is relevant given that only five SVCs are utilized.

At the time of the EWIC outage, the active power generation in the Irish system is greater than the demand. Hence, the bus voltage is increased by the SVC with LFC and PIFC to increase power consumption. On the other hand, the CSVC without frequency control ensures the best voltage control. Due to the limits imposed in the VFC, voltage fluctuations remain within the maximum operating range (1.1 pu). Even though PIFC provides the minimum steady-state error, the overall transient response of the frequency is better when the LFC is used.

TABLE 7.5 Parameters of the SVCs and PLLs [56].

Name	Values		
CSVC	$K_r = 25$	$T_r = 0.01$	
LFC	$K_g = 75$	$T_g = 0.005$ s	$db = 0.015$ Hz
PIFC	$K_p = 1.5$	$K_i = 50$	$db = 0.015$ Hz
PLL	$T_{1,\,LF} = 0.2$ s	$T_{2,\,LF} = 2$ s	$T_f = 0.01$ s

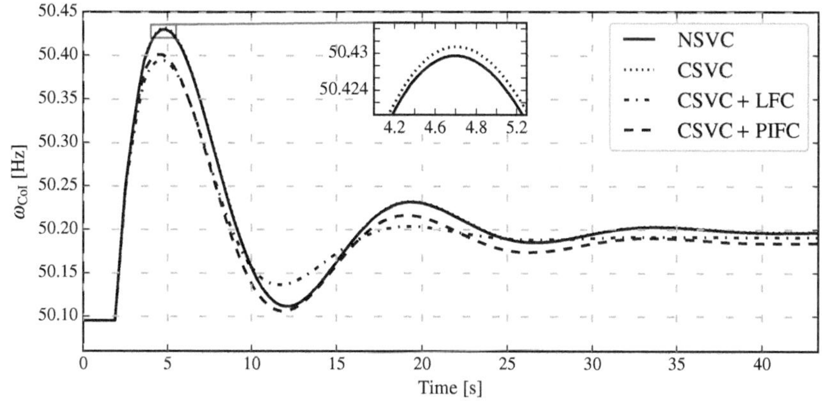

FIG. 7.28 Response of the frequency after the loss of the EWIC [56].

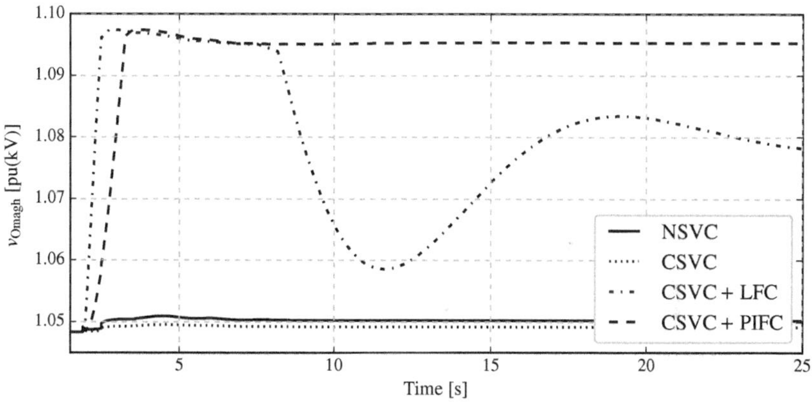

FIG. 7.29 Response of the voltage at bus Omagh Main after the loss of the EWIC [56].

7.4.1.2 Remarks

The following remarks on the SVC-based VFC are relevant.

- Despite using a relatively small amount of energy and only a few SVC, the examples discussed in the section clearly show that the VFC is promising. Note also that the more the loads that participate to the VFC, the lower their power variation to achieve the same amount of regulation.
- The VFC does not compromise the voltage profile at the load buses. This, again, is a particularly promising result that can help convince distribution system operators, prosumers, and aggregators to participate to VFC.
- Overall, the control strategy with the best dynamic response is the PIFC. The parameters of such a controller, which in the examples discussed previously are obtained with a simple trial and error technique, have proved to be particularly robust to changes in the loading level and network topology.

7.4.2 Smart transformer

This section introduces STs and their frequency support capability. Here, we envisage the STs being implemented as a three-stage solid-state transformer [61], placed at the interface between the HV transmission system and the MV distribution system. The three stages of the ST therefore consist of an HVAC-HVDC rectifier stage, an HVDC-MVDC DC-DC converter stage, and an MVDC-MVAC inverter stage, as shown in Fig. 7.30.

The rectifier connects to the transmission system using a PLL to maintain synchronization. The control in this stage applies outer-power, inner-current control with the aim of regulating the HVDC voltage to a reference value. The DC-DC converter has a medium-frequency transformer to step down the voltage, and its control aims to regulate the MVDC voltage. The inverter converts the MVDC voltage back to the MVAC voltage that supplies the distribution system so that in this stage the control is used for MVAC voltage regulation. The voltages at each port are independent.

The capacitors in the HVDC and MVDC isolate the reactive power from HVAC to MVAC so that the ST only transfers the active power and decouples the reactive power. Due to the decoupled voltage and reactive power regulation in the ST, the MVAC side can independently control its connected demand in response to frequency variations [62] similarly to the SVC in Section 7.4.1. Meanwhile, the HVAC side can support the HVAC voltage by reactive power compensation. In this way, the ST can separately and independently provide frequency and voltage support to the system.

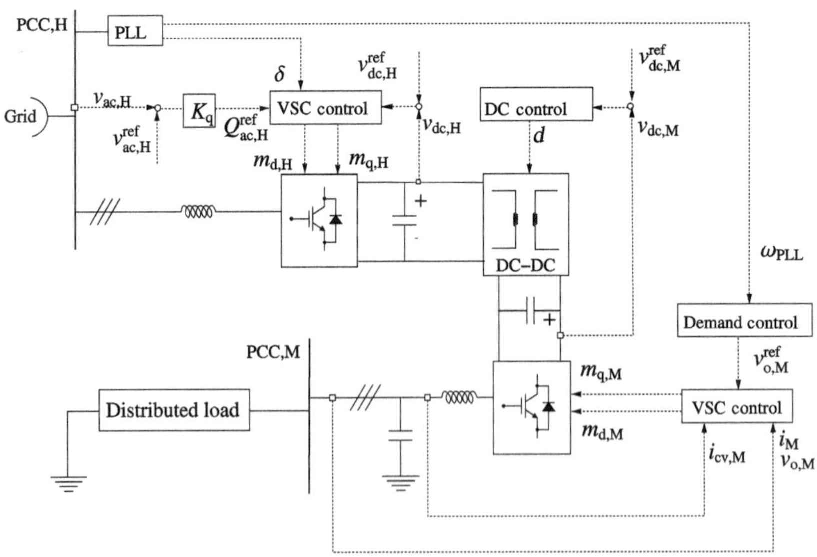

FIG. 7.30 Basic three-stage ST topology and its control scheme of ST [62].

Fig. 7.30 shows the topology and control scheme of the ST. In this section, the focus is on frequency support aspects only. The full model of the ST is given in [63, 64]. The reactive power compensation in the HVAC side uses the droop control as follows:

$$q_{\mathrm{ac,H}}^{\mathrm{ref}}(t) = K_{\mathrm{q}}\left(v_{\mathrm{ac,H}}^{\mathrm{ref}} - v_{\mathrm{ac,H}}(t)\right) + q_{\mathrm{ac,H},o}, \tag{7.13}$$

where $q_{\mathrm{ac,H}}^{\mathrm{ref}}$ is the reactive power reference to the outer power inner current control of the rectifier, $q_{\mathrm{ac,\ H},\ o}$ is the initial reactive power set point, K_{q} is the droop gain, $v_{\mathrm{ac,H}}^{\mathrm{ref}}$ is the voltage set point, and $v_{\mathrm{ac,H}}$ is the bus voltage of the connected grid.

The regulation of the MVAC voltage achieves demand control by making use of the voltage dependency of the demand. Although generally the load power is positively linked to its supply voltage, increasing use of power electronic loads may weaken this relationship as power electronic interfaced loads typically work in a constant power regime, where decreasing the voltage increases current and losses in the system. Thus, it is preferable to identify the load voltage sensitivity before regulating the voltage; otherwise, the effect on demand and hence on frequency may be negative. Voltage sensitivity is a method utilized to identify the load voltage sensitivity, defined as the percentage of the load power change Δp in response to a 1% voltage reduction Δv as Eq. (7.14) [65]. $S_v > 0$ means that a voltage reduction reduces the load consumption; $S_v < 0$ means that a voltage reduction increases the load consumption. According to the value of S_v, the MVAC voltage is regulated in response to the grid frequency as in Eq. (7.15):

$$S_v(t) = \frac{\Delta p(t)}{\Delta v(t)}, \tag{7.14}$$

$$v_{\mathrm{o,M}}^{\mathrm{ref}}(t) = \frac{-S_v(t)}{|S_v(t)|}\left(K_{\mathrm{t}}\Delta\dot{\omega}_{\mathrm{g}}(t) + K_{\mathrm{d}}\Delta\omega_{\mathrm{g}}(t)\right) + v_{\mathrm{o,M},o}, \tag{7.15}$$

where $v_{\mathrm{o,\ M}}^{\mathrm{ref}}$ is the reference voltage of the outer-voltage, inner-current controller of the inverter; $-(S_v/|S_v|)$ is the relationship (positive or negative) between the voltage and loading; K_{t} is the RoCoF gain; K_{d} is the droop gain for the frequency deviation; $\Delta\omega_{\mathrm{g}}$ is the grid frequency deviation with respect to the nominal frequency; and $v_{\mathrm{o,\ M},\ o}$ is the initial MVAC voltage set point. In the remainder of this section, it is assumed that the maximum allowable voltage variation is 0.1 pu, the allowable frequency deviation is 2 Hz (0.04 pu), RoCoF is 0.5 Hz/s (0.01 pu/s), and $K_{\mathrm{t}} = 2.5$ and $K_{\mathrm{d}} = 10$, corresponding to the maximum voltage variation in response to the maximum frequency deviation and RoCoF.

Reference [62] points out that the available loading power used for frequency support from a typical distribution system is around 6%–10%, and the reactive power used for voltage support is greater than 60% of the ST capacity.

7.4.2.1 Example

In this example, the level of frequency support achievable by placing STs at the interface between the transmission system and loads is investigated. The IEEE 39-bus test system (see Fig. 7.18) is again used for the case study. Different levels of penetration of STs in the system are investigated by gradually interfacing various loads, starting initially from bus 3 and ending with the loads at buses 3, 4, 7, 8, 12, 15, 16, 18, 20, 21, 23, and 25–27 all interfaced through an ST. This latter case represents the situation where 68.7% of the total loading in the system is interfaced through an ST. In each case, the HV rectifier connects to the transmission system bus and the MV inverter is directly connected to the load. The load voltage sensitivity is set to be 1.6 for the active power and 3 for the reactive power with a maximum allowable variation range ± 0.1 pu. The initial operating point for each ST is set equal to the initial loading. The case study verifies the frequency support available from the ST and discusses the system frequency stability with the increasing quantity of the load regulated by ST. The contingency is the outage of generator Gen1.

Fig. 7.31A shows the trajectories of the frequency of the CoI after the contingency, for the varying levels of ST interfaced load penetration, where the bottom curve represents the case for no STs. It can be seen that the inclusion of the STs with demand control can enhance the system frequency stability in terms of the frequency deviation and RoCoF.

Fig. 7.31B shows the voltage magnitude at bus 3, and again the bottom curve shows the lowest voltage dip which represents the case where no ST is present. As it can be seen, the reactive power compensation provided by the ST raises the voltage at the bus so that even with only one ST, the voltage dip decreases from approximately 0.025 to 0.02 pu. Note that on the MV load side the controlled voltage dip is approximately 0.06 pu corresponding to a 0.016 pu frequency deviation and 0.0015 pu/s RoCoF. As the ST penetration increases, the voltage stability improves due to the reactive power compensation.

7.4.2.2 Remarks

Based on the simulation results discussed previously, the following remarks are relevant.

- The ST has decoupled voltage regulation and reactive power regulation between its primary and secondary sides, and this can be used to independently provide frequency and voltage support in the system.
- Provided that the load has an adequate voltage dependency, significant frequency support can be provided through the ST dynamic manipulation of the load voltage. Increased use of STs to interface to loads enhances the level of frequency support available. Simultaneously, the ST can be used to provide voltage support.

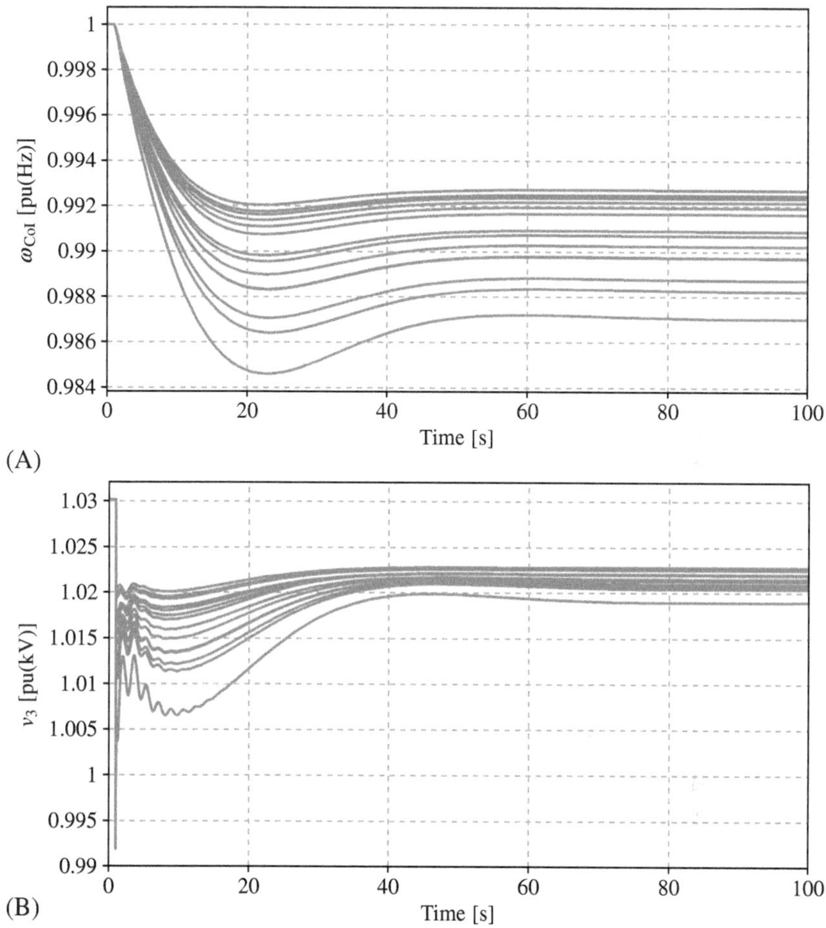

FIG. 7.31 Response of the IEEE 39-bus system with different penetrations of ST following the generator Gen1 outage. The lowest trajectory is the scenario where no ST applied, and the highest trajectory is the scenario where the penetration of STs in the system is 68.7% [62]. (A) CoI frequency; (B) voltage magnitude at bus 3.

References

[1] O. Anaya-Lara, F. Hughes, N. Jenkins, G. Strbac, Contribution of DFIG-based wind farms to power system short-term frequency regulation, IEE Proc. Gener. Transm. Distrib. 153 (2) (2006) 164–170.

[2] P. Moutis, A. Vassilakis, A. Sampani, N. Hatziargyriou, DC switch driven active power output control of photovoltaic inverters for the provision of frequency regulation, IEEE Trans. Sustain. Energy 6 (4) (2015) 1485–1493.

[3] Á. Ortega, F. Milano, Modeling, simulation, and comparison of control techniques for energy storage systems, IEEE Trans. Power Syst. 32 (3) (2017) 2445–2454, https://doi.org/10.1109/TPWRS.2016.2602211.

[4] D. Ochoa, S. Martínez, Fast-frequency response provided by DFIG-wind turbines and its impact on the grid, IEEE Trans. Power Syst. 32 (5) (2017) 4002–4011.

[5] J. Cerqueira, F. Bruzzone, C. Castro, S. Massucco, F. Milano, Comparison of the dynamic response of wind turbine primary frequency controllers, in: Proceedings of the IEEE PES General Meeting, July, 2017, pp. 1–5.

[6] S. Ruthe, C. Rehtanz, S. Lehnhoff, Towards frequency control with large scale virtual power plants, in: Proceedings of the IEEE PES Innovative Smart Grid Technologies Europe (ISGT Europe), October, 2012, ISSN: 2165-4816, pp. 1–6.

[7] P. Moutis, N.D. Hatziargyriou, Decision trees-aided active power reduction of a virtual power plant for power system over-frequency mitigation, IEEE Trans. Ind. Inform. 11 (1) (2015) 251–261.

[8] N. Etherden, V. Vyatkin, M.H.J. Bollen, Virtual power plant for grid services using IEC 61850, IEEE Trans. Ind. Inform. 12 (1) (2016) 437–447.

[9] H. Beck, R. Hesse, Virtual synchronous machine, in: 2007 9th International Conference on Electrical Power Quality and Utilisation, October, 2007, pp. 1–6.

[10] S. D'Arco, J. Suul, O.B. Fosso, A virtual synchronous machine implementation for distributed control of power converters in smartgrids, Electr. Power Syst. Res. 122 (2015) 180–197.

[11] Q. Zhong, G. Weiss, Synchronverters: inverters that mimic synchronous generators, IEEE Trans. Ind. Electron. 58 (4) (2011) 1259–1267.

[12] L. Huang, H. Xin, Z. Wang, K. Wu, H. Wang, J. Hu, C. Lu, A virtual synchronous control for voltage-source converters utilizing dynamics of DC-link capacitor to realize self-synchronization, IEEE J. Emerg. Sel. Top. Power Electron. 5 (4) (2017) 1565–1577.

[13] M. Liserre, G. Buticchi, M. Andresen, G. De Carne, L.F. Costa, Z. Zou, The smart transformer: impact on the electric grid and technology challenges, IEEE Ind. Electron. Mag. 10 (2) (2016) 46–58.

[14] A.Q. Huang, M.L. Crow, G.T. Heydt, J.P. Zheng, S.J. Dale, The future renewable electric energy delivery and management (FREEDM) system: the energy internet, Proc. IEEE 99 (1) (2011) 133–148.

[15] L. Xie, P.M.S. Carvalho, L.A.F.M. Ferreira, J. Liu, B.H. Krogh, N. Popli, M.D. Ilic, Wind integration in power systems: operational challenges and possible solutions, Proc. IEEE 99 (1) (2011) 214–232, https://doi.org/10.1109/JPROC.2010.2070051.

[16] G. Ramtharan, J.B. Ekanayake, N. Jenkins, Frequency support from doubly fed induction generator wind turbines, IET Renew. Power Gener. 1 (1) (2007) 3–9, https://doi.org/10.1049/iet-rpg:20060019.

[17] J.M. Mauricio, A. Marano, A. Gómez-Expósito, J.L.M. Ramos, Frequency regulation contribution through variable-speed wind energy conversion systems, IEEE Trans. Power Syst. 24 (1) (2009) 173–180, https://doi.org/10.1109/TPWRS.2008.2009398.

[18] J. Ekanayake, N. Jenkins, Comparison of the response of doubly fed and fixed-speed induction generator wind turbines to changes in network frequency, IEEE Trans. Energy Convers. 19 (4) (2004) 800–802.

[19] J. Morren, S.W.H. de Haan, W.L. Kling, J.A. Ferreira, Wind turbines emulating inertia and supporting primary frequency control, IEEE Trans. Power Syst. 21 (1) (2006) 433–434, https://doi.org/10.1109/TPWRS.2005.861956.

[20] J. Morren, J. Pierik, S.W.H de Haan, Inertial response of variable speed wind turbines, Electr. Power Syst. Res. 76 (11) (2006) 980–987, https://doi.org/10.1016/j.epsr.2005.12.002.

[21] I.D. Margaris, S.A. Papathanassiou, N.D. Hatziargyriou, A.D. Hansen, P. Sorensen, Frequency control in autonomous power systems with high wind power penetration, IEEE Trans. Sustain. Energy 3 (2) (2012) 189–199, https://doi.org/10.1109/TSTE.2011.2174660.

[22] Á. Ortega, F. Milano, Impact of frequency estimation for VSC-based devices with primary frequency control, in: IEEE PES Innovative Smart Grid Technologies Conference Europe (ISGT-Europe), September, 2017, pp. 1–6, https://doi.org/10.1109/ISGTEurope.2017.8260199.
[23] P.W. Sauer, M.A. Pai, Power System Dynamics and Stability, Prentice Hall, Upper Saddle River, NJ, 1998.
[24] J.G. Slootweg, H. Polinder, W.L. Kling, Dynamic modelling of a wind turbine with doubly fed induction generator, in: IEEE PES Summer Meeting, July, vol. 1, 2001, pp. 644–649.
[25] F. Milano, Power System Modelling and Scripting, Springer, London, 2010.
[26] M. Singh, S. Santoso, Dynamic models for wind turbines and wind power plants, Tech. Rep., NREL, 2011. https://doi.org/10.2172/1028524, http://www.osti.gov/scitech/servlets/purl/1028524.
[27] F. Milano, R. Zárate-Miñano, A systematic method to model power systems as stochastic differential algebraic equations, IEEE Trans. Power Syst. 28 (4) (2013) 4537–4544.
[28] Á. Ortega, F. Milano, A. Musa, L. Toma, D. Preotescu, Definition of frequency under high dynamic conditions—Deliverable 2.1, Tech. Rep., European Commission, 2017. www.re-serve.eu.
[29] B. Tamimi, C. Cañizares, K. Bhattacharya, Modeling and performance analysis of large solar photo-voltaic generation on voltage stability and inter-area oscillations, in: 2011 IEEE Power and Energy Society General Meeting, July, 2011, pp. 1–6, https://doi.org/10.1109/PES.2011.6039797.
[30] H.S. Ko, G.G. Yoon, W.P. Hong, Active use of DFIG-based variable-speed wind-turbine for voltage regulation at a remote location, IEEE Trans. Power Syst. 22 (4) (2007) 1916–1925, https://doi.org/10.1109/TPWRS.2007.908475.
[31] F. Fernández-Bernal, L. Rouco, P. Centeno, M. González, M. Alonso, Modelling of photovoltaic plants for power system dynamic studies, in: 2002 Fifth International Conference on Power System Management and Control Conf. Publ. No. 488, April, 2002, pp. 341–346, https://doi.org/10.1049/cp:20020059.
[32] Electrical Energy Storage, White Paper, IEC, 2011.
[33] M. Beaudin, H. Zareipour, A. Schellenberglabe, W. Rosehart, Energy storage for mitigating the variability of renewable electricity sources: an updated review, Energy Sustain. Dev. 14 (2010) 302–313.
[34] P.F. Ribeiro, B.K. Johnson, M.L. Crow, A. Arsoy, Y. Liu, Energy storage systems for advanced power applications, Proc. IEEE 89 (12) (2001) 1774–1756.
[35] F. Milano, Á. Ortega, Converter-Interfaced Energy Storage Systems—Context, Modelling and Dynamic Analysis, Cambridge University Press, Cambridge, UK, 2019.
[36] Á. Ortega, F. Milano, Combined frequency and RoCoF control of converter-interfaced energy storage systems, in: IFAC Workshop on Control of Smart Grid and Renewable Energy Systems (CSGRES), Jeju, South Korea, June, 2019, pp. 1–7.
[37] F. Milano, R. Zárate-Miñano, Study of the interaction between wind power plants and SMES systems, 12th Wind Integration Workshop, October, London, UK 2013.
[38] Á. Ortega, F. Milano, Design of a control limiter to improve the dynamic response of energy storage systems, in: Power Engineering Society General Meeting, IEEE, July, Denver, Colorado, USA, 2015.
[39] S.K.M. Kodsi, C.A. Cañizares, Modeling and simulation of IEEE 14-bus system with FACTS controllers, University of Waterloo, Tech. Rep., 2003-3. Waterloo, 2003.
[40] A. Chauhan, R. Saini, Statistical analysis of wind speed data using Weibull distribution parameters, in: 1st International Conference on Non Conventional Energy (ICONCE 2014), January, Kalyani, India, 2014, pp. 160–163.

[41] J. Rocabert, A. Luna, F. Blaabjerg, P. Rodríguez, Control of power converters in AC microgrids, IEEE Trans. Power Electron. 27 (11) (2012) 4734–4749.

[42] L. Zhang, L. Harnefors, H. Nee, Power-synchronization control of grid-connected voltage-source converters, IEEE Trans. Power Syst. 25 (2) (2010) 809–820.

[43] X. Wang, Y.W. Li, F. Blaabjerg, P.C. Loh, Virtual-impedance-based control for voltage-source and current-source converters, IEEE Trans. Power Electron. 30 (12) (2015) 7019–7037.

[44] J. Chen, T. O'Donnell, Parameter constraints for virtual synchronous generator considering stability, IEEE Trans. Power Syst. 34 (3) (2019) 2479–2481.

[45] J. Chen, M. Liu, T. O'Donnell, Replacement of synchronous generator by virtual synchronous generator in the conventional power system, IEEE PES General Meeting, August, 2019, pp. 1–5.

[46] J. Chen, M. Liu, C. O'Loughlin, F. Milano, T. O'Donnell, Modelling, simulation and hardware-in-the-loop validation of virtual synchronous generator control in low inertia power system, in: Power Systems Computation Conference (PSCC), June, 2018, pp. 1–7.

[47] J. Liu, Y. Miura, T. Ise, Comparison of dynamic characteristics between virtual synchronous generator and droop control in inverter-based distributed generators, IEEE Trans. Power Electron. 31 (5) (2016) 3600–3611.

[48] J. Chen, M. Liu, F. Milano, T. O'Donnell, Placement of virtual synchronous generator controlled electric storage combined with renewable generation, in: IEEE PowerTech, June, 2019, pp. 1–6.

[49] I. Cvetkovic, D. Boroyevich, R. Burgos, C. Li, M. Jaksic, P. Mattavelli, Modeling of a virtual synchronous machine-based grid-interface converter for renewable energy systems integration, in: IEEE 15th Workshop on Control and Modeling for Power Electronics (COMPEL), June, 2014, pp. 1–7.

[50] J. Alipoor, Y. Miura, T. Ise, Power system stabilization using virtual synchronous generator with alternating moment of inertia, IEEE J. Emerg. Sel. Top. Power Electron. 3 (2) (2015) 451–458.

[51] D. Li, Q. Zhu, S. Lin, X.Y. Bian, A self-adaptive inertia and damping combination control of VSG to support frequency stability, IEEE Trans. Energy Convers. 32 (1) (2017) 397–398.

[52] F. Wang, L. Zhang, X. Feng, H. Guo, An adaptive control strategy for virtual synchronous generator, IEEE Trans. Ind. Appl. 54 (5) (2018) 5124–5133.

[53] A. Ballanti, L.N. Ochoa, K. Bailey, S. Cox, Unlocking new sources of flexibility: CLASS: the world's largest voltage-led load-management project, IEEE Power Energy Mag. 15 (3) (2017) 52–63.

[54] A.J. Collin, G. Tsagarakis, A.E. Kiprakis, S. McLaughlin, Development of low-voltage load models for the residential load sector, IEEE Trans. Power Syst. 29 (5) (2014) 2180–2188, https://doi.org/10.1109/TPWRS.2014.2301949.

[55] N. Mohan, T.M. Undeland, W.P. Robbins, Power Electronics: Converters, Applications and Design, third ed., John Wiley & Sons, New York, NY, 2003.

[56] M.A.A. Murad, G. Tzounas, M. Liu, F. Milano, Frequency control through voltage regulation of power system using SVC devices, in: IEEE PES General Meeting, August, 2019, pp. 1–5.

[57] H.M. Ayres, I. Kopcak, M.S. Castro, F. Milano, V.F. da Costa, A didactic procedure for designing power oscillation dampers of FACTS devices, Simul. Model. Pract. Theory 18 (6) (2010) 896–909, https://doi.org/10.1016/j.simpat.2010.02.007.

[58] M.A.A. Murad, Á. Ortega, F. Milano, Impact on power system dynamics of PI control limiters of VSC-based devices, in: 2018 Power Systems Computation Conference (PSCC), June, 2018, pp. 1–7, https://doi.org/10.23919/PSCC.2018.8442549.

[59] J. Egan, P. O'Rourke, R. Sellick, P. Tomlinson, B. Johnson, S. Svensson, Overview of the 500 MW EirGrid East-West Interconnector, considering System Design and execution-phase issues, Universities Power Engineering Conference (UPEC), September, 2013, pp. 1–6.

[60] EirGrid, SONI, All-Island Ten Year Transmission Forecast Statement 2016, Tech. Rep., 2017.

[61] X. She, A.Q. Huang, R. Burgos, Review of solid-state transformer technologies and their application in power distribution systems, IEEE J. Emerg. Sel. Top. Power Electron. 1 (3) (2013) 186–198.

[62] J. Chen, R. Zhu, M. Liu, G. De Carne, M. Liserre, F. Milano, T. O'Donnell, Smart transformer for the provision of coordinated voltage and frequency support in the grid, in: IECON—44th Annual Conference of the IEEE Industrial Electronics Society, October, 2018, pp. 5574–5579.

[63] A.A. Milani, M.T.A. Khan, A. Chakrabortty, I. Husain, Equilibrium point analysis and power sharing methods for distribution systems driven by solid-state transformers, IEEE Trans. Power Syst. 33 (2) (2018) 1473–1483.

[64] Y. Tu, J. Chen, H. Liu, T. O'Donnell, Smart transformer modelling and hardware in-the-loop validation, in: IEEE 10th International Symposium on Power Electronics for Distributed Generation Systems, June, 2019, pp. 1–7.

[65] G. De Carne, M. Liserre, C. Vournas, On-line load sensitivity identification in LV distribution grids, IEEE Trans. Power Syst. 32 (2) (2017) 1570–1571.

Chapter 8

Control of power electronics-driven power sources

Taoufik Qoria and Xavier Guillaud

Laboratory of Electrical Engineering and Power Electronics, École Centrale de Lille, Lille, France

8.1 Introduction

Power electronic converters have been developed for several tenths of years with many types of applications. One of the major applications is the drive of electrical machines mainly used in the beginning for industrial applications, and now propagated in many household appliances. The beginning of the renewable energy story had a moderate influence on the power electronic converters market. Indeed, in the old small wind turbines, the mechanical power was converted to electrical power thanks to induction machines directly connected to the grid. Very quickly, the power of the wind turbine increased and more rules were imposed, for example, voltage ride-through capability, injection of reactive power [1]. The connection of wind turbines with power electronic converters became compulsory also to optimize the aerodynamic power conversion. At the same time, photovoltaic energy started also to grow with an obvious mandatory direct current (**DC**)/alternating current (AC) conversion, which requires a power electronic interfacing. Beside renewables, the development of many projects of high-voltage direct current (HVDC) links also increases the impact of power converters on the grid.

The influence of power electronic converters in the grid cannot be neglected anymore; also many studies confirm the good behavior of the new grids comprising a high rate of these new devices. Moreover, it is likely that the type of connection currently used will not be sufficient in the future. Indeed, the power electronic converters are connected to the grid as current injectors (i.e., they are controlled based on the grid-following concept). It assumes that the grid is strong enough to behave as a voltage source at the point of common coupling of the converter. However, with the increase of these power electronic converters, this becomes less true. It is likely that some converters will need to switch from current source mode to a voltage source mode. This is the concept of grid-forming control.

Converter-Based Dynamics and Control of Modern Power Systems
https://doi.org/10.1016/B978-0-12-818491-2.00008-0

The general idea of this chapter is to start from the fundamental concepts of active power management in the grid with a voltage source to propose a classification of different types of controls for power electronic converters based on fundamental thinking.

Some general ideas are first recalled for the grid-following converter, but more focus is made on the grid-forming control. This approach is valid for the two main topologies found on the grid: Two-level voltage source converter (VSC) and modular multilevel converters (MMC).

8.2 Main topologies used for the power electronic converters connected to the grid

8.2.1 Two-level voltage source converter

As mentioned previously, VSC has already been used for many years mainly for the drive of electrical machines. The first question to ask is why "voltage" source converter and not "Current" source converter? Indeed, before using transistors for the drive of DC motor or synchronous motor, the Thyristors based-power converters were connected to a current source. Owing to the operating mode of the Thyristors, the current source operation mode was compulsory. When switching to transistor based-power converters, both choices, that is, current source or voltage source were available. This choice had an influence on the type of switches. Indeed, a current source converter (CSC) supposes to use a switch composed of a transistor in series with a diodes whereas the VSC needs a transistor in parallel with a diode, as shown in Fig. 8.1.

In case of a CSC, the two switches in series induce more losses than in the case of a VSC. The VSC has been chosen for a matter of efficiency and because it is easier to manage a voltage source with a capacitor than a current source with an inductor.

A converter is always a double modulator of electrical quantities. In the case of a VSC, the DC bus voltage is modulated to generate the modulated voltage on the AC side (v_{ma}, v_{mb}, v_{mc}), as illustrated in Fig. 8.2, but at the same time, the grid currents are modulated to generate the DC current i_m. Thus, this topology could be either called "DC VSC" or "AC CSC." If the modulation is intrinsically symmetric, the control applied on this converter is not symmetric. Indeed, only one modulation is used to achieve a given functionality. Most of time, the voltage modulation is used in order to control the AC current. This is the reason why this converter is named "VSC," since the modulation of the DC voltage source is much more used than the modulation of the grid currents.

8.2.2 Modular multilevel converter

For the high power applications, the DC bus voltage is limited by the technology of the cables. For several years, the nominal voltage for the DC cable

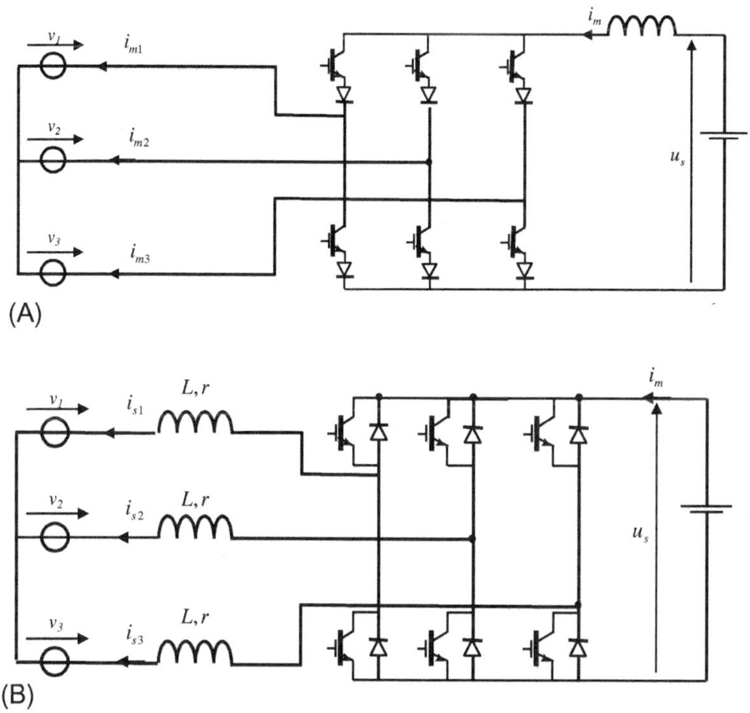

FIG. 8.1 Current source converter vs voltage source converter. (A) Current source converter (CSC) and (B) Voltage source converter (VSC).

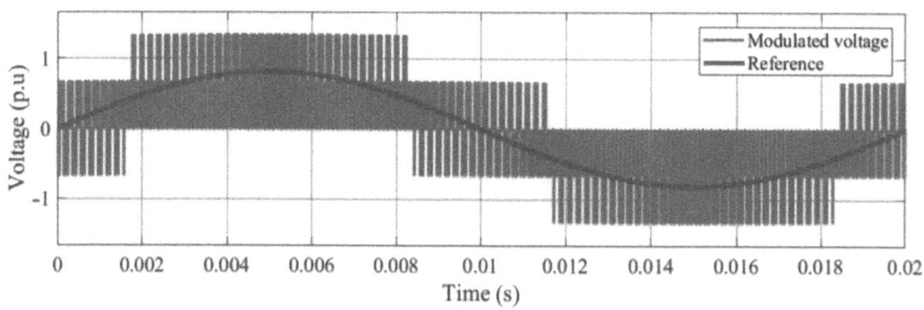

FIG. 8.2 Voltage source converter waveforms (v_{ma}).

has to been set to 320 kV between one pole and the ground. Now, some new projects are expected with higher voltage 400 kV, up to 525 kV. In all cases, this voltage supposes to put hundreds of transistors in series. Switching hundreds of transistors at the same time is extremely challenging [2], the concept of MMC is a solution to this issue. It is based on a series association of elementary submodules, which do not switch at the same time. Submodules may use a half-bridge, full-bridge, or hybrid topologies [3]. Fig. 8.3 depicts the half-bridge submodule, which is the most used application in today's industrial applications [4].

Each submodule may have two states depending on the state of the transistor and the current direction. One state is called "active state": the phase current is flowing into the capacitor.

Two possible situations:	$i>0$ and T_{2j} is off; D_{1j} is conducting the current $i<0$ and T_{1j} is on; T_{1j} is conducting the current

The other state is called "inactive state": the submodules are short circuited.

Two possible situations:	$i>0$ and T_{2j} is on; T_{2j} is conducting the current $i<0$ and T_{1j} is off; D_{2j} is conducting the current

Fig. 8.4 illustrates the four different situations. A stack of N submodules is called an arm, each arm is connected to an inductor ($L_{arm}R_{arm}$) and its aim is to limit the derivative of the current in case of a DC fault. Each phase comprise of two identical arms. One is connected to the positive pole of the DC bus and the other is connected to the negative pole of the DC bus. The association of two arms in the same phase is called a leg. The MMC is composed of three legs (see Fig. 8.5).

The series association of hundreds of elementary submodules results in a quasisinusoidal modulated voltage waveform, as shown in Fig. 8.6. As the phase currents (i_u, i_l) are flowing into the capacitors, the voltage across the capacitors has to be balanced. This is the aim of the balancing control algorithm (BCA). Many papers already have been published on this topic [5, 6].

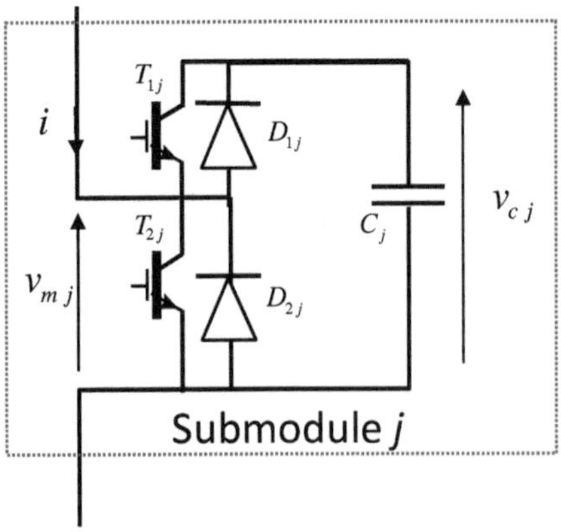

FIG. 8.3 Topology of an elementary submodule used in a MMC.

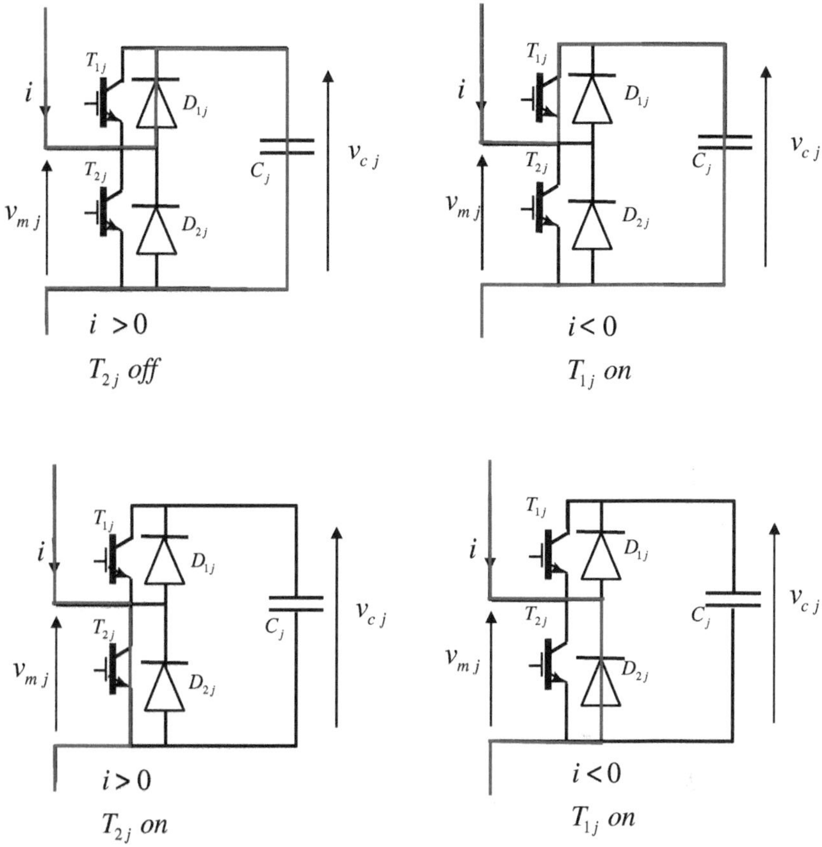

FIG. 8.4 The four situations of a MMC submodule.

8.3 General considerations about power control in a voltage source converter

A VSC connected on a grid may have various roles, but in all cases, an active power control has to be implemented. The general organization of the control depends on the choice made for the type of active power control. To develop this idea, a single phase quasistatic model of the AC side of the VSC is given in Fig. 8.7. In the sequel, the resistors are neglected.

Let us define the phasor variables:

Grid voltage:	$\overline{V}_g = V_g e^{j\delta_g}$
Modulated voltage:	$\overline{V}_m = V_m e^{j\delta_m}$
Grid current:	$\overline{I}_g = I_g e^{j\theta_i} = I_g e^{j(\delta_g - \phi)} = \overline{I}_P + j\overline{I}_Q$

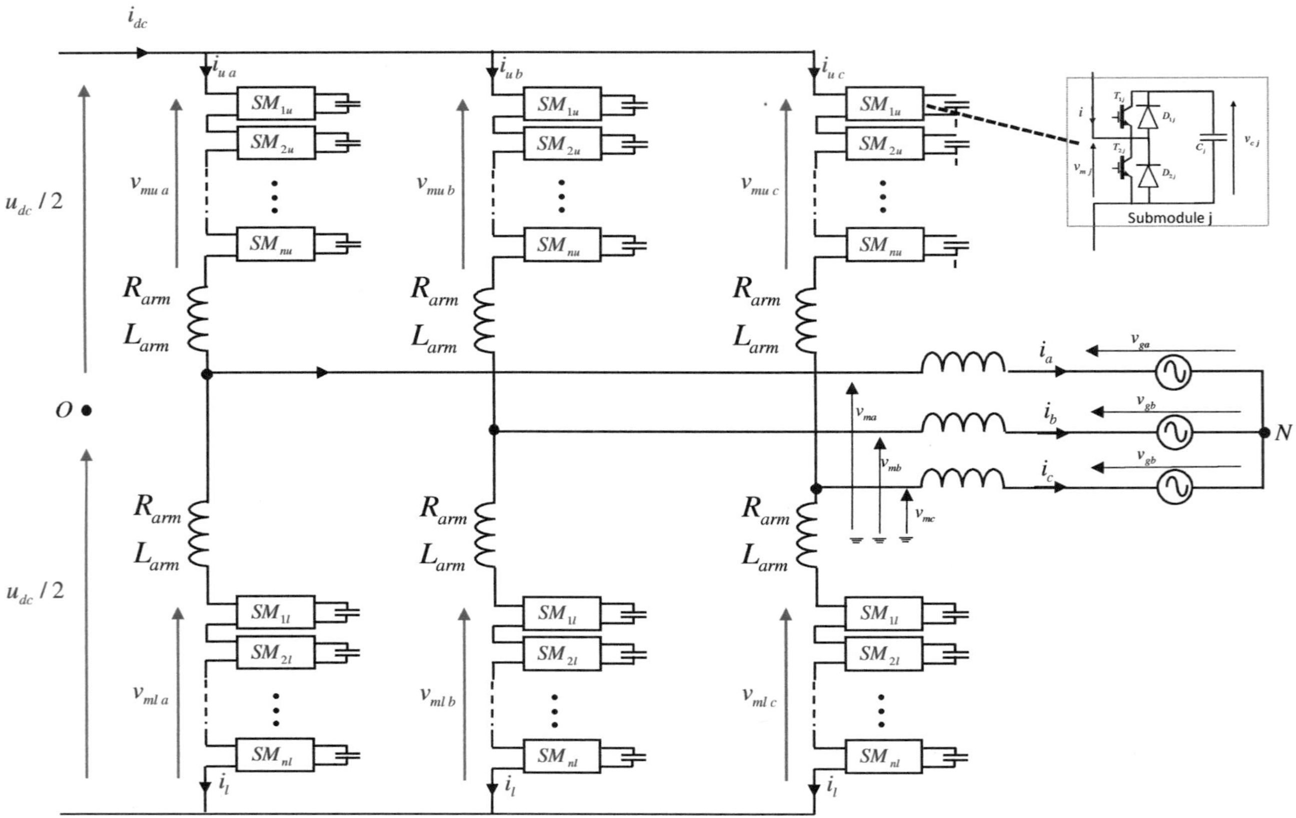

FIG. 8.5 MMC topology.

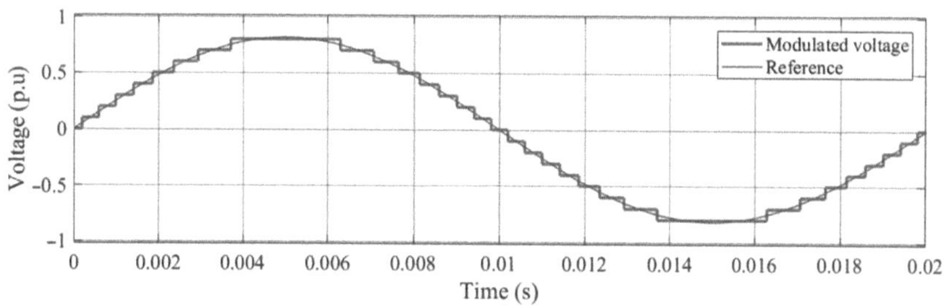

FIG. 8.6 AC voltage waveform by the MMC (v_{ma}).

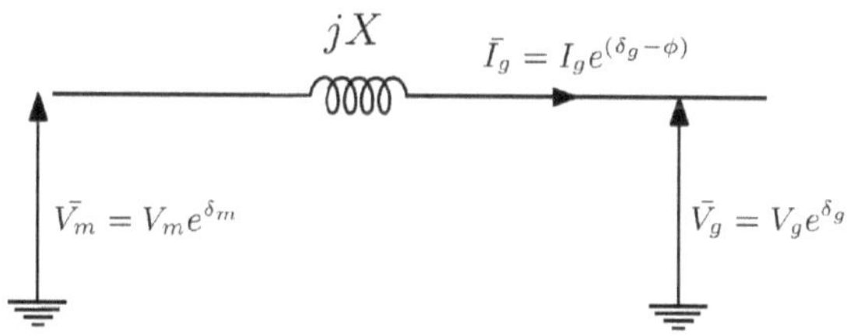

FIG. 8.7 Single phase quasi static model of the VSC connected to the grid.

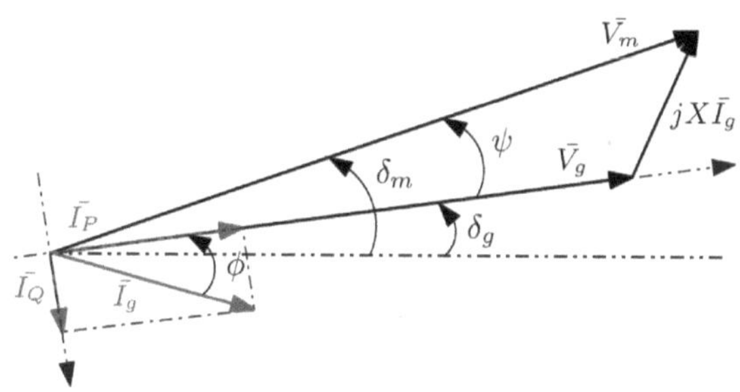

FIG. 8.8 Phasor representation of the basic system.

In quasistatic operation, the frequency is supposed to be nearly constant, unique in the whole system and equal to the grid frequency ω_g.The grid current can be decomposed into two components:

$\overline{I}_P$ is the active current and $\overline{I}_Q$ is the reactive current. Fig. 8.8 shows the phasor representation of this simple system.

The active power at the point of common coupling is:

$$P = \mathrm{Re}\left(\overline{V}_g \overline{I_g}^*\right) = V_g I_g \cos(\phi) = V_g I_P \tag{8.1}$$

In steady state, another formulation, based on the voltage, can be derived from this expression. Neglecting the resistor, a Kirchhoff law applied on this simple system yields:

$$\overline{V}_m = jX\overline{I} + \overline{V}_g \tag{8.2}$$

It results in:

$$P = \mathrm{Re}\left(\overline{V}_g\left(\frac{\overline{V}_m - \overline{V}_g}{jX}\right)^*\right) = \frac{V_g V_m}{X}\sin(\psi) = \frac{V_g V_m}{X}\sin(\delta_m - \delta_g) \quad (8.3)$$

The formulation of the power based on the current is the origin of the current controlled-VSC. The second formulation is the origin of the voltage controlled-VSC.

As already mentioned, one of the aim of the converter is to control the active power exchange. Let's define P^* as the power set point. When using the first formulation of the active power referring to the current, it can be deduced that the reference for the active current $I_{P_{ref}}$ is obtained through a division by the root mean square (RMS) voltage at the point of common coupling (PCC) (Fig. 8.9A). A current loop is implemented, and in steady state, it is assumed that:

$$I_P = I_P^* \quad (8.4)$$

A more robust control for the power will be achieved with a closed-loop control (Fig. 8.9B). The inevitable inaccuracies of the model will be compensated. Moreover, if an integral control is embedded in the controller, in steady state, it yields:

$$P = P^* \quad (8.5)$$

Controlling the active current I_P means that the phase of the grid voltage is known in order to align $\overline{I}_P$ on $\overline{V}_g$, as shown in Fig. 8.8.

A first consequence can be drawn: for this control, an estimation of the grid angle is compulsory.

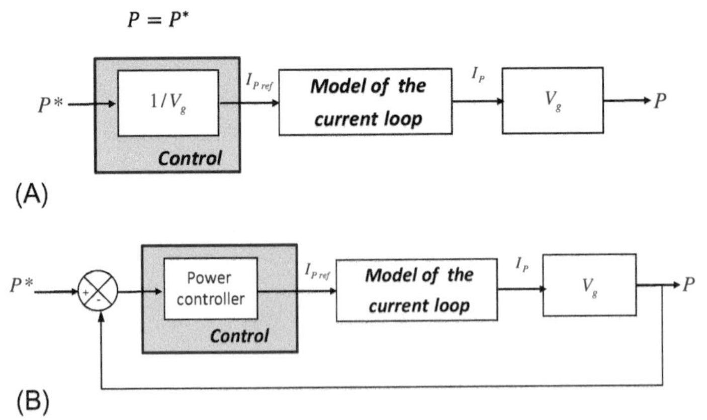

FIG. 8.9 Principle of the power control with the current. (A) Open loop power control, (B) Closed loop power control.

Some similar considerations could be done on the reactive power Q: a reactive power reference Q^* defines a reference for the reactive I_Q^* and in steady state:

$$I_Q = I_Q^* \tag{8.6}$$

To sum up, given P^* and Q^*, the current $\overline{I}_g$ is known. As $\overline{V}_g$ is an external input for the system, it can be deduced that, in case of grid voltage variation, the voltage $\overline{V}_m$ "follows" the grid voltage $\overline{V}_g$ to maintain the grid current constant $\overline{I_g}$ in phase and magnitude. This is the origin of the so-called grid-following control. Actually, in the literature, the grid-following terminology can also be found under the name "grid-feeding".

As the control is achieved in the time-domain simulation, the control has to generate instantaneous current references. Let us define the notations:

Time-domain angle:	Grid voltage angle:	θ_g
	Estimate of θ_g:	$\tilde{\theta}_g$

In time-domain, the general organization of a grid-following control is given in Fig. 8.11. It is decomposed in three main parts:

- Average value control of the active and reactive power. The reactive power loop can be replaced by an RMS voltage loop.
- Estimation of the grid angle.
- Current control. In Fig. 8.10, the current control is implemented in (abc) frame, whereas in Fig. 8.11 the current control is implemented in $(d\text{-}q)$ frame. This does not change the general organization. In any case, an

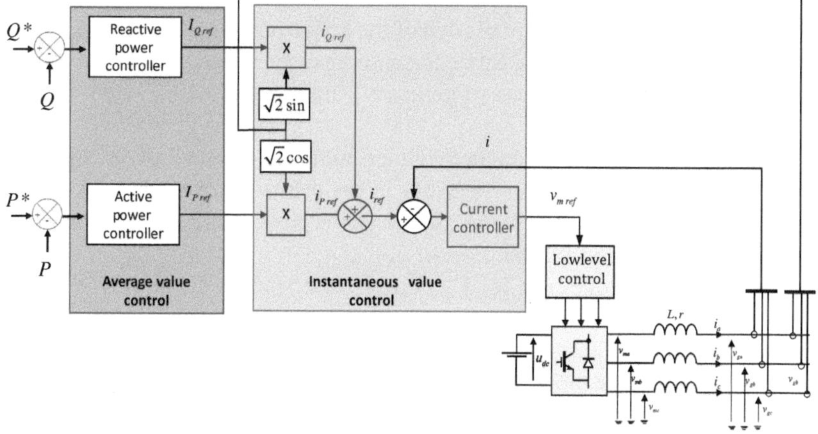

FIG. 8.10 General organization of the grid-following control in (abc) frame.

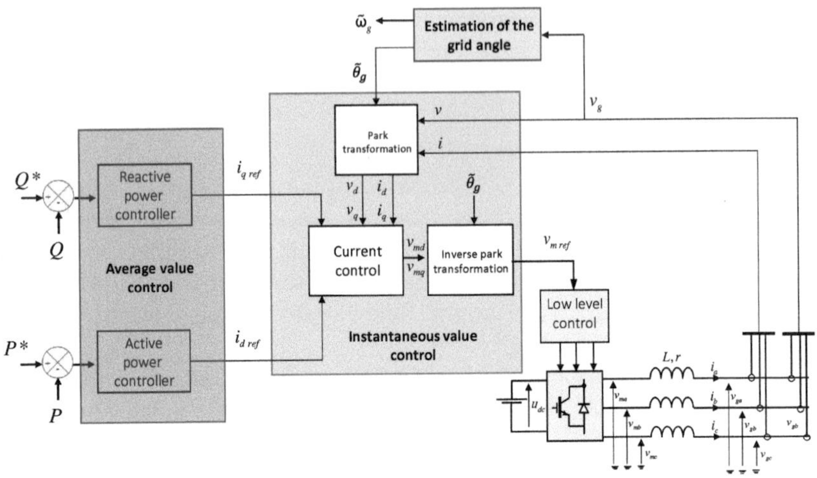

FIG. 8.11 General organization of the grid-following control in (*d-q*) frame.

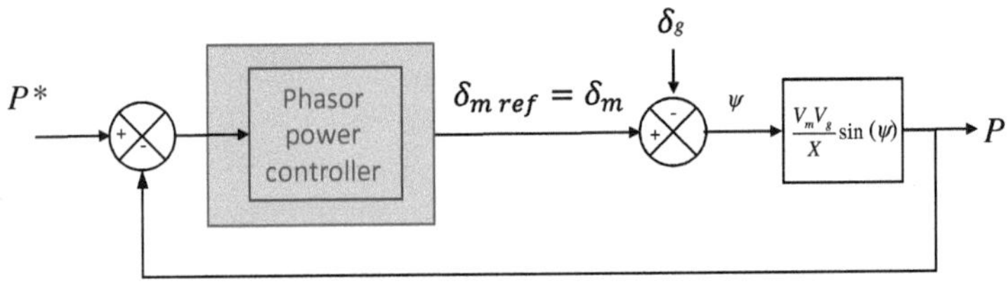

FIG. 8.12 Principle of the closed-loop power control with the voltage controlled mode.

estimate of the grid angle is needed either to generate the three-phase sinusoidal references or the angle to be used in the park transform.

Let us now use the second formulation of the active power to explain the basics of the second type of control. Since the active power is linked with the modulated angle δ_m, the control has to generate a reference for this angle $\delta_{m_{ref}}$, as shown in Fig. 8.12.

It is assumed that the modulated voltage angle $\delta_{m_{ref}}$ is equal to δ_m.No information on the grid angle is compulsory in the control. However, as it can be seen, the grid angle δ_g acts as a disturbance on the power control. This important point will be discussed further.

The magnitude reference for V_m, $V_{m_{ref}}$ is independent of the active power control, it can be modified with respect to a voltage or reactive power loop. Hence, contrary to the previous control, a variation on the grid voltage has no direct effect on the modulated voltage since the active power only generates an angle reference and not the magnitude. To modify the difference of angle between the modulated voltage angle δ_m and the grid voltage angle δ_g, the control has to modify slightly the modulated voltage frequency. An internal frequency has to be defined: ω_m.This frequency is equal to the grid frequency in steady-state.

The real control is implemented in time domain, so the proposed control has to generate a time domain angle. Let us define the time-domain angles associated with the phasor angles:

$$\theta_g = \omega_g t + \delta_g \quad \theta_m = \omega_m t + \delta_m \tag{8.7}$$

With:

ω_g: grid frequency

ω_m: modulated voltage frequency

As for the previous control, this control can be implemented either in (*abc*) frame (Fig. 8.13A) or in (*d-q*) frame (Fig. 8.13B).

Contrary to the grid-feeding/following control, the angle used in this control is linked with the modulated voltage v_m and not with the grid voltage anymore.

Since the control generates a voltage reference, it is called grid-forming control in the literature. Note that this voltage control can also be used in

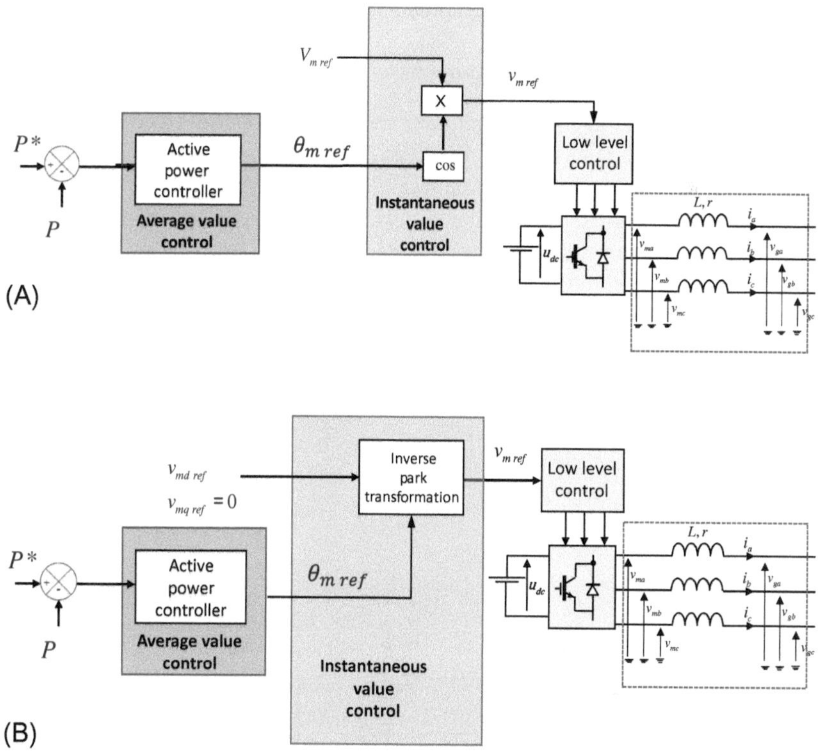

FIG. 8.13 General organization of the grid-forming control. (A) (*abc*) frame, (B) (*d-q*) frame.

stand-alone. In this case, the active power is imposed by the load and not the control anymore.

This section has presented the main principles of the two types of controls for the VSC. In the next section, both controls are detailed with a major focus on the grid-forming control since the grid-following control has already been greatly documented in many publication and hand books.

8.4 Current control of a VSC—Grid-following control

8.4.1 Introduction

The previous section has presented the main functionalities of grid-following control. In this section the different functionalities are detailed for the two main topologies found on the grid: two level converters and MMC. The synchronization to the grid is the same for both converters but the MMC needs one more control due to possible internal current circulation. In the last part of this section, some considerations will be given on the ancillary services that this kind of converters can provide to the grid.

8.4.2 Synchronization to the grid

As explained before, this control requires the knowledge of the grid angle. This functionality is not always easy to achieve especially in case of unbalanced three-phase voltages induced by a fault on the AC grid. Many papers and handbooks have been published on this topic. Here, the classical three-phase phase locked loop (PLL) also called synchronous reference frame phase-locked loop (SRF-PLL) is presented. The usual structure of the SRF-PLL is shown in Fig. 8.14 and the major principle is recalled in the following lines.

A Park transformation is applied on the three-phase voltage at the point of common coupling to generate the two (d-q) components v_{gd}, v_{gq}.

$$\begin{bmatrix} v_{gd} \\ v_{gq} \end{bmatrix} = \sqrt{\frac{2}{3}} \begin{bmatrix} \cos\left(\tilde{\theta}_g\right) & \cos\left(\tilde{\theta}_g - \frac{2\pi}{3}\right) & \cos\left(\tilde{\theta}_g - \frac{4\pi}{3}\right) \\ -\sin\left(\tilde{\theta}_g\right) & -\sin\left(\tilde{\theta}_g - \frac{2\pi}{3}\right) & -\sin\left(\tilde{\theta}_g - \frac{4\pi}{3}\right) \end{bmatrix} \begin{bmatrix} v_{ga} \\ v_{gb} \\ v_{gc} \end{bmatrix} \tag{8.8}$$

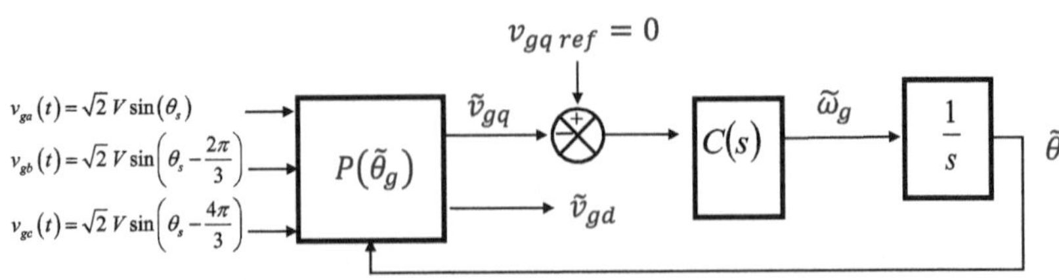

FIG. 8.14 Structure of the SRF-PLL.

This expression can be simplified:

$$\begin{bmatrix} v_{gd} \\ v_{gq} \end{bmatrix} = \sqrt{3}V_G \begin{bmatrix} \cos\left(\theta_g - \tilde{\theta}_g\right) \\ \sin\left(\theta_g - \tilde{\theta}_g\right) \end{bmatrix} \tag{8.9}$$

If the v_{gq} component is canceled, it means that $\tilde{\theta}_g = \theta_g$.This is the aim of the closed loop system presented in Fig. 8.14. The v_{gq} component is compared with a null reference and the difference is applied on a PI control which generates an estimate of the grid frequency $\tilde{\omega}_g$. This frequency is integrated to give an estimate of the grid frequency $\tilde{\theta}_g$. A block diagram of the system is presented in Fig. 8.15A. It highlights the nonlinearity of the system. It is possible to linearize the system around an operating point, by assuming a small difference between the grid phase angle and the estimated one. The nonlinearity can be neglected ($\sin\left(\theta_g - \tilde{\theta}_g\right) \approx \left(\theta_g - \tilde{\theta}_g\right)$). On the basis of this assumption a linear model is obtained, as shown in Fig. 8.15B.

To avoid the steady state error between the voltage reference $v_{gq\ ref}$ and the measured quadratic voltage v_q, a proportional integral (PI) controller is required:

$$C(s) = k_{p_{PLL}} + \frac{k_{i_{PLL}}}{s} \tag{8.10}$$

where, $k_{p_{PLL}}$and $k_{i_{PLL}}$ are the proportional control gain and integrator gain, respectively.

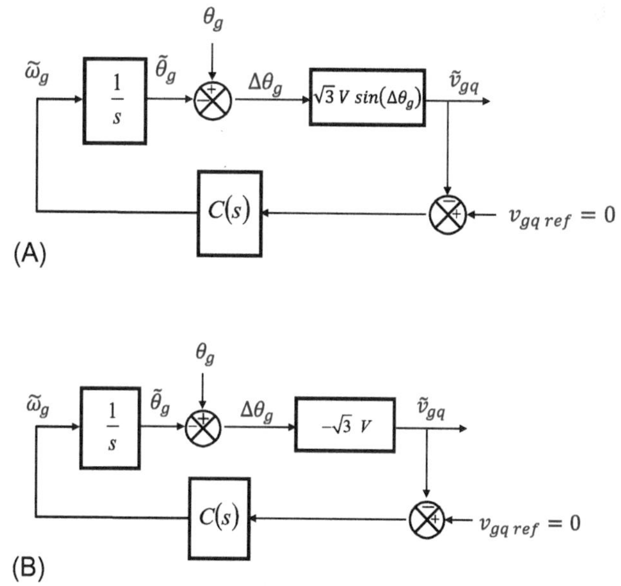

FIG. 8.15 Block diagram of the system to control (A) nonlinear model, (B) linear model.

From the control structure in Fig. 8.15B, the characteristic polynomial $P_c(s)$ of the system is:

$$P_c(s) = 1 - \frac{k_{p_{PLL}}}{\sqrt{3}Vk_{i_{PLL}}}s - \frac{1}{\sqrt{3}Vk_{i_{PLL}}}s^2 \tag{8.11}$$

It is then possible to identity this polynomial to a reference second order polynomial $P_{ref}(s)$.

$$P_{ref}(s) = 1 + \frac{2\xi_{ref}}{\omega_{nref}}s + \frac{1}{\omega_{nref}^2}s^2 \tag{8.12}$$

ξ_{ref} and $\omega_{n\ ref}$ are characteristic of the 2nd order polynomial. Choosing a value of 0.7 for the damping, yields to the relation between the 5% response times $T^R_{5\%}\omega_{n_{ref}}$:

$$T^R_{5\%} \cong \frac{3}{\omega_{nref}} \tag{8.13}$$

In most cases, the PLL response time is chosen between 10 ms $< T^R_{5\%} <$ 100 ms. It is then possible to calculate $\omega_{n\ ref}$ and then the value of the PI controller:

$$k_{p_{PLL}} = -\frac{2\xi_{ref}\omega_{nref}}{\sqrt{3}V_G}, k_{i_{PLL}} = -\frac{\omega_{nref}^2}{\sqrt{3}V_G} \tag{8.14}$$

8.4.3 AC current loop

As presented in Figs. 8.10 and 8.11, two solutions exist for the control of the grid current. One is implemented in the (*abc*) frame. The currents are following some sinusoidal references. The resonant controllers [7] are well adapted for this kind of control: since an infinite gain at a given frequency is induced by the resonant element of the controller the error between to sinusoidal variables can be canceled in steady state.

However, most of the current controllers are implemented in a (*d-q*) frame. Indeed, with an adequate choice for the frame frequency the variables can be constant in steady state. A simple PI controller is sufficient for a good accuracy of the current control.

Fig. 8.16 presents the block diagram of the current control and the model of the system to be controlled. Two (*d-q*) frame are defined, one for the grid (frequency ω_g, angle δ_g), the other for the control (frequency $\tilde{\omega}_g$, angle $\tilde{\delta}_g$). In steady state, both (*d-q*)frames are supposed to be superimposed. The design of the control is based on this assumption. Thanks to the compensation and decoupling actions in the control, it is as if the output of the two controllers $v_{rld\ ref}$, $v_{rlq\ ref}$ were directly driven the input of the *R*, *L* filter v_{rld}, v_{rlq}. Since the resistance can, most of the time be neglected, the design of the PI controller becomes very simple since the system to control can be assimilated to a PI

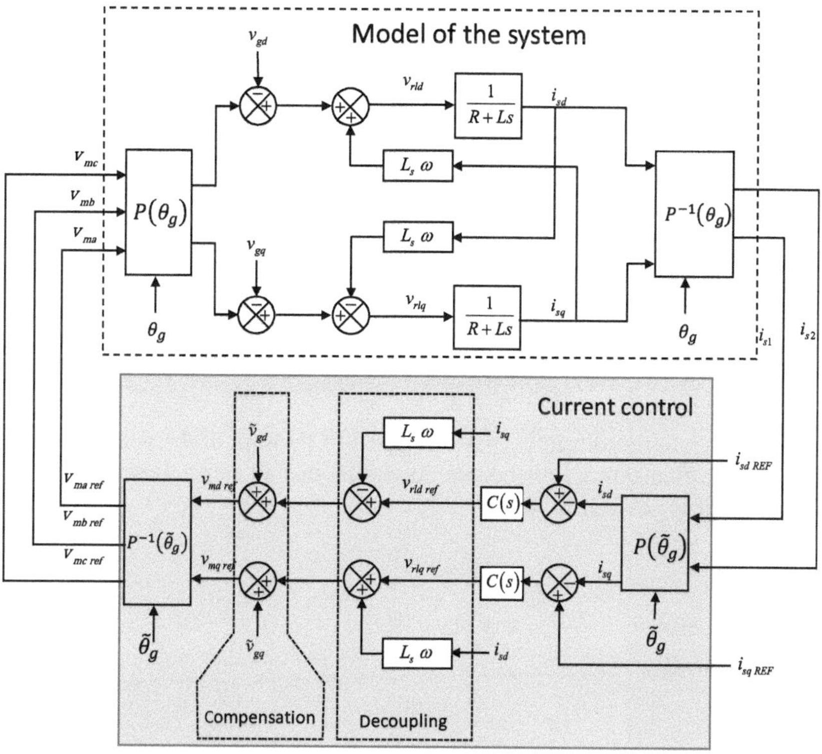

FIG. 8.16 Block diagram of the current control in (d-q) frame.

controller in series with an integrator $^1/_{Ls}$. The same method as for the PLL can be used to derive the parameters of the controller.

This method works properly when the grid is strong. In case of weak grid, the parameters of the PI controller have to be adjusted. This have been explained in some publications [8].

8.4.4 MMC control

To design the high-level control, a simplified model is needed for each arm. Indeed, it is possible to define an equivalent model for each arm. The following demonstration is valid for any phase and the upper and lower arm. To simplify the notations, the phase index is removed, the index "u/l" is used to refer to the upper and lower arm in the same time.

Let's define v_{ctot} as the sum of all the N elementary submodule voltages of an arm:

$$v_{ctotu/l} = \sum_{i=1}^{N} v_{ci} \tag{8.15}$$

Then:

$$\frac{C}{N}\frac{dv_{ctotu/l}}{dt}=\frac{C}{N}\sum_{i=1}^{N}\frac{dv_{ci}}{dt} \tag{8.16}$$

For each of the active submodules, it is possible to write:

$$i_{u/l}=C\frac{dv_{ci}}{dt} \tag{8.17}$$

For the nonactive submodules, the variation of voltage is null.
Let us define $n_{u/l}$ the number of active submodules in an arm. It yields:

$$\frac{C}{N}\frac{dv_{ctotu/l}}{dt}=\frac{n_{u/l}}{N}i_{u/l} \tag{8.18}$$

The modulated voltage $v_{mu/l}$ is composed of the sum of the voltage modulated by the elementary submodules. Assuming that all elementary voltages v_{ci} are equal, it is possible to write:

$$v_{mu/l}=\frac{n_{u/l}}{N}v_{ctot} \tag{8.19}$$

Let us define $i_m=\frac{n_{u/l}}{N}i_{u/l}$ and $m_{u/l}=\frac{n_{u/l}}{N}$.
It results in:

$$i_m=m_{u/l}i_{u/l} \tag{8.20}$$

$$v_{mu/l}=m_{u/l}v_{ctot} \tag{8.21}$$

An arm of the MMC can be assimilated to a chopper with a modulation ratio $m_{u/l}$, as shown Fig. 8.17. This model is called Average Arm model [9].
Let us suppose the voltage $v_{NO}=0$, the Kirchhoff laws applied for the phase i (Fig. 8.5), gives the following equations:

$$\frac{u_{dc}}{2}=v_{mu}+R_{arm}i_u+L_{arm}\frac{di_u}{dt}+Ri+L\frac{di_g}{dt}+v_g \tag{8.22}$$

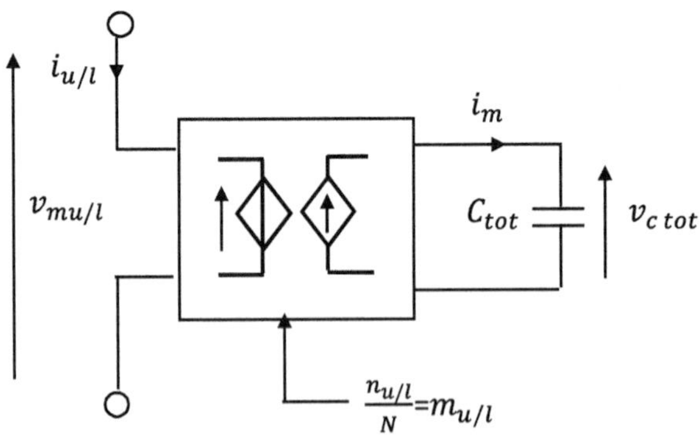

FIG. 8.17 Average arm model of an arm of a MMC.

$$\frac{u_{dc}}{2} = v_{ml} + R_{arm} i_l + L_{arm} \frac{di_l}{dt} - Ri - L\frac{di_g}{dt} - v_g \tag{8.23}$$

Adding and subtracting these two equations results in:

$$u_{dc} = v_{mu} + v_{ml} + R_{arm}(i_u + i_l) + L_{arm} \frac{d(i_u + i_l)}{dt} \tag{8.24}$$

$$0 = v_{mu} - v_{ml} + R_{arm} i_g + L_{arm} \frac{di_g}{dt} + 2Ri_g + 2L\frac{di_g}{dt} + 2v_g \tag{8.25}$$

Let us define: $v_{diff} = \frac{v_{mu} + v_{ml}}{2}$, $v_m = \frac{-v_{mu} + v_{ml}}{2}$, $i_{diff} = \frac{i_u + i_l}{2}$
It yields:

$$v_m - v_g = \left(\frac{R_{arm}}{2} + R\right) i_g + \left(\frac{L_{arm}}{2} + L\right) \frac{di_g}{dt} \tag{8.26}$$

$$\frac{u_{dc}}{2} - v_{diff} = R_{arm} i_{diff} + L_{arm} \frac{di_{diff}}{dt} \tag{8.27}$$

Eqs. (8.17), (8.26), (8.27) represent the MMC model. The block diagram of the model is presented at the top of Fig. 8.18. From this model, it is possible to derive the control of the MMC. Eq. (8.26) shows that it is possible to control the AC current with the set of three-phase voltage $v_{m_{abc}}$. The same kind of (*d-q*) control as in a two-level VSC can be implemented to control the grid current.

The current i_{diff} is flowing into the DC bus or circulating between the arms. Since this circulating current is not desired, a circulating current suppression controller is implemented [10]. It has been demonstrated that this current is a three phase current with a $-2\omega_g$ frequency. A specific (*d-q*) frame with this frequency is used to set this current to 0 in steady state.

The bottom of Fig. 8.18 is giving the general organization of the MMC control. It includes the AC current control and the circulating current suppression controller. The output of this controller are the reference variables for v_{mabc} and $v_{diff_{abc}}$. From these variables, the reference for the modulated voltage $v_{mu_{abc}}$ and $v_{ml_{abc}}$ are calculated by inverting Eqs. (8.26), (8.27). Finally the 6 modulated ratios $m_{u_{abc}}$ and $m_{l_{abc}}$ are derived by dividing the previous variables by the DC bus voltage.

The main difference between MMC and two-level VSC is about the management of the differential current. However, in term of active and reactive power control, the AC current loop in an MMC is similar to what is implemented in a two-level VSC. Simulation results of 20-level MMC are illustrated in Fig. 8.19.

8.4.5 Ancillary services with grid-following converters

Owing to the grid codes imposed to the generators and to the HVDC link connected to the grid, all these devices are asked to provide ancillary services to the grid. Two main functionalities are mentioned: frequency and voltage support.

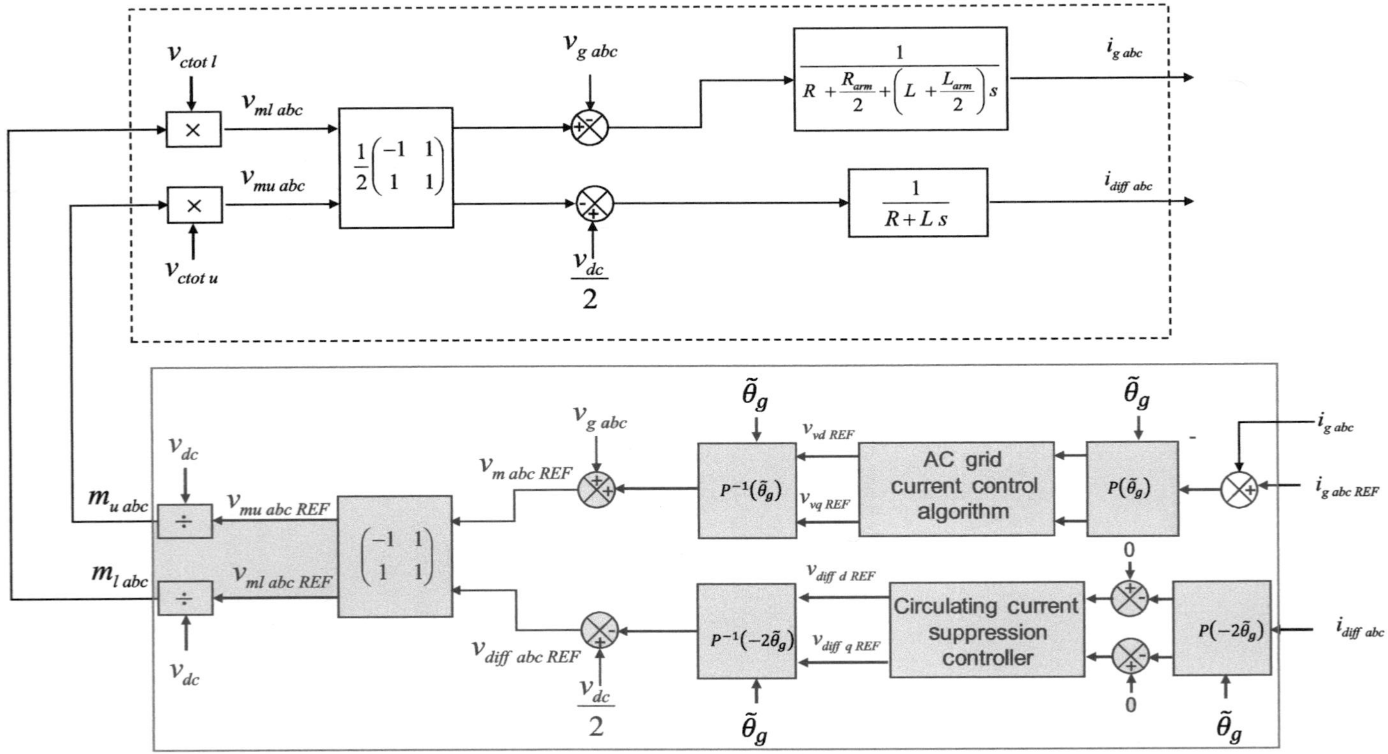

FIG. 8.18 MMC model and control including the AC current control and the circulating current suppression control.

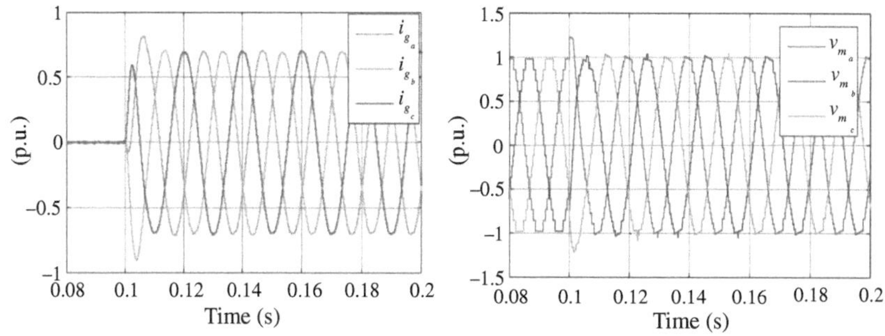

FIG. 8.19 Current and voltage dynamics in response to power step of $P^* = 0.5$ p.u.

The voltage support is linked with the reactive power. Depending on the application, a reactive power control or voltage control is generating the reactive current reference. The active current reference is linked with the frequency support. Since there is an active power control, it is possible to add an outer loop for the frequency support as presented in Fig. 8.20.

As for the synchronous machine control, it is possible to add a dead band Δf_{db}, it is also possible to limit the frequency support to a given amount of power ΔP_{max}. These limitations are very important to limit the solicitations of the primary source.

An inertial effect can be added to this control [11] but it supposes to derivate the frequency. To avoid some unwanted noises in the control a filter on the frequency has to be added. However, it limits the effectiveness of the inertial effect.

8.5 Voltage control of an ideal VSC based grid-forming control

8.5.1 Introduction

As explained in Section 8.3, it is possible to control the active power with the modulated voltage angle δ_m. Eq. (8.3) gives the relation between the power and the angle. However, this a static equation hides some large oscillations which can appear during the transient [12, 13] and which have to be damped. The beginning of this section is first focused on the dynamic model between the active power and the angle and then the integration of a damping resistance to improve the dynamic characteristics.

The basics of the power control with a voltage source are then explained and different ways of implementing the droop control are proposed and compared for an ideal voltage source. The introduction of a LC filter at the output of the VSC is studied. This section ends by some considerations about the possible ancillary services which can be provided by a grid-forming converter.

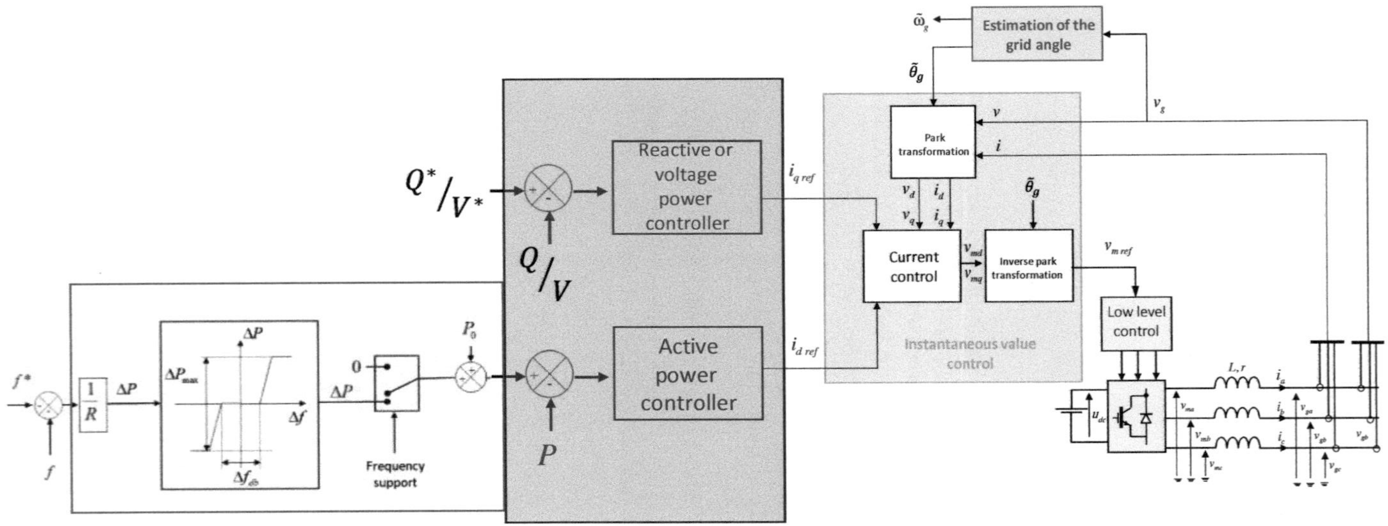

FIG. 8.20 Ancillary services with power electronic converters.

In the sequel, all the models are considered in per unit. The base for the active power is the nominal active power. The base for the voltage is the nominal voltage. For the frequency, the base is the nominal frequency ω_b.

8.5.2 Principle of the power control with the voltage

The system to study is depicted in Fig. 8.21. It is composed of two sets of three-phase voltages (one is an external voltage source which represents the AC grid. The other one is the modulated voltage $v_{m_{abc}}$ controlled by the converter).

The external voltage can be either the voltage at PCC $v_{g_{abc}}$ or the Thevenin equivalent voltage source $e_{g_{abc}}$. Depending on the choice the expression of the power is:

$$P = \frac{V_g V_m}{L\omega_g} \sin\left(\delta_m - \delta_g\right) \tag{8.28}$$

$$P = \frac{E_g V_m}{\left(L + L_g\right)\omega_g} \sin\left(\delta_m\right) \tag{8.29}$$

In this system, the resistors are neglected so the active power is the same at each node of the grid. $e_{g_{abc}}$ is considered as the voltage angle reference.

A dynamic model between δ_m and the active power P delivered by the converter is developed in the following section. In the proposed analysis, only the connection inductance L is taken into consideration in the model. Let us start from the electrical equations of this system expressed in a synchronous (d-q) frame where the frequency is assumed to be equal to the grid frequency. The dynamic equations of this circuit can be written:

$$\frac{L}{\omega_b}\frac{di_{g_d}}{dt} = v_{md} - v_{gd} - R\, i_{g_d} + \omega_g L\, i_{g_q} \tag{8.30}$$

$$\frac{L}{\omega_b}\frac{di_{g_q}}{dt} = v_{mq} - v_{gq} - R\, i_{g_q} - \omega_g L\, i_{g_d} \tag{8.31}$$

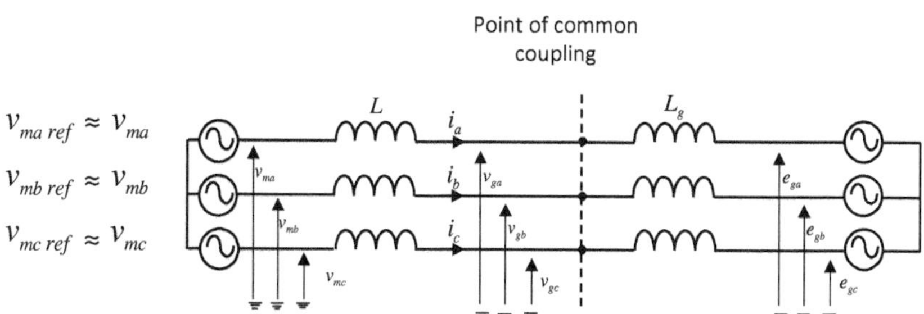

FIG. 8.21 Exchange of power between three-phase voltage sources.

The d-axis is supposed to be aligned with the AC source V_g and taken into account the notations for the modulated voltage v_m, the previous equations becomes:

$$\frac{L}{\omega_b}\frac{di_{gd}}{dt} = V_m \cos(\delta_m) - V_g - R\, i_{g_d} + \omega_g L\, i_{g_q} \tag{8.32}$$

$$\frac{L}{\omega_b}\frac{di_{g_q}}{dt} = V_m \sin(\delta_m) - R\, i_{g_q} - \omega_g L\, i_{g_d} \tag{8.33}$$

These nonlinear equations can be linearized around an operating point for the modulated voltage (V_{m0}, δ_{m0}). The grid frequency is supposed to be constant $\omega_g = \omega_0 = 1$ p. u. The linear expressions of $V_m\cos(\delta_m)$ and $V_m\sin(\delta_m)$ around an operating point x_0 is given by:

$$V_m \cos(\delta_m) = -V_{m0} \sin(\delta_{m0})\Delta\delta_m + \Delta V_m \cos(\delta_{m0}) \tag{8.34}$$

$$V_m \sin(\delta_m) = V_{m0} \cos(\delta_{m0})\Delta\delta_m + \Delta V_m \sin(\delta_{m0}) \tag{8.35}$$

where, the symbol "Δ" and "0" denote respectively the small derivation and the initial value of variables. The instantaneous power delivered by the converter is:

$$P = v_{md} i_d + v_{mq} i_q \tag{8.36}$$

It can be linearized around on operating point:

$$\Delta P = \Delta v_{md} i_{gd0} + v_{md0} \Delta i_{g_d} + \Delta v_{mq} i_{gq0} + v_{mq0} \Delta i_{g_q} \tag{8.37}$$

After some calculation, it yields:

$$\Delta P = \underbrace{\frac{N_{p\delta}}{\left(\frac{L}{\omega_b}s + R\right)^2 + (\omega_g L)^2}}_{T_{p\delta}} \Delta\delta_m + \underbrace{\frac{N_{pV}}{\left(\frac{L}{\omega_b}s + R\right)^2 + (\omega_g L)^2}}_{T_{pV}} \Delta V_m \tag{8.38}$$

With:

$$N_{p\theta} = \frac{V_{m0}\left(\frac{L}{\omega_b}\right)^2 \left(V_g\left(R\sin(\delta_{m0}) + L\omega_g \cos(\delta_{m0})\right) - V_{m0} L\,\omega_g\right)}{(R)^2 + (\omega_g L)^2} s^2$$

$$+ \frac{2R\left(\frac{L}{\omega_b}\right) V_{m0}\left(V_g\left(R_c \sin(\delta_{m0}) + L\omega_g \cos(\delta_{m0})\right) - V_{m0} L\omega_g\right)}{(R)^2 + (\omega_g L)^2} s$$

$$+ \ldots V_{m0} V_g \left(R\sin(\theta_{m0}) + L\omega_g \cos(\theta_{m0})\right)$$

$$N_{pV} = \frac{\left(\frac{L}{\omega_b}\right)^2 \left(V_g\left(-R\cos(\theta_{m0}) + L\omega_g \sin(\theta_{m0})\right) + R\right)}{(R)^2 + (\omega_g L)^2} s^2$$

$$+ \left[\frac{2R_c \frac{L}{\omega_b}\left(V_g\left(-R_c\cos(\theta_{m0}) + L\omega_g \sin(\theta_{m0})\right) + V_{m0}R_c\right)}{(R)^2 + (\omega_g L)^2} + V_{m0}\frac{L}{\omega_b} \right] s$$

$$+ \ldots 2R_c V_{m0} + V_g\left(-R_c\cos(\theta_{m0}) + L_c\omega_g \sin(\theta_{m0})\right)$$

In steady-state, and neglecting the resistance, the expression of the active power becomes:

$$\Delta P = \frac{V_{m0}V_g}{\omega_g L}\cos(\delta_{m0})\Delta\delta_m + \frac{V_{m0}V_g}{\omega_g L}\sin(\delta_{m0})\Delta V_m \tag{8.39}$$

In case the grid angle δ_{m0} is small, the well-known expression of the power is retrieved:

$$\Delta P = \frac{V_{m0}V_g}{\omega_g L}\Delta\delta_m \tag{8.40}$$

The linearized model has been validated by a comparison with the nonlinear dynamic models (see Fig. 8.22). A step is applied on the angle $\Delta\delta_m = 0.15$ rad. The initial active power transfer between sources is set to $p_{ac0} = 0$ p.u.

This simulation result highlights a poorly damped system. The denominator of the transfer function is a second order and the conjugated poles are:

$$s_{1-2} = -\frac{R}{L}\omega_b \pm \omega_g\omega_b \tag{8.41}$$

In a transmission system, the resistance is always extremely low which explains a low damping coefficient and then the large oscillations obtained in the simulation results.

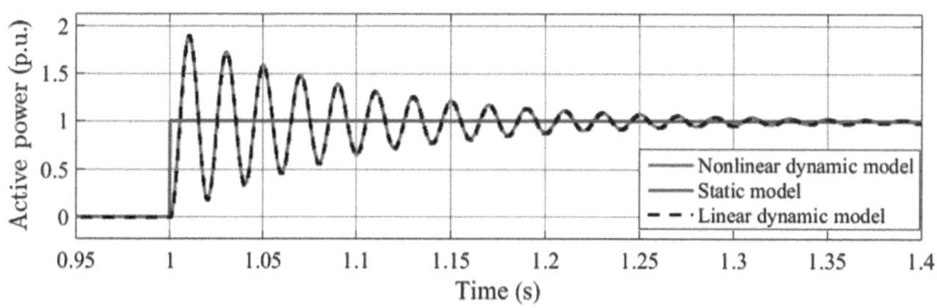

FIG. 8.22 Comparison between, static, nonlinear dynamic and linear dynamic models.

8.5.3 Effect of adding a virtual transient damping resistance

A solution to the oscillations noticed in the previous section is to add a virtual damping resistance (VDR). The magnitude of the modulated voltage is not constant any more. It is adjusted with respect to the current such as:

$$v_{md} = v'_{m_d} - R_v i_d \tag{8.42}$$

$$v_{mq} = v'_{m_q} - R_v i_q \tag{8.43}$$

R_v is defined as a virtual resistance. $v_m{}'$ is a new variable in the control. Eqs. (8.30), (8.31) becomes

$$\frac{L}{\omega_b}\frac{di_{g_d}}{dt} = v'_{m_d} - v_{gd} - (R_c - R_v)i_{g_d} + \omega_g L\, i_{g_q} \tag{8.44}$$

$$\frac{L}{\omega_b}\frac{di_{g_q}}{dt} = v'_{m_q} - v_{gq} - (R_c - R_v)i_{g_q} - \omega_g L\, i_{g_d} \tag{8.45}$$

Based on that, the new poles of the system are:

$$s_{1-2} = \underbrace{-\frac{R+R_v}{L}\omega_b}_{\Re} \pm \underbrace{j\omega_g\omega_b}_{\Im} \tag{8.46}$$

The damping coefficient is:

$$\zeta_{1-2} = \frac{R+R_v}{\sqrt{(R+R_v)^2 + (L\omega_g)^2}} \tag{8.47}$$

From this expression, it is possible to calculate the value of R_v to get a given damping. Since this damping resistance is only useful during the transient, a washout filter is added in order to cancel the effect during the steady state.

$$v_{md} = v'_{md} - R_v \frac{s}{s+\omega_{VDR}} i_d,\, v_{mq} = v'_{mq} - R_v \frac{s}{s+\omega_{VDR}} i_q \tag{8.48}$$

Fig. 8.23 illustrates a voltage control for the VSC embedding a virtual damping resistance. As shown in Fig. 8.22, a step on the angle $\Delta\delta_m = 0.15$ rad is applied. It can be noticed from Fig. 8.24 a much more damped behavior due to the damping resistance.

To confirm, the effect of the damping resistance, a comparison between both Bode diagrams of the transfer functions between the active power and the angle is presented in Fig. 8.25. The Bode diagram of $T_{p\delta}$ without VDR shows a high peak of resonance, while, the Bode diagram of $T_{p\delta}$ with VDR shows that this peak of resonance has disappeared due to the introduction of the damping resistance. The gain is the nearly constant on a wide range of frequency, it can be concluded that the static gain can be used for the design of the controller. This is the aim of the following section.

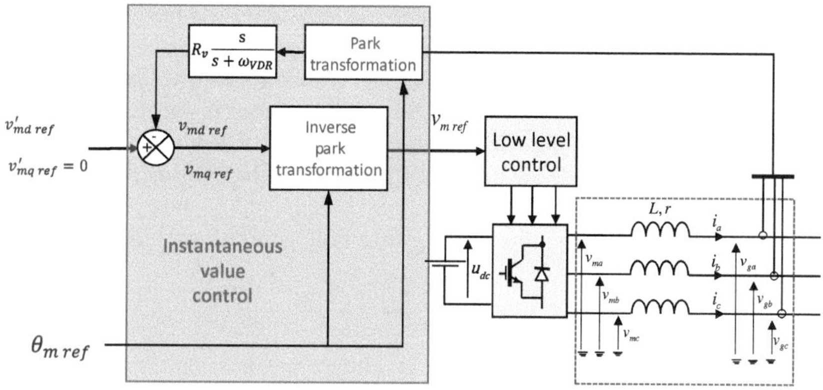

FIG. 8.23 Voltage control for the VSC including a damping resistance.

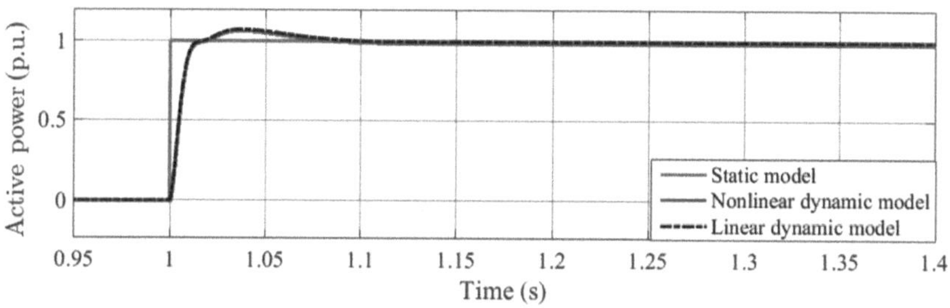

FIG. 8.24 Dynamic behavior of the active power with a damping resistance.

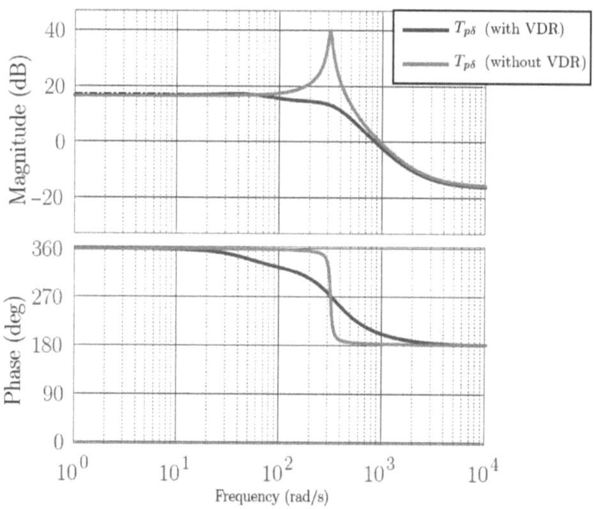

FIG. 8.25 Bode diagrams of $T_{p\delta}$ with or without damping resistance.

The control is based on the quasistatic model which links the angle $\Delta\delta_m$ to the active power. In this model, the grid angle δ_g appears as a disturbance. It can be compensated if a PLL is implemented to estimate this grid angle ($\tilde{\delta}_g$). It should be noticed that the role of the PLL is not the same as in the grid-following control. It is not used in the Park transformation (Fig. 8.26).

An integrator is included in the controller to cancel the static error. $\tilde{\delta}_g = \delta_g$ in steady state.

It can be concluded that the active power is equal to its reference in steady state:

$$P = P^* \tag{8.49}$$

A filter is also included in the loop to mitigate the noise on the power measurement. However, it may also have another role as it is shown in the second solution to control the power which is presented in Fig. 8.27.

Contrary to the previous control, the estimated frequency is used instead of the estimated angle. In steady-state, it gives the same results as previously, the active power P is equal to its reference P^*. However, in transient, the behavior is different. With the second type of control, it can be written:

$$\omega_m = \frac{\omega_c}{\omega_c + s}\left[\tilde{\omega}_g + m_p(P^* - P)\right] \tag{8.50}$$

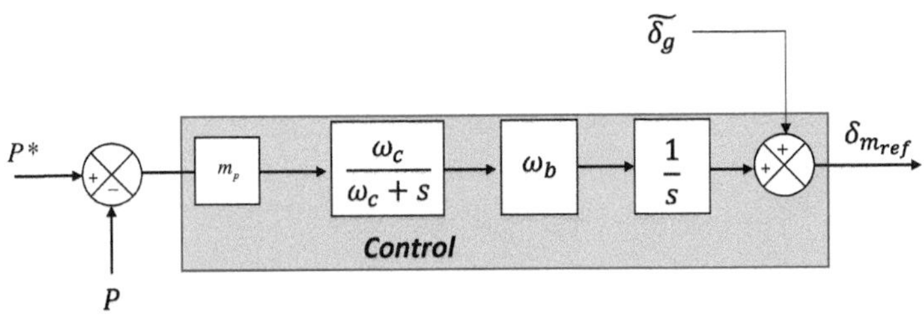

FIG. 8.26 Block diagram of the first solution for the active power control with the simplified model of the system.

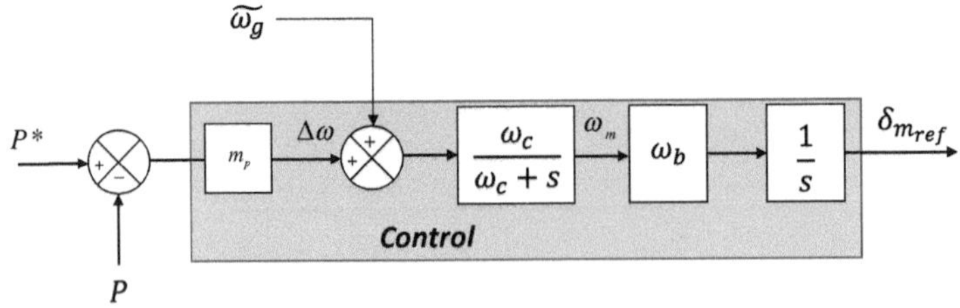

FIG. 8.27 Block diagram of the second solution for the active power control.

It can be rearranged in:

$$s\omega_m = \omega_c\left(\widetilde{\omega_g} - \omega_m\right) + \omega_c m_p (P^* - P) \tag{8.51}$$

In time domain, it yields:

$$\omega_m = \int_0^t \left[\omega_c\left(\widetilde{\omega_g} - \omega_m\right) + \omega_c m_p (P^* - P)\right] dt \tag{8.52}$$

In case of a frequency variation on the grid, ω_m is not modified instantaneously. This is a kind of inertial effect. To enhance this inertial effect, it is possible to present the control in another way (Fig. 8.28). This formulation is often referred in the literature as a VSM [14].

The modulated voltage frequency is calculated with the equation:

$$s\omega_m = \frac{1}{2H}\left[K\left(\widetilde{\omega}_g - \omega_m\right) + (P^* - P)\right] \tag{8.53}$$

As it can be noticed, the form of Eqs. (8.51), (8.53) is the same. It demonstrates the equivalence between both controls. By identification, it results in:

$$\omega_c = \frac{K}{2H} \quad m_p = \frac{1}{K} \tag{8.54}$$

For the first control, the modulated angle is given by the following equations:

$$\delta_m = \widetilde{\delta}_g + \frac{\omega_b}{s}\left(m_p\left(\frac{\omega_c}{\omega_c + s}\right)(P^* - P)\right) \tag{8.55}$$

$$\omega_m = \widetilde{\omega}_g + \left(m_p\left(\frac{\omega_c}{\omega_c + s}\right)(P^* - P)\right) \tag{8.56}$$

In case of a frequency variation on the grid, the PLL detects the variation and modifies ω_m. The inertial effect is not effective anymore with this type of control. To illustrate this phenomenon, a comparison of the dynamics of the system

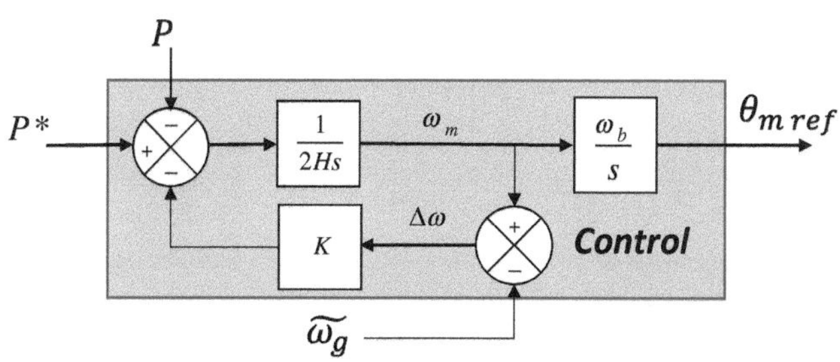

FIG. 8.28 A variant of presentation for the second solution of control.

with both types of control is presented in case of a frequency variation on the voltage source (Fig. 8.29).

8.5.4 Control without PLL

As in every types of control, the compensation of the disturbance is never compulsory. The output of the PLL can be replaced by a constant frequency set to the nominal value of the frequency (Fig. 8.30). The new equation to derive the modulated frequency is:

$$\omega_m = \left(\frac{\omega_c}{\omega_c + s}\right)\left(\omega_{set} + m_p(P^* - P)\right) \tag{8.57}$$

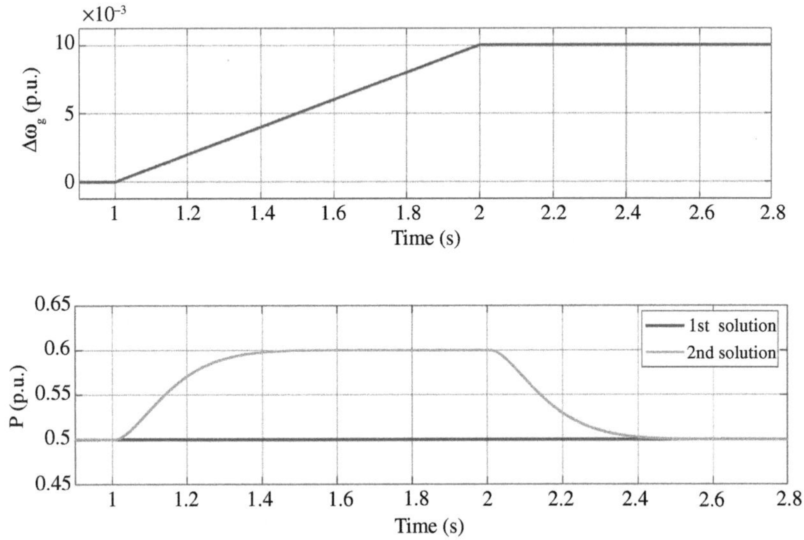

FIG. 8.29 System dynamics with both types of control in case of a frequency variation.

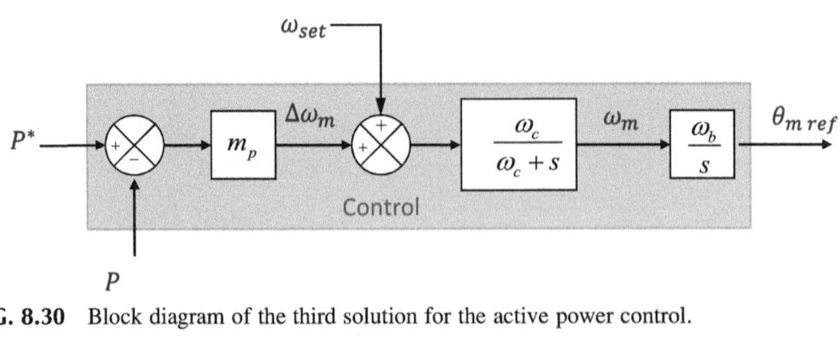

FIG. 8.30 Block diagram of the third solution for the active power control.

The general organization of the control is unchanged. As the previous control, this type of control also brings an inertial effect. However, the static equation is modified. Since $\omega_m = \omega_g$ in steady-state, it can be deduced:

$$P = P^* + \frac{1}{m_p}\left(\omega_{set} - \omega_g\right) \tag{8.58}$$

This equation is similar to a frequency droop control. It means that, in case of a frequency variation, there is an inherent frequency support with this kind of control. The coefficient m_p characterizes the frequency support.

8.5.5 Introduction of a LC filter in the grid-forming converter

In certain conditions, a LC filter has to be placed at the output of a two-level VSC. The modulated voltage v_m is not applied anymore to the connection inductance but it is replaced by the voltage across the capacitor e_g, as shown in Fig. 8.31.

A current and voltage loops have to be implemented to control the state variables of the LC filter. The design of the controller has been described in [15]. The general organization of the control is unchanged. However, the active power control does not deliver the angle for the modulated voltage anymore but for the reference of voltage across the capacitor.

It can be shown that the overall dynamics of the system is nearly unchanged. The simulation in Fig. 8.32 compares the simulation results with or without the LC filter when a step is applied on the active power reference. A very slight

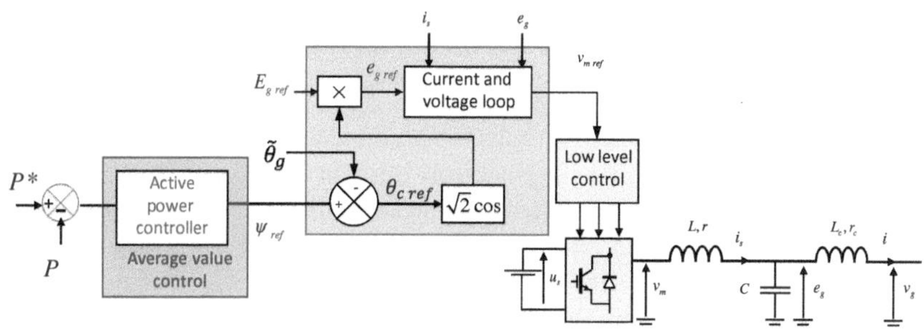

FIG. 8.31 Grid-forming converter with a LC filter.

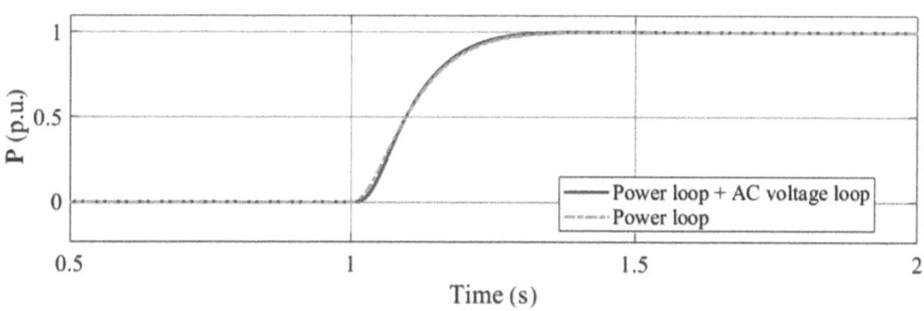

FIG. 8.32 Comparison of the simulation results with or without a LC filter at the output of the VSC.

difference can be noticed on the time response for the power including the AC current and voltage loop or not. Since the results with or without LC filter are very similar, the LC filter will not be considered further.

8.5.6 Ancillary services with grid-forming converters

As with the grid-following converters, the grid-forming converters can participate to the ancillary services for the grid (Fig. 8.33). Since the grid-forming control is a voltage control, a voltage regulation can be easily implemented with this type of control. However, it is also possible to implement a reactive power control if needed.

In term of frequency support, the choice operated on the type of droop control may have a major consequence. Indeed, the droop control without PLL embeds an inherent frequency support. Then, it is not possible to add an outer loop for the same functionality. Hence, with this kind of control, it is not possible to add any dead-band or power limitation to the frequency support. This can be a major issue for the primary source which provides the energy.

It is possible to add an outer loop to the two controls including the PLL. The difference between both controls is in term of inertial effect. Indeed, the second control provides an inertial effect whereas the first control, with the PLL on the angle does not provide this inertia. The second control has the advantage of integrating an inertial effect without any derivation on the frequency.

8.6 Test of grid-forming converters behaviors in various situations

In this section, some simulation results are presented to highlight the properties of the three different grid-forming controls introduced previously.

The external system is represented by a Thevenin equivalent, as shown in Fig. 8.34. The short circuit ratio (SCR) can be modified by adjusting the grid inductance (L_g), the resistance (r_g) is considered as negligible. The magnitude of the voltage source is constant. In the first step, the grid frequency is considered as constant.

The simulation parameters are listed in Table 8.1.

The aim of the first solution presented in Fig. 8.26 is only to control the active power. The value of the gain m_p can be modified to adjust the dynamics of the system, as shown in Fig. 8.35, where a step is applied on the active power reference.

The value of ω_c has been chosen to 31.4 rad/s. This control reveals an excellent robustness against a grid impedance variation as it has been demonstrated in [16]. As it is highlighted in Fig. 8.36, this mitigates the sensitivity of the grid-following converter to the variation of SCR which has been reported in the literature [17, 18].

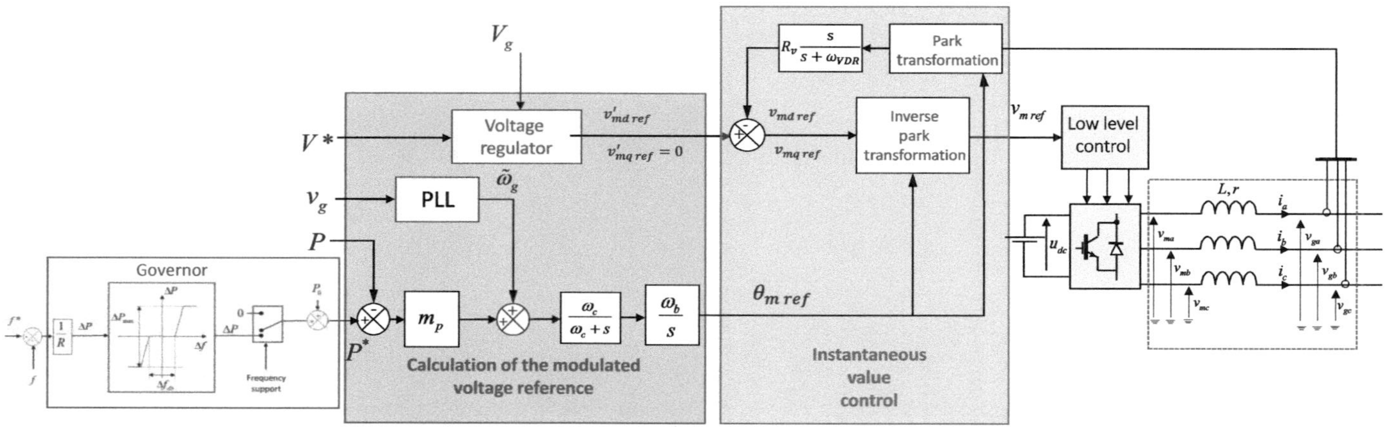

FIG. 8.33 Ancillary services with a grid-forming control.

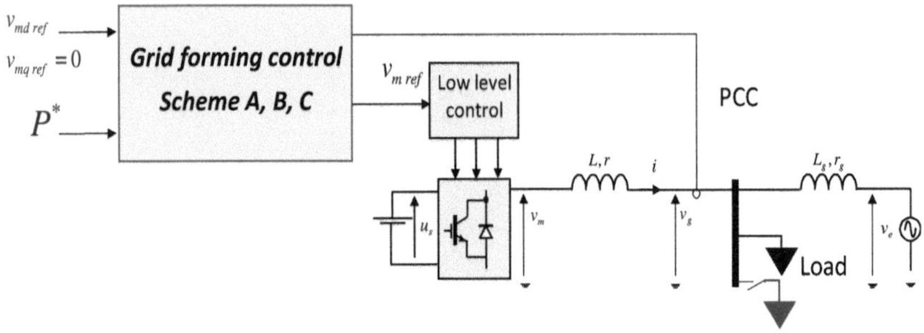

FIG. 8.34 Connection of the grid-forming converter to a Thevenin equivalent.

TABLE 8.1 List of the parameters for the simulation.

Symbol	Value	Symbol	Value
P_b	1 GW	f_{sw}	5 kHz
Cos ϕ	0.95	u_{dc}	640 kV
Fb	50 Hz	U_b	320 kV
$L=L_c$	0.15 p.u.	$R=R_c$	0.005 p.u.
C	0.066 p.u.	SCR	20
General controller parameters			
k_{pv}	0.52 p.u.	k_{iv}	1.16 p.u.
k_{iv}	0.73 p.u.	k_{pc}	1.19 p.u.
R_v	0.09 p.u.	ω_{VDR}	60 rad/s
First solution parameters: generic values			
m_p	0.0048 p.u.	ω_c	31.4 rad/s
Second solution parameters: generic values			
m_p	0.003 p.u.	ω_c	33.3 rad/s
Third solution parameters: generic values			
m_p	0.04 p.u.	ω_c	31.4 rad/s

With the second solution, an inertial effect is provided. The gain m_p cannot be chosen with respect to the dynamics anymore since the product $\omega_c m_p$ is relative to the inertial effect as already reported. Hence, to maintain a given inertial effect, the value of ω_c has to be adjusted when modifying m_p. Fig. 8.37 shows the results for a value of 5 s. It shows that the increase of m_p may induce large overshoot. It is possible to find a value for m_p which results in a 5% overshoot.

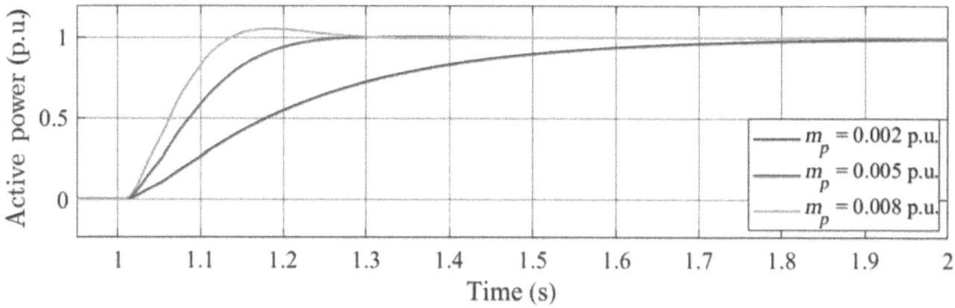

FIG. 8.35 Dynamic of a grid-forming control (Scheme A) for various values of m_p.

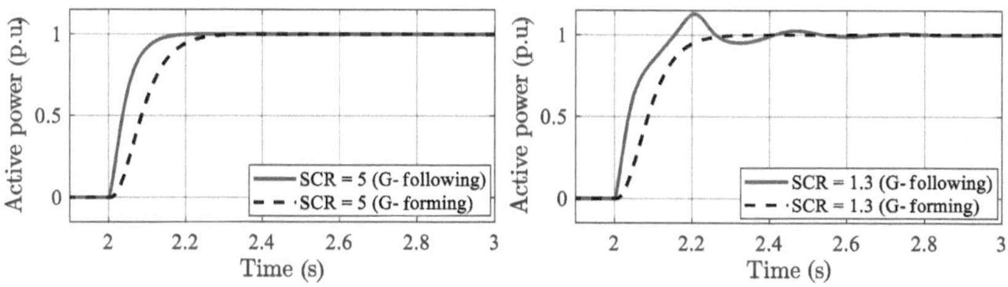

FIG. 8.36 Comparison of grid-following and grid-forming (first solution) for two SCR values.

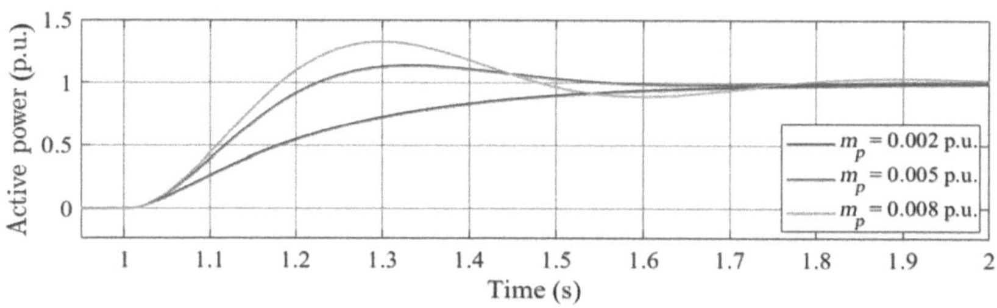

FIG. 8.37 Effect of the choice of m_p on the system dynamics (second solution) $H = 5$ s.

However, the active power time response may be slightly slower than with the first solution.

With the third solution, the gain m_p is relative to the frequency support. As explained previously, its numerical value has to be coordinated with the other sources connected to the same grid in order to provide a good load sharing in case of frequency support. Increasing the inertial effect is possible only by decreasing ω_c value. This has a large effect on the stability as highlighted in Fig. 8.38. It is the reason why this type of control has often been referred as "zero inertia" (VSMOH) [19]. However, a derivative effect can be added on the active power in order to stabilize the system [20].

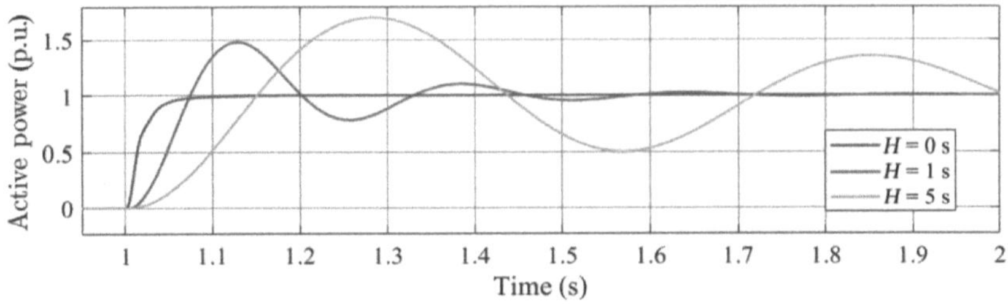

FIG. 8.38 Effect of the increase of H in the stability of the system (third solution).

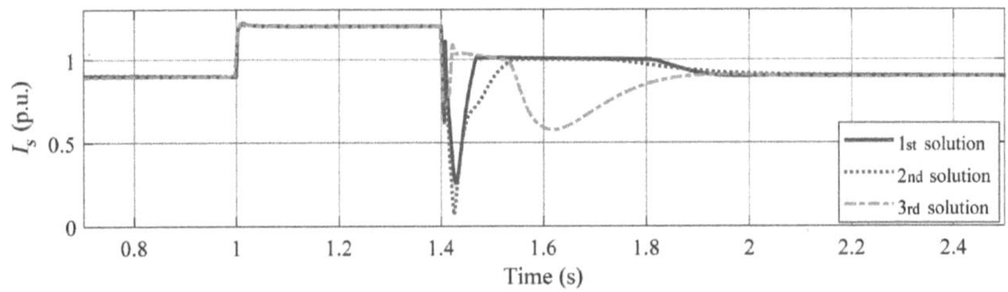

FIG. 8.39 Short circuit at the PCC.

The transient stability of the grid-forming control has been tested of various events which would induce an overcurrent in the converter. The current limitation is obtained first with saturation on the current reference and then the introduction of virtual impedance which increases the critical clearing time [21]. After the fault, the resynchronization is operated with no external signals. As soon as the current decreases below a given limit, the virtual impedance is removed.

In Fig. 8.39, the different solutions for grid-forming VSC are compared when a bolted fault occurs at the PCC.

To highlight the inertial effect, a small variable frequency system has been implemented. The voltage source frequency (ω_e) is defined by the system presented in Fig. 8.40. It represents an equivalent electrical system with inertia ($H_{eq} = 5\,\text{s}$) and a dynamics represented by a lead lag filter ($T_N = 1\,\text{s}$, $T_D = 1.5\,\text{s}$). A frequency support is implemented with a droop control.

In a first step, the first and second solutions are compared in terms of inertial effect with the generic parameters. The outer loop for frequency support is not activated for the grid-forming control. To modify the grid frequency, a load variation is applied at the node located at the PCC (Fig. 8.34). As it can be noticed in Fig. 8.41, the first solution does not provide any inertial effect whereas the power increases with the second solution to counteract the frequency variation.

In a second step, the second and third solutions are compared in term of inertial effect. Since a droop effect is inherently embedded in the third solution, the outer frequency loop is activated in the second solution to have the same

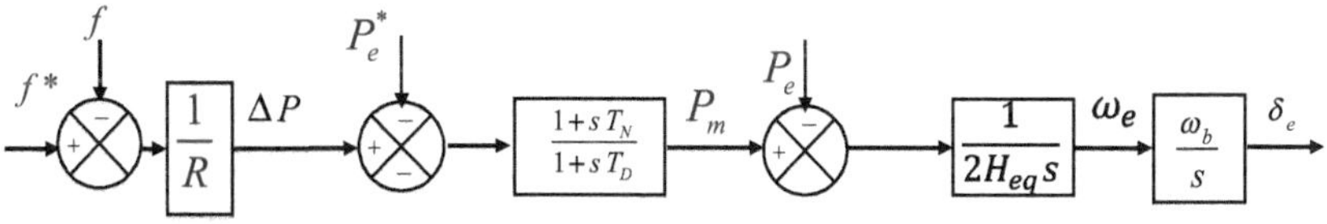

FIG. 8.40 Equivalent system for the definition of the voltage source frequency.

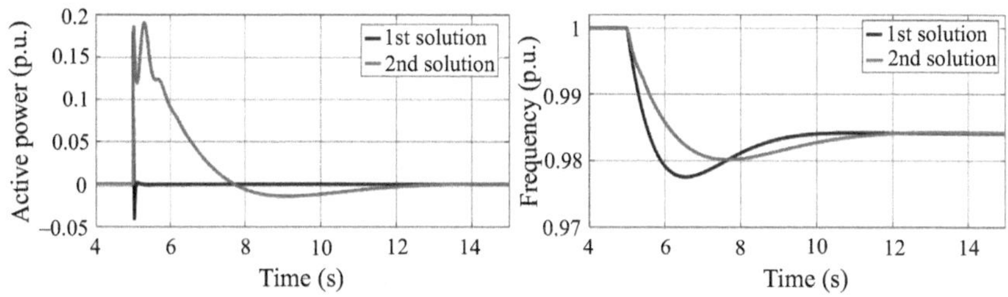

FIG. 8.41 Comparison between first and second solution in case of grid frequency variation.

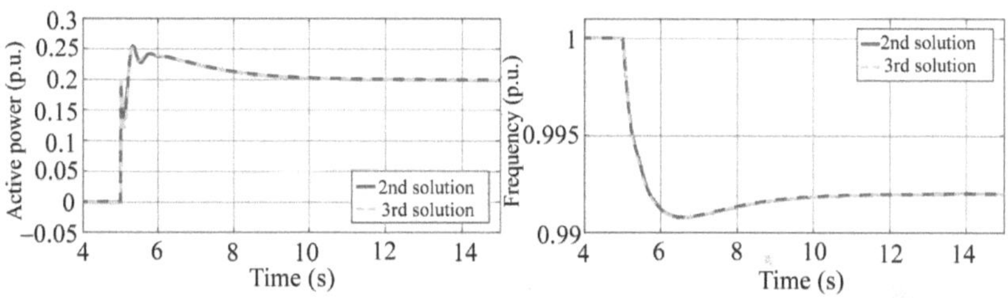

FIG. 8.42 Comparison between second and third solution in case of grid frequency variation.

behavior. In Fig. 8.42, the simulation results are presented for both controls with the same value of inertia and the same droop. These results show that the dynamic behavior is akin.

8.7 Proposal of a classification for the main types of grid-forming controls

In the literature, lots of vocabularies have been defined about the different types of control for the power electronic converters connected to the grid [22] but as it has been shown in the previous sections, there are only two fundamental concepts for the control of the converter [23]. Either, the control is based on the injection of an active/reactive current in a voltage source, or the drive of the modulated voltage with respect to an active power reference. In the first solution, called in the literature, grid-feeding or grid-following control, a closed loop current control is implemented. However, some few authors propose open-loop current control [24, 25]. It does not change the fundamental concept.

However, a PLL is required to inject the current with the right phase with respect to voltage at the point of common coupling.

In the second case, called in the literature "grid-forming control [22]", the fundamental concept of the power control with the voltage supposes to control the angle between the modulated voltage and the voltage at the point of common coupling. Three different solutions have been proposed. The first solution provides the most accurate active power control. With the third control solution, the active power control is lighter control since the active power delivered by the converter is depending on an external power reference and the grid frequency. The second solution of control is intermediary. An inertial effect is brought with this control but the active power, in steady state is not depending on the frequency. In the literature, the second and third solutions are developed with various names. The Synchronverter [25] shown in Fig. 8.43 can be considered as a variant of these solutions.

Indeed, generation of the angle is very similar with what has already been presented. However, a simplified model of the rotor and the stator of the synchronous machine is implemented in real-time. The output of a reactive power regulation is delivering the value of a virtual inductor current i_f which is used in Eqs. (8.59), (8.60). It defines the torque, the RMS value of the modulated voltage, and the reactive power.

$$T_e = M_f i_f \sum_{j=1}^{3} i_j \sin\left(\theta_{m\,ref} - \frac{2\pi}{3}(j-1)\right) \tag{8.59}$$

$$Q = M_f i_f \omega_m \sum_{j=1}^{3} i_j \cos\left(\theta_{m\,ref} - \frac{2\pi}{3}(j-1)\right) \tag{8.60}$$

$$V_{m\,ref} = M_f i_f \omega_m \tag{8.61}$$

A more important variant can be considered when a current control is embedded in the grid-forming control. Indeed, the grid-forming control does not embed any current limitation in the algorithm [26]. A possibility is to derive a current reference from the voltage reference and then recalculate the voltage. A solution proposed in [27] is to generate a current reference from a quasistatic inductance model (r, l). The voltage regulation provides a kind of virtual EMF $(e_{md_{ref}}, e_{mq_{ref}})$. This solution is illustrated in Fig. 8.44.

When neglecting the resistance, the $(d\text{-}q)$ current references are derived thanks to the following equations:

$$i_{qref} = \frac{v_{gd} - e_{md\,ref}}{l_s \omega_m} \quad i_{dref} = \frac{-e_{md\,ref}}{l_s \omega_m} \tag{8.62}$$

Then a current loop is implemented to generate the modulated voltage:

$$v_{md} = v_{gd} - l_s \omega_m i_q + C(s)\left(i_{d\,ref} - i_d\right) \tag{8.63}$$

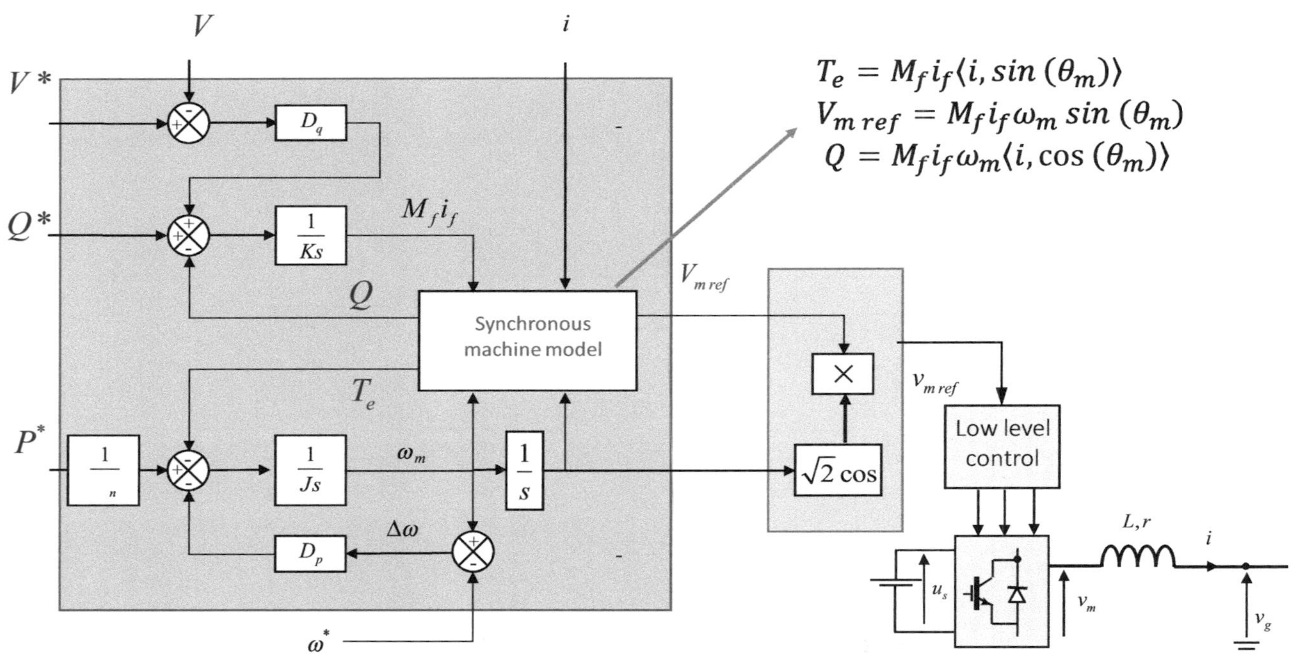

FIG. 8.43 Synchronverter.

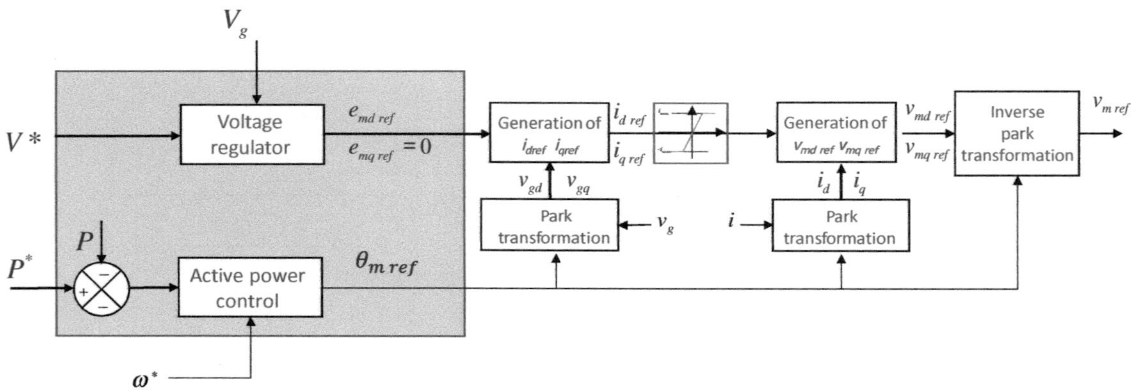

FIG. 8.44 Grid-forming control including a current reference.

$$v_{mq} = v_{gq} + l_s \omega_m i_d + C(s)\left(i_{qref} - i_q\right) \tag{8.64}$$

$C(s)$ is the representing the transfer function of a controller.

Other solutions to generate a current reference in order to limit the current has been proposed in [28], it has been called "VISMA," another solution called by the author as "Power synchronization method" has also been presented in [13]. The way to generate the current references and calculate the modulated voltage is not exactly the same but the principle is similar.

Doing so, it is possible to include a current saturation in the control algorithm to limit the current in case of an event on the grid that could induce an over current in the converter.

8.8 Conclusion

This chapter has presented the main functionalities of grid-feeding/following converters. Table 8.2 is a summary of the main properties for the various types of controls which have been presented.

TABLE 8.2 Classification of the different types of control.

	Grid feeding/ following	Grid forming		
Type of control	Open loop or close loop current control	Solution 1	Solution 2	Solution 3
PLL	Yes	Yes: estimation of the grid voltage angle	Yes: estimation of the grid voltage frequency	No
Active power tracking	Yes	Yes	Yes but influence of the inertial effect on the active power	No
Inertial effect	Yes with the derivative of the frequency	Yes with the derivative of the frequency	Yes	Yes
Primary frequency support	Yes/optional	Yes/optional	Yes/optional	Yes/ compulsory

Different topics have been identified to compare this control. The PLL is a very important point which can be somewhat confused. Indeed, the use of a PLL is often associated with a poor robustness in case of a weak grid. It can be true in case of grid-following control but not for grid-forming control. Indeed, the aim of the PLL for both types of control is not the same. The grid following control is using an estimate of the grid angle since the injected current has to be referenced to the grid voltage. In the grid forming control, it is an internal angle reference. The information about the grid angle or frequency is helpful only to reject the grid disturbance and to achieve a better active power control. Indeed, this kind of information is of highest importance on the functionalities which are achieved by a power converter in the power system.

An accurate active power tracking is an important topic mainly for HVDC or storage applications. Grid following and grid forming—(solution 1 and 2) are both good solutions to provide this functionality. If the grid following can give some better response time, grid forming—solution 1 reveals an excellent robustness against a grid impedance variation. It's also possible to impose a reference for the active power in solution 3 of the grid-forming control. However, the active power is modified as soon as the frequency is varying due to the droop control.

The inertia effect is inherent to grid forming solution 2 and 3. It is possible to add an inertial effect with grid following control. However, a derivative of the frequency is needed. Primary frequency support is optional for every type of control, except for the third solution, where a compulsory frequency support has been highlighted. In high power application, this can be an important drawback.

References

[1] N.R. Ullah, T. Thiringer, D. Karlsson, Voltage and transient stability support by wind farms complying with the E.ON Netz grid code. IEEE Trans. Power Syst. 22 (4) (Nov. 2007) 1647–1656, https://doi.org/10.1109/TPWRS.2007.907523.

[2] S.J. Finney, G.P. Adam, B.W. Williams, D. Holliday, I.A. Gowaid, Review of dc–dc converters for multi-terminal HVDC transmission networks. IET Power Electron. 9 (2) (Feb. 2016) 281–296, https://doi.org/10.1049/iet-pel.2015.0530.

[3] R. Zeng, L. Xu, L. Yao, B.W. Williams, Design and operation of a hybrid modular multilevel converter. IEEE Trans. Power Electron. 30 (3) (Mar. 2015) 1137–1146, https://doi.org/10.1109/TPEL.2014.2320822.

[4] HVDC PLUS (VSC)—HVDC | IGBT, MMC | HVDC (High Voltage Direct Current)|Siemens Global Website. https://new.siemens.com/global/en/products/energy/high-voltage/high-voltage-direct-current-transmission-solutions/hvdc-plus.html (Accessed March 23, 2020).

[5] F. Deng, Z. Chen, A control method for voltage balancing in modular multilevel converters. IEEE Trans. Power Electron. 29 (1) (2014) 66–76, https://doi.org/10.1109/TPEL.2013.2251426.

[6] H. Saad, X. Guillaud, J. Mahseredjian, S. Dennetiere, S. Nguefeu, MMC capacitor voltage decoupling and balancing controls. IEEE Trans. Power Deliv. 30 (2) (2015) 704–712, https://doi.org/10.1109/TPWRD.2014.2338861.

[7] C. Xie, X. Zhao, K. Li, J. Zou, J.M. Guerrero, A new tuning method of multiresonant current controllers for grid-connected voltage source converters. IEEE J. Emerg. Sel. Top. Power Electron. 7 (1) (2019) 458–466, https://doi.org/10.1109/JESTPE.2018.2833806.

[8] A. Egea-Alvarez, F. Hassan, Advanced vector control for voltage source converters connected to weak grids, IEEE Trans. Power Syst. 30 (6) (2015) 3072–3081.

[9] P. Delarue, F. Gruson, X. Guillaud, Energetic macroscopic representation and inversion based control of a modular multilevel converter. in: 2013 15th European Conference on Power Electronics and Applications (EPE), Lille, France, Sep. 2013, pp. 1–10, https://doi.org/10.1109/EPE.2013.6631859.

[10] T. Qingrui, Z. Xu, L. Xu, Reduced switching-frequency modulation and circulating current suppression for modular multilevel converters. IEEE Trans. Power Deliv. 26 (3) (Jul. 2011) 2009–2017, https://doi.org/10.1109/TPWRD.2011.2115258.

[11] T. Pieter, H. Pierre, C. Stijn, Penetration of Renewables and Reduction of Synchronous Inertia in the European Power System—Analysis and Solutions, Asset(Nov. 2018).

[12] T. Qoria, Q. Cossart, C. Li, X. Guillaud, F. Gruson, X. Kestelyn, Deliverable 3.2: Local Control and Simulation Tools for Large Transmission Systems, (2018) 89.

[13] L. Zhang, L. Harnefors, H. Nee, Power-synchronization control of grid-connected voltage-source converters. IEEE Trans. Power Syst. 25 (2) (May 2010) 809–820, https://doi.org/10.1109/TPWRS.2009.2032231.

[14] H. Beck, R. Hesse, Virtual synchronous machine. in: 2007 9th International Conference on Electrical Power Quality and Utilisation, Oct. 2007, pp. 1–6, https://doi.org/10.1109/EPQU.2007.4424220.

[15] T. Qoria, F. Gruson, F. Colas, X. Guillaud, M. Debry, T. Prevost, Tuning of cascaded controllers for robust grid-forming voltage source converter. in: 2018 Power Systems Computation Conference (PSCC), Jun. 2018, pp. 1–7, https://doi.org/10.23919/PSCC.2018.8443018.

[16] E. Rokrok, T. Qoria, A. Bruyere, B. Francois, X. Guillaud, Effect of using PLL-based grid-forming control on active power dynamics under various SCR. in: IECON 2019—45th Annual Conference of the IEEE Industrial Electronics Society, 1 Oct. 2019, pp. 4799–4804, https://doi.org/10.1109/IECON.2019.8927648.

[17] J.Z. Zhou, H. Ding, S. Fan, Y. Zhang, A.M. Gole, Impact of short-circuit ratio and phase-locked-loop parameters on the small-signal behavior of a VSC-HVDC converter. IEEE Trans. Power Deliv. 29 (5) (Oct. 2014) 2287–2296, https://doi.org/10.1109/TPWRD.2014.2330518.

[18] A. Adib, B. Mirafzal, X. Wang, F. Blaabjerg, On stability of voltage source inverters in weak grids. IEEE Access 6 (2018) 4427–4439, https://doi.org/10.1109/ACCESS.2017.2788818.

[19] M. Yu, et al., Use of an inertia-less virtual synchronous machine within future power networks with high penetrations of converters. in: 2016 Power Systems Computation Conference (PSCC), Jun. 2016, pp. 1–7, https://doi.org/10.1109/PSCC.2016.7540926.

[20] T. QORIA, F. GRUSON, F. COLAS, G. Denis, T. PREVOST, G. Xavier, Inertia effect and load sharing capability of grid forming converters connected to a transmission grid, in: The 15th IET international conference on AC and DC Power Transmission, UK, Jan. 2019. p. 6.

[21] T. Qoria, F. Gruson, F. Colas, G. Denis, T. Prevost, X. Guillaud, Critical clearing time determination and enhancement of grid-forming converters embedding virtual impedance as current limitation algorithm. IEEE J. Emerg. Sel. Top. Power Electron. (2019) 1, https://doi.org/10.1109/JESTPE.2019.2959085.

[22] J. Rocabert, A. Luna, F. Blaabjerg, P. Rodríguez, Control of power converters in AC microgrids. IEEE Trans. Power Electron. 27 (11) (Nov. 2012) 4734–4749, https://doi.org/10.1109/TPEL.2012.2199334.

[23] T. Qoria, T. Prevost, G. Denis, F. Gruson, F. Colas, X. Guillaud, Power converters classification and characterization in power transmission systems. in: 2019 21st European Conference on Power Electronics and Applications (EPE '19 ECCE Europe), Genova, Italy, Sep. 2019, pp. 1–9, https://doi.org/10.23919/EPE.2019.8914783.

[24] I. Erlich, Control challenges in power systems dominated by converter interfaced generation and transmission technologies, in: NEIS 2017; Conference on Sustainable Energy Supply and Energy Storage Systems, Sep. 2017, pp. 1–8.

[25] I. Erlich, et al., New control of wind turbines ensuring stable and secure operation following islanding of wind farms. IEEE Trans. Energy Convers. 32 (3) (Sep. 2017) 1263–1271, https://doi.org/10.1109/TEC.2017.2728703.

[26] T. Qoria, C. Li, K. Oue, F. Gruson, F. Colas, X. Guillaud, Direct AC voltage control for grid-forming inverters. J. Power Electron. 20 (1) (Jan. 2020) 198–211, https://doi.org/10.1007/s43236-019-00015-4.

[27] O. Mo, S. D'Arco, J.A. Suul, Evaluation of virtual synchronous machines with dynamic or quasi-stationary machine models. IEEE Trans. Ind. Electron. 64 (7) (Jul. 2017) 5952–5962, https://doi.org/10.1109/TIE.2016.2638810.

[28] Y. Chen, R. Hesse, D. Turschner, H.-P. Beck, Comparison of methods for implementing virtual synchronous machine on inverters. Renew. Energy Power Qual. J. (Apr. 2012) 734–739, https://doi.org/10.24084/repqj10.453.

Chapter 9

Converter-based swing dynamics

David Raisz and Antonello Monti

Institute for Automation of Complex Power Systems, RWTH Aachen University, Aachen, Germany

9.1 Introduction

The increasing share of renewable energy sources and energy storage systems leads to a gradual replacement of classical synchronous generators by converter-interfaced generators. As a consequence, the rotating mass (or inertia) in the power system is decreasing over time. It results in a different dynamic behavior as compared to classical power systems: in case of disturbances or a power imbalance, the frequency excursions become larger, the rate of change of frequency (ROCOF) or the gradient of the frequency changes is increasing (see Section 3.2.4). These phenomena might lead to instability problems or loss of synchronism [1, 2].

As the control of power converters allows for a wide range of functionalities, several control strategies have been proposed, which aim at mimicking synchronous generators' properties such as synchronization, voltage, frequency regulation, and black-start. Converters providing such capabilities are called "grid forming" units, as opposed to "grid feeding" or "grid supporting" converters [3].

The state-of-the-art solution to mitigate the above-mentioned drawbacks related to inertia-less systems is the emulation of inertial behavior by additional control schemes. In the literature, several solutions have been proposed to implement such schemes—these solutions are widely referred to as virtual synchronous machines (VSM), virtual synchronous generators, or synthetic inertia schemes.

Some of the approaches implement a higher (third to seventh) order dynamic model of synchronous machines. A second order model with an intrinsic reactive power control dynamics is used in the so-called Synchronverter [4]. The "frequency-power response" based topology emulates the release of kinetic energy based on the derivative of the measured frequency [5]. An even simpler approach uses a power-frequency droop to provide frequency regulation. A recent approach implements a virtual induction machine instead of a virtual

Converter-Based Dynamics and Control of Modern Power Systems
https://doi.org/10.1016/B978-0-12-818491-2.00009-2

synchronous machine [6], and in a new class of methods called virtual oscillator control (VOC), a nonlinear oscillator is implemented within the converter controller to achieve a fast and globally asymptotically stable synchronization mechanism [7]. These methods can be assessed based on properties like

- complexity vs ease of implementation,
- numerical stability or the need for a frequency derivative calculation,
- reliance on a PLL which is known to cause instability in weak grids,
- transient response,
- intrinsic overcurrent limitation capability.

This chapter focuses on the so-called swing-equation based approach because—despite its simplicity—it captures the dynamic behavior that is used to model traditional power system dynamics, and therefore it is a prevalent candidate to provide inertia emulation in future power systems.

9.1.1 Dynamics of the swing equation

As introduced in Section 3.2.3, the swing equation (SE) is often used as a low order approximation a synchronous generator's (SG) dynamic behavior.

Let us consider the single-machine-infinite-bus (SMIB) system in Fig. 9.1, where a SG is connected to a network, which is represented by its Thévenin equivalent. The network equivalent model comprises a constant voltage source E and an impedance. (The impedance can be approximated by a series reactance in transmission systems.)

The dynamic behavior of this system can be described by the following equations [8]:

$$\dot{\omega} = \frac{1}{M}\left(P_{ref} - P - D\dot{\delta}\right) \tag{9.1}$$

$$\dot{\delta} = \omega - \omega_s \tag{9.2}$$

$$P = V^2 G - EV(B \sin\delta + G \cos\delta) \tag{9.3}$$

where M is the inertia constant, ω is the rotational speed, ω_s represents the synchronous speed, P_{ref} denotes the reference ("mechanical") power, P is the actual power output, $\overline{Y} = G + jB$ is the admittance of the equivalent system, V is the

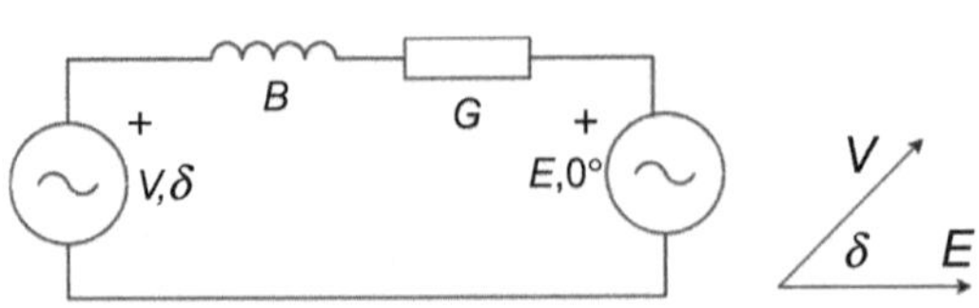

FIG. 9.1 Simple model for demonstration of the SE.

SG terminal voltage, δ denotes the angle between E and V, and D represents the damping constant.

The SE can be linearized around an operating point to simplify the stability analysis. The obtained "small-signal model" is valid for small disturbances around that operating point, and captures system stability information for this small region only.

By linearizing Eqs. (9.1)–(9.3) around an operating point (for an explanation see, e.g., [8]) we obtain the small-signal model Eq. (9.4):

$$\begin{bmatrix} \Delta\dot{\delta} \\ \Delta\dot{\omega} \end{bmatrix} = \begin{bmatrix} 0 & 1 \\ \frac{EV}{M}(B\cos\delta - G\sin\delta) & -\frac{D}{M} \end{bmatrix} \begin{bmatrix} \Delta\delta \\ \Delta\omega \end{bmatrix} \tag{9.4}$$

The Eigenvalues of this system can be obtained as

$$\lambda_{1,2} = -\gamma \pm \sqrt{\gamma^2 + \frac{EV}{M}(B\cos\delta - G\sin\delta)} \tag{9.5}$$

with

$$\gamma = \frac{D}{2M} \tag{9.6}$$

From Eqs. (9.3)–(9.5) it can be observed, that the dynamics of this system depends on the actual operating point (P or δ). It can also be seen that this dependency results from the nonlinearity in Eq. (9.3) between the power P and the angle δ.

9.1.2 Linear swing dynamics

By introducing the concept of linear and uniform swing dynamics (LSD), our first goal is to eliminate the nonlinearity, that is, the dependence of the dynamics on the actual operating point. This nonlinearity originates in the power transfer equation. LSD provides a linear SE dynamics—this makes it possible to extend the validity range of small-signal analysis for the large part of (in some cases the whole) state space, that is, for small as well as larger disturbances. The application of LSD results in the same, fixed swing-related Eigenvalues across a large (in some cases the whole) power range. This can make it easier for system operators to assess the dynamics and stability of the system, without the need to perform modal analysis at different operating points.

The second goal of developing the LSD concept is to ensure a uniform swing dynamic performance, that is, the same dynamics for all generators in a multi-machine system. For conventional power systems there are well-established concepts for assessing stability. However, as the power system approaches ever-larger percentages of RES, the number of power electronic-based generating units is increasing by several orders of magnitude. It is imperative to have coherent requirements or design rules for converters (including their controllers), otherwise power system dynamic behavior would be very hard to analyze and predict. LSD, however, provides uniform Eigenvalues of the system, regardless of the number of generators actually supplying the grid, provided that they all apply the LSD principle.

The expected location of the Eigenvalues (or, in other words, the values of inertia constants and damping) can be determined based on required dynamic performance, as shown, e.g., in Ref. [9].

With the LSD concept presented in this chapter our goal is to contribute to the creation of the requirements for the converters providing VSM behavior in the future power systems.

In this chapter we focus on disturbances that do not force the inverter to enter its current-limited operating regime, e.g., changes in power reference or disturbances of system frequency. These are the disturbances usually considered in small-signal stability analysis [10], however, as mentioned earlier, we do not assume these disturbances to be "small."

Throughout the chapter we assume a timescale-separation between the dynamic processes shown here and other dynamics like that of current, voltage, or DC link control loops, a PLL, or power line dynamics.

The chapter summarizes results published in Refs. [11–15] and largely builds on content in Ref. [16].

9.2 LSD concepts for single-machine-infinite-bus systems

For the sake of introducing the LSD concept, we first focus on a SMIB system, as shown in Fig. 9.2.

Here the voltage V represents the output voltage of the controllable power electronic converter. The voltage E of the infinite bus, and the admittance $\overline{Y} = G + jB$ of the grid are assumed to be constant.

Power electronic converter control provides several degrees of freedom to implement the LSD concept. In the following Chapters we derive several methods for the control of the converter so that the swing-related dynamics of the system comprise constant Eigenvalues, independent of the operating point.

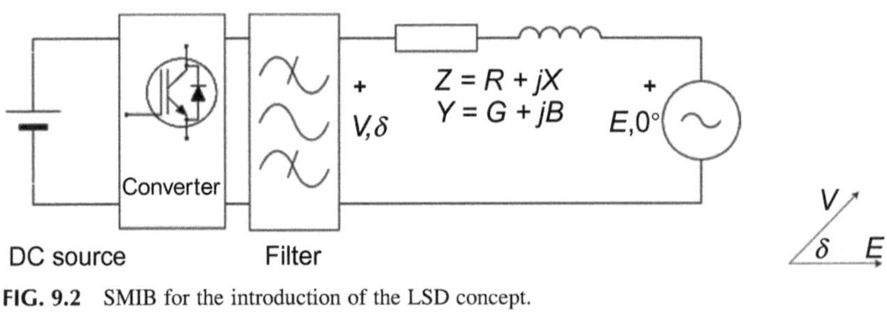

FIG. 9.2 SMIB for the introduction of the LSD concept.

9.2.1 Voltage control-based LSD

In order to derive the Voltage-based LSD concept, first let us assume an inductive network ($G=0$) so that Eq. (9.3) becomes

$$P=-EBV\sin\delta \tag{9.7}$$

In this chapter we make the nonlinear $P(\delta)$ relation (9.7) behave linearly, in order to eliminate the dependence of the Eigenvalues Eq. (9.5) on the actual operating point or δ.

Since E and B are constant in Eq. (9.7), this linearization can only be achieved by an appropriate control of $V\sin\delta$. One has to keep in mind, when considering the control of the output voltage of the inverter that the voltages in a power system have to be kept within a tolerance band around the nominal voltage. This tolerance band is usually 5% for transmission systems and 10% for distribution systems. We aim at achieving a linear $P(\delta')$ relationship as in Eq. (9.8):

$$P=-EB(1-\varepsilon)E\delta' \tag{9.8}$$

where ε is the voltage tolerance, usually 0.05 pu or 0.1 pu. We use here the notation δ' instead of δ, because—in order to obtain a linear power-angle characteristic—the angle that belongs to a certain power must be different from the angle that results from a constant voltage V.

The solution of Eq. (9.8) will yield

$$\delta'(P)=\frac{P}{-(1-\varepsilon)E^2B} \tag{9.9}$$

The voltage control law can now be expressed from Eqs. (9.7), (9.8) as

$$V(\delta')=\frac{(1-\varepsilon)E}{\sin\delta'}\delta' \tag{9.10}$$

It can be shown that for $\varepsilon=0.05$ and by using Eqs. (9.9), (9.10) a linear $P(\delta')$ relationship is achieved in the angle range [0, 44 degrees], the corresponding power range [0, 0.73] and voltage range [0.95, 1.05] (Fig. 9.3).

For $\varepsilon=0.1$ and by using Eqs. (9.9), (9.10) the $P(\delta')$ relationship can be made linear in the angle range [0, 62 degrees], the corresponding power range [0, 0.97] and voltage range [0.9, 1.1] (Fig. 9.4).

This means that by controlling the voltage, but keeping it within prespecified limits, the power transfer equation can be made linear for up to 97% of the original transfer capacity.

From Eq. (9.8), the linearization of P with respect to δ will yield $\frac{\partial P}{\partial \delta}=-E^2B$ $(1-\varepsilon)$. This means, that by using Eqs. (9.9), (9.10) the VSM dynamics will be defined by:

$$\begin{bmatrix}\Delta\dot{\delta}\\ \Delta\dot{\omega}\end{bmatrix}=\begin{bmatrix}0 & 1\\ -\frac{(1-\varepsilon)S_{sc}}{M} & -\frac{D}{M}\end{bmatrix}\begin{bmatrix}\Delta\delta\\ \Delta\omega\end{bmatrix} \tag{9.11}$$

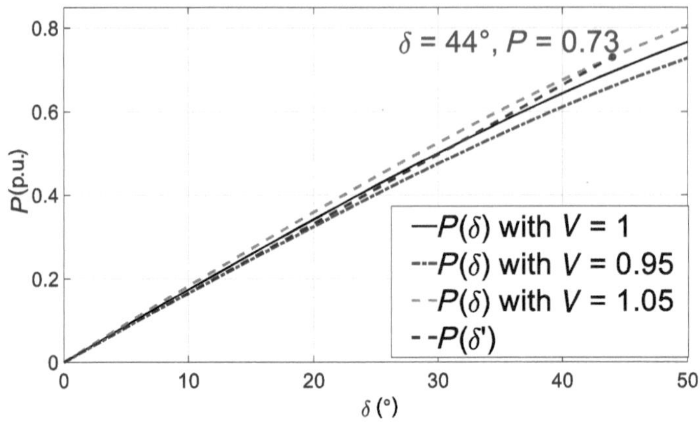

FIG. 9.3 Linearization with $\varepsilon = 0.05$.

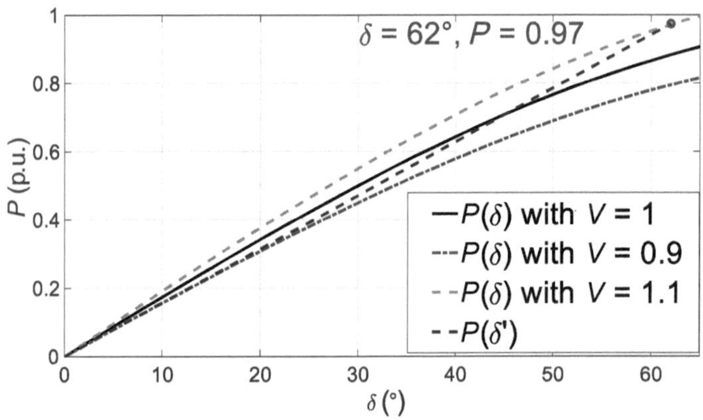

FIG. 9.4 Linearization with $\varepsilon = 0.1$.

where $S_{sc} = -E^2B$ is the short-circuit power at the grid connection point of the converter. The Eigenvalues of Eq. (9.11) are

$$\lambda_{1,2} = -\gamma \pm \sqrt{\gamma^2 - (1-\varepsilon)S_{sc}/M}. \tag{9.12}$$

It can be observed from Eqs. (9.12), (9.6), that

- The Eigenvalues are independent from the actual operating point.
- The two constants M and D can be used for selecting the required Eigenvalues in Eq. (9.12).The values for M and D can eventually be limited by the capacity or rating of the VSM (i.e. battery capacity or inverter rating), but these constraints are not considered at this stage.

The disadvantages of the Voltage-based LSD method are:

- it couples frequency control and voltage control,

- the Eigenvalues depend on E and B (and in case of resistive networks, on G),
- and an estimate of E and B (and G) is required.

9.2.1.1 *Stability analysis*

Here we give a short formal proof that the voltage-control based LSD provides an asymptotically stable system at its steady-state operating points.

Section 6.3.2 in Ref. [8] presents a Lyapunov function for stability analysis, for the system described by Eqs. (9.1)–(9.3). We follow here the same approach for the system described by Eqs. (9.1), (9.2) with (9.8)–(9.10). The only non-trivial equilibrium point of the system is:

$$\delta_s = -\frac{P}{(1-\varepsilon)\,E^2B}$$

$$\Delta\omega = 0 \tag{9.13}$$

Let TE denote the total energy function of the system, comprising the kinetic energy E_k and the potential energy E_p:

$$TE = E_k + E_p \tag{9.14}$$

where

$$E_k = \frac{1}{2}M\Delta\omega^2 \tag{9.15}$$

and

$$E_p = \frac{1}{2}b\left(\delta^2 - \delta_s^2\right) - P(\delta - \delta_s) \tag{9.16}$$

with

$$b = -E^2B(1-\varepsilon) \tag{9.17}$$

The energy function TE satisfies the definition of a Lyapunov function if and only if

(i) it has stationary point at Eq. (9.13)
(ii) it is positive definite in the vicinity of Eq. (9.13)
(iii) its derivative is nonpositive.

We show below that all three conditions hold. Condition (i) can be proven by calculating

$$\text{grad}\, TE = \begin{bmatrix} \dfrac{\partial TE}{\partial \Delta\omega} \\ \dfrac{\partial TE}{\partial \delta} \end{bmatrix} = \begin{bmatrix} \dfrac{\partial E_k}{\partial \Delta\omega} \\ \dfrac{\partial E_p}{\partial \delta} \end{bmatrix} = \begin{bmatrix} M\Delta\omega \\ -(P - b\delta) \end{bmatrix} \tag{9.18}$$

The gradient is zero at Eq. (9.13).

Condition (ii) can be proven by determining the Hessian matrix:

$$H = \begin{bmatrix} \frac{\partial^2 TE}{\partial \Delta\omega^2} & \frac{\partial^2 TE}{\partial \Delta\omega\, \partial\delta} \\ \frac{\partial^2 TE}{\partial\delta\, \partial\Delta\omega} & \frac{\partial^2 TE}{\partial\delta^2} \end{bmatrix} = \begin{bmatrix} M & 0 \\ 0 & b \end{bmatrix} \tag{9.19}$$

Since $M > 0$ and $b > 0$ always, therefore the Hessian H is positive definite. Condition (iii) can be proven by determining

$$\frac{dTE}{dt} = \frac{dE_k}{dt} + \frac{dE_p}{dt} \tag{9.20}$$

Since (using Eqs. 9.1, 9.2)

$$\frac{dE_k}{dt} = \left[M\frac{d\Delta\omega}{dt}\right]\Delta\omega = [P - b\delta]\Delta\omega - D\Delta\omega^2 \tag{9.21}$$

and

$$\frac{dE_p}{dt} = \frac{\partial E_p}{\partial\delta}\frac{d\delta}{dt} = -[P - b\delta]\Delta\omega \tag{9.22}$$

it follows that

$$\frac{dTE}{dt} = -D\Delta\omega^2 \tag{9.23}$$

which is always negative.

We have shown that $TE(\delta, \Delta\omega)$ is a Lyapunov function, and the equilibrium point Eq. (9.13) is asymptotically stable.

Asymptotic stability does not imply, that the voltage time function will always be within the limits at $1 \pm \varepsilon$ p.u., but this is not a problem from a practical operation point of view, because for short time intervals these limits can be exceeded. The relation of the magnitude and the time span of voltage sags or swells is given e.g., by CBEMA or ITIC curves [17].

9.2.1.2 *Resistive-inductive network*

In this chapter we derive the Voltage-based LSD concept for SMIB systems, where the grid is represented by a resistive-inductive series impedance, instead of a purely reactive impedance, as shown in Fig. 9.2.

The power transfer across the network equivalent impedance can now be written as

$$P = V^2 G - EV(B\sin\delta + G\cos\delta) \tag{9.24}$$

Consistently with the purely reactive grid case, we use the following expressions to define the expected behavior of the linearized system:

$$V = E(1 - \varepsilon) \tag{9.25}$$

$$\sin\delta \approx \delta' \tag{9.26}$$

In Eqs. (9.8), (9.26) the linear approximation of the sine function was achieved by using the first term of its Fourier series. However, this approach is not feasible in case of the cosine function: using

$$\cos\delta \approx 1 \tag{9.27}$$

the $P(\delta)$ relation will be inappropriate, i.e., not in between the "lower" und "upper" limit curves determined by $(1-\varepsilon)$ and $(1+\varepsilon)$, as shown in Ref. [12]. Therefore, we will use the approximation (Fig. 9.5)

$$\cos\delta \approx 1 - \frac{abs(\delta')}{\pi/2} \tag{9.28}$$

This way Eq. (9.24) becomes

$$P = -E^2(1-\varepsilon)\left(B\delta' - G\frac{2}{\pi}abs(\delta') + G\varepsilon\right) \tag{9.29}$$

For $E=1$, $R/X=1$ (which is typical for many distribution systems) and $|G+jB|=1$ this results in an approximation as shown in an example below (Fig. 9.6), for $\delta > 0$:

For high R/X ratios (above approx. 7), it is possible that the LSD voltage controller will produce a voltage reference lower than $(1\text{-}\varepsilon)$ pu, as for example in the power interval between $P=1.1$ and 1.6 in Fig. 9.7.

In such cases (for R/X ratios above 7), the linear curve can be adjusted. We have shown in Ref. [12], that it is possible to construct a linear curve between the $P(\delta)$ curves that belong to the voltage limits, even for a purely resistive network, for $\varepsilon=$ 10%. This is not a limitation, since in distribution systems, (where R/X is typically high) usually 10% voltage tolerance is allowed. In transmission

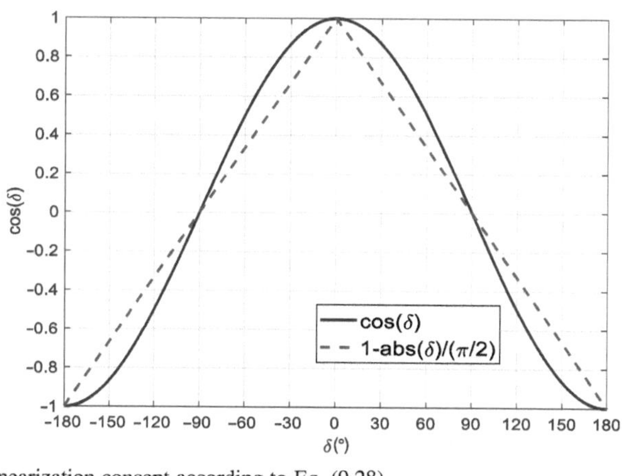

FIG. 9.5 Linearization concept according to Eq. (9.28).

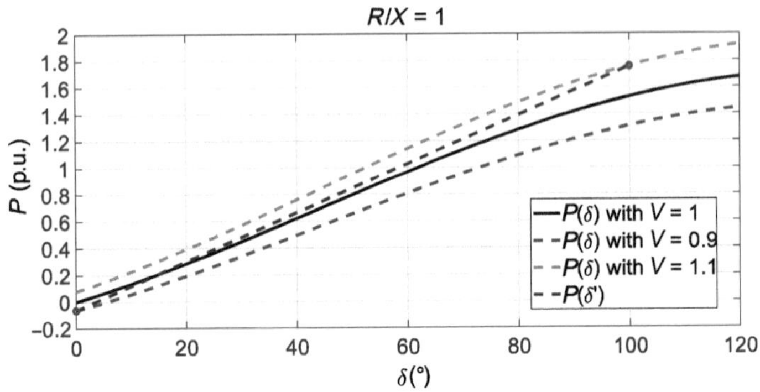

FIG. 9.6 Linearization for $R/X=1$ and using Eq. (9.28).

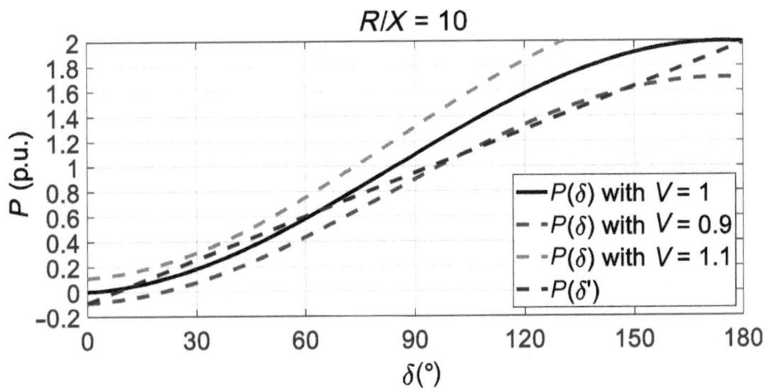

FIG. 9.7 Linearization for $R/X=10$ and using Eq. (9.28).

systems, $\varepsilon=5\%$ is usually applied, but in these systems R/X is typically much smaller than 1, and therefore Eqs. (9.9), (9.10) can be used.

For the ease of further analysis, we will assume $R/X<7$ and use Eqs. (9.25), (9.26), (9.28).

Solving Eq. (9.29) for δ we obtain

$$\delta'(P)=-\frac{P+GE^2\varepsilon(1-\varepsilon)}{E^2(1-\varepsilon)\left(B\pm G\frac{2}{\pi}\right)} \tag{9.30}$$

and then the solution for V gives:

$$V(\delta')=E\left(K\pm\sqrt{K^2+(1-\varepsilon)\left(\frac{2}{\pi}abs(\delta')-\rho\,\delta'+\varepsilon\right)}\right) \tag{9.31}$$

with

$$\mathrm{K}=\frac{1}{2}(\cos\delta'+\rho\,\sin\delta') \tag{9.32}$$

and

$$\rho = B/G \tag{9.33}$$

From Eq. (9.29) it follows that

$$\frac{\partial P}{\partial \delta'} = -E^2(1-\varepsilon)\left(B - G\frac{2}{\pi}\operatorname{sign}(\delta')\right) \tag{9.34}$$

and therefore the system Eigenvalues become

$$\lambda_{1,2} = -\gamma \pm \sqrt{\gamma^2 - (1-\varepsilon)E^2/M\left(B - \frac{2}{\pi}G\operatorname{sign}(\delta')\right)}. \tag{9.35}$$

with $\gamma = D/2M$.

This means, that although the Eigenvalues are not fixed at all operating points, they only take a finite number of possible values, depending on the actual sign of the angles. In case of a SMIB system, there are four possible Eigenvalues, instead of two.

9.2.2 Adaptive voltage control-based LSD

As it can be observed from Eq. (9.12), the Eigenvalues of the system—in case of the voltage control-based LSD—are dependent on system parameters, namely on S_{sc} which in turn depends on E and B. If the system parameters change, the system dynamics can change as well.

In order to keep both Eigenvalues at previously defined values regardless of the actual system parameters, the SE constants M and D must be adapted to changes in those system parameters, as follows. If the short-circuit power at the grid connection point is known or measured, γ is chosen according to RoCoF and nadir requirements, and M and D are updated using Eqs. (9.36), (9.37) when system parameters change.

$$M = M_0\, S_{sc}/S_{sc,0} \tag{9.36}$$

and since γ is to be held constant,

$$D = D_0\, S_{sc}/S_{sc,0} \tag{9.37}$$

This way both Eigenvalues of the system can be kept at $\lambda_{1,2} = -\gamma \pm \sqrt{\gamma^2 - (1-\varepsilon)S_{sc,0}/M_0}$ regardless of the system parameters.

9.2.3 Inertia-based LSD

Power electronic devices are more flexible than traditional synchronous generators in the sense that their inertia constant and damping constant are not fixed, but can be controlled. We made use of this in the previous chapter. In what follows, we seek to control the parameters M and D, so that the LSD behavior can be provided without coupling the voltage control and the SE control loops, and thus mitigate the main drawback of the voltage-based LSD method.

We focus again at a general class of SMIB systems with inductive-resistive grid. Let us denote the admittance magnitude by

$$Y = \sqrt{B^2 + G^2} \tag{9.38}$$

In order to obtain fixed Eigenvalues, we can write from Eq. (9.5) the following control rules for the inertia and damping variables:

$$\begin{aligned} M &= \frac{M_0}{Y}(B\cos\delta - G\sin\delta) \\ D &= \frac{D_0}{Y}(B\cos\delta - G\sin\delta) \end{aligned} \tag{9.39}$$

Assuming constant inverter voltage V, the steady state is found at (δ_0, ω_0) which solves Eq. (9.40):

$$\begin{aligned} P_{ref} - V^2 G &= -EV(B\sin\delta_0 + G\cos\delta_0) \text{ and} \\ 0 &= \omega_0 - \omega_n \end{aligned} \tag{9.40}$$

Let $f(\delta, \omega)$ denote the right-hand side of Eq. (9.3), then

$$\left.\frac{df}{d\delta}\right|_{\delta_0,\omega_0} = \frac{Y\left(\left(P_{ref} - V^2 G\right)(B\sin\delta_0 + G\cos\delta_0) + EVY^2\right)}{M_0(B\cos\delta_0 - G\sin\delta_0)^2} \tag{9.41}$$

Using Eq. (9.40), and noting that

$$(B\sin\delta_0 + G\cos\delta_0)^2 + (B\cos\delta_0 - G\sin\delta_0)^2 = Y^2 \tag{9.42}$$

we obtain

$$\left.\frac{df}{d\delta}\right|_{\delta_0,\omega_0} = \frac{EVY}{M_0} \tag{9.43}$$

Therefore the small-signal approximation of Eqs. (9.1)–(9.3) with (9.39) yields

$$\begin{bmatrix} \Delta\dot{\delta} \\ \Delta\dot{\omega} \end{bmatrix} = \begin{bmatrix} 0 & 1 \\ \frac{EVY}{M_0} & -\frac{D_0}{M_0} \end{bmatrix} \begin{bmatrix} \Delta\delta \\ \Delta\omega \end{bmatrix} \tag{9.44}$$

The Eigenvalues of the system are.

$$\lambda_{1,2} = -\gamma \pm \sqrt{\gamma^2 + EVY/M_0} \text{ with } \gamma = \frac{D_0}{2M_0} \tag{9.45}$$

The disadvantages of this method are:

- the Eigenvalues depend on E, V, and Y,
- an estimate of B and G is required in Eq. (9.39).

9.2.4 Adaptive inertia-based LSD

In this section we show that the dependence of the Eigenvalues on E, V and Y can be eliminated. Let γ be an arbitrary positive number and Ω an arbitrary nonnegative number. In order to obtain constant Eigenvalues, we can intuitively write from Eq. (9.5) the following control rules for the inertia and damping variables:

$$M = \frac{EV(B\cos\delta - G\sin\delta)}{-(\gamma^2+\Omega)} \text{ and} \quad D = 2M\gamma \tag{9.46}$$

Similarly to Eq. (9.41), we can calculate

$$\left.\frac{df}{d\delta}\right|_{\delta_0,\omega_0} = \frac{-(\gamma^2+\Omega)\left(\left(P_{ref}-V^2G\right)(B\sin\delta_0+G\cos\delta_0)+EVY^2\right)}{EV(B\cos\delta_0-G\sin\delta_0)^2} \tag{9.47}$$

and with Eqs. (9.40), (9.47) we obtain

$$\left.\frac{df}{d\delta}\right|_{\delta_0,\omega_0} = -\left(\gamma^2+\Omega\right) \tag{9.48}$$

We can also calculate

$$\left.\frac{df}{d\omega}\right|_{\delta_0,\omega_0} = -2\gamma \tag{9.49}$$

Therefore the small-signal approximation of Eqs. (9.1)–(9.3) with (9.46) yields

$$\begin{bmatrix}\Delta\dot{\delta}\\ \Delta\dot{\omega}\end{bmatrix} = \begin{bmatrix}0 & 1\\ -(\gamma^2+\Omega) & -2\gamma\end{bmatrix}\begin{bmatrix}\Delta\delta\\ \Delta\omega\end{bmatrix} \tag{9.50}$$

The Eigenvalues of the system are

$$\lambda_{1,2} = -\gamma \pm j\sqrt{\Omega} \tag{9.51}$$

for arbitrarily chosen positive γ and nonnegative Ω.

The disadvantages of this method are:

- an estimate of B, G and E is required in Eq. (9.46).

9.2.5 Reverse approach: The delta-based SE

The traditional swing equation can be replaced by a structurally similar, modified SE, that—as we will see later—produces more accurate LSD behavior. In order to derive such a modified SE, we start from defining the desired Eigenvalues as in Eq. (9.51). We observe that a small-signal behavior as in Eq. (9.50) will exactly define this desired dynamics. We realize, that

$$\dot{\omega} = (\gamma^2 + \Omega)\left(\delta_{ref} - \delta\right) - 2\gamma\dot{\delta}$$
$$\dot{\delta} = \omega - \omega_n \tag{9.52}$$

when linearized, will yield Eq. (9.50), and at the same time the steady-state values of the state variables will be.

$$\delta = \delta_{ref}, \omega = \omega_n \tag{9.53}$$

We call Eq. (9.52) the delta-based SE [15]. We assume constant inverter voltage V. The reference angle can be calculated by solving

$$P_{ref} = V^2G - EV\left(B\,\sin\delta_{ref} + G\,\cos\delta_{ref}\right) \tag{9.54}$$

for δ_{ref}, so that in steady state $P = P_{ref}$ will hold. The disadvantages of this method are:

- the estimate of B, G and E is required in Eq. (9.54).

9.2.6 Comparison of SMIB LSD concepts

In this section we compare the LSD concepts designed for SMIB systems, described in the previous section, by presenting and analyzing time-domain simulation results. The simulation model is a simplified LV network with typical values. It consists of two controlled voltage sources, one representing the infinite bus with $E = 231V_{rms}/0$ degrees and the other representing the inverter (in cases other than the voltage-control based LSD, a constant magnitude of $V = 231\ V_{rms}$ is used). The impedance between them is purely reactive, assuming a LV line of 0.5 km length and 0.37 Ω/km specific reactance, resulting in $B = -5.4054$ S. Despite the fact that for LV systems $R \approx X$ is typical, we neglect the line resistances here, so that the effects to be demonstrated be more clearly visible. Adding resistive component would make the swings more similar and the Eigenvalues less dependent on the operating point. This is clear if we observe Eq. (9.5): in case of $R = X$ the term $(B\cos\delta - G\sin\delta)$ varies in the range $[-2.7, -3.8]$ whereas without resistance it varies in the much wider range $[-2.7, 0]$, as δ goes from 0 to 90 degrees. (Nevertheless, the resistive-inductive case yields results that are consistent with the ones presented below.) The damping and inertia constants are defined as $D = 15{,}000\ Ws$ and $M = 2000\ Ws^2$. From Eqs. (9.44), (9.50) the parameters γ and Ω were determined as $\gamma = \frac{D}{2M}$ and $\Omega = -\left(\gamma^2 + \frac{EVB}{M}\right)$.

For the voltage-based LSD, $\varepsilon = 0.1$ was used, and in order to obtain the same dynamics as the other LSD methods, the values of D and M were both multiplied by $(1 - \varepsilon)$ so that Eq. (9.12) be the same as Eqs. (9.45), (9.51).

Fig. 9.8 shows time domain simulation results for this system, for four cases: No LSD, voltage-control based LSD, Inertia based LSD and Delta-based LSD. (The Adaptive Voltage-control based method gives exactly the same results as

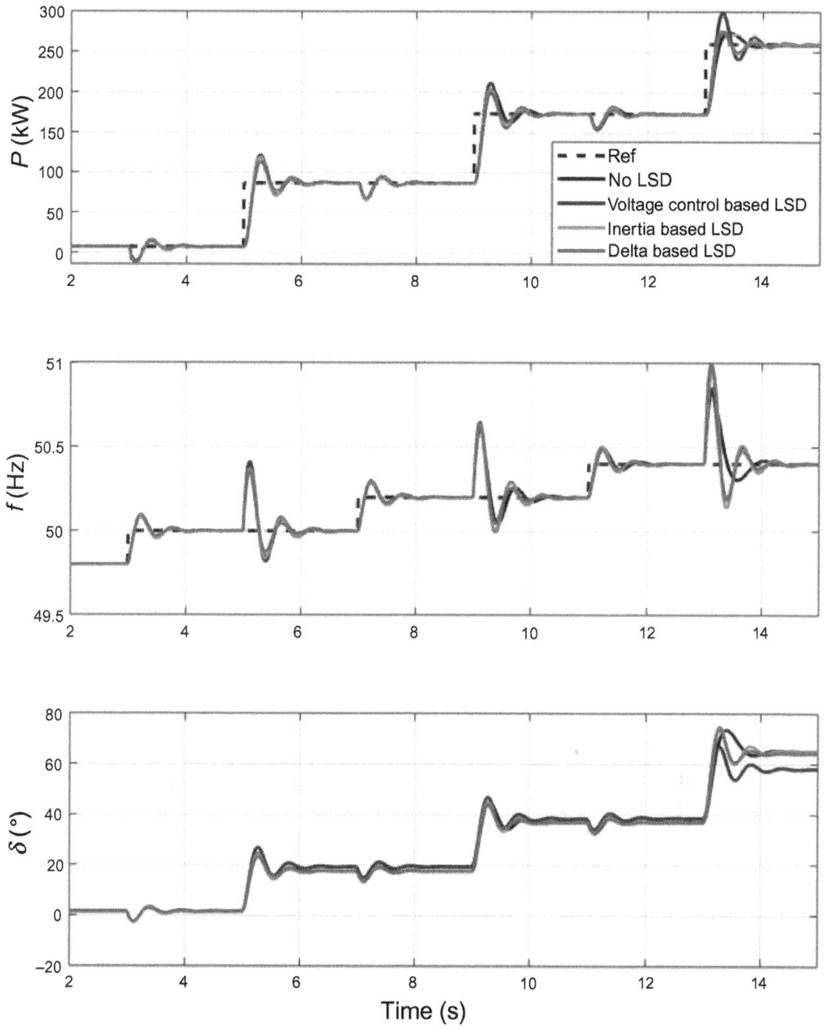

FIG. 9.8 Time domain simulation results using various SMIB LSD concepts.

the Voltage-control based method in case of constant system parameters. The Adaptive inertia based LSD method was also simulated, and it gave exactly identical results with the Inertia based LSD method using the above parameters, therefore it is not shown here separately.)

At $t=\{1,5,9\}$ s the reference power was changed from 2.5% to $\{30\%,60\%,90\%\}$ of the maximal power $EVB=288.4$ kW. At $t=\{3,7,11\}$ s the reference frequency was changed from 49.8 Hz to $\{50,50.2,50.4\}$ Hz. The goal of these scenarios is to cover almost the full range of the power capability (giving a certain safety margin for oscillations below the maximal power).

Another goal is to ensure that the dynamics is observed at various operating points that are different from each other not only in one of the state variables (δ) but also in the other state variable (frequency).

Analyzing the results, it is possible to determine that the angle δ of the voltage V converges to the same steady state values in all cases except the voltage-control based LSD. This is the behavior we expect from theory as explained under Eq. (9.8). The dynamic behavior in case of the three LSD methods is very similar, and significantly different from the "No LSD" case. Coherently with the theory, the No-LSD case is different for each operating point, whereas the LSD methods produce very similar transients regardless of the operating point.

A more accurate analysis of the transients is carried out below. The frequency transients from the intervals (3..5) s, (7..9) s, and (11..13) s are extracted and plotted in Fig. 9.9 after subtracting from them the respective reference values.

It can be observed, that the three transients are identical in case of the delta-based LSD, however there are slight differences between the three transients in case of other LSD methods. For all these transients we used a curve-fitting approach to experimentally determine the Eigenvalues of the system [18].

We determined the parameters p_i of Eq. (9.55) for each method and each transient period:

$$\widetilde{\Delta f}\,(t) = p_1 e^{p_2 t} \cos\left(p_3 t + p_4\right) \tag{9.55}$$

and from p_2 and p_3 we determined the estimated Eigenvalues. The results are shown in Tables 9.1 and 9.2 with four digits accuracy (usually the Eigenanalysis is carried out with 4–7 significant digits [19]).

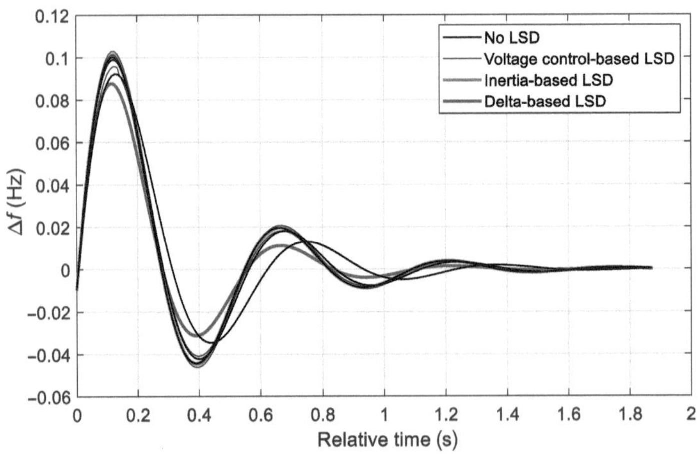

FIG. 9.9 Frequency transients at three different operation points and various LSD methods (three curves for each LSD method and also for the No LSD case).

TABLE 9.1 **Estimated Eigenvalues for different methods at various operation points.**

λ	(3..5) s	(7..9) s	(11..13) s
No LSD	−3.0068±11.5727i	−3.0476±11.2639i	−3.1737±10.1987i
Voltage	−3.0514±11.4055i	−3.0039±11.5449i	−2.9394±11.4954i
Inertia	−3.0070±11.5779i	−3.0124±11.5445i	−3.0187±11.5090i
Delta	−3.7499±11.4086i	−3.7499±11.4086i	−3.7494±11.4085i

TABLE 9.2 **Estimated Eigenvalues for different methods at various operation points.**

λ	(5..7) s	(9..11) s	(13..15) s
No LSD	−3.0413±11.2952i	−3.1734±10.2942i	−3.7450±7.6698i
Voltage	−2.9953±11.5722i	−2.9266±11.5230i	−2.7416±11.3608i
Inertia	−3.0105±11.5898i	−3.0177±11.5437i	−3.0297±11.5239i
Delta	−3.7483±11.4077i	−3.7489±11.4078i	−3.7488±11.4078i

The expected Eigenvalues of the system are—based on the design parameters: −3.7500±11.4086i from Eq. (9.51). It can be observed, that the Delta-based LSD delivers the expected Eigenvalues at all operating points. The Voltage control and Inertia-based LSD methods, however, produce dynamics with Eigenvalues that are different from the expected one, and the Eigenvalues are also slightly dependent on the operating point. These deviations have several reasons.

First, in Eq. (9.3) we neglected the fact that in the lossless case $B = -1/(\omega L)$ depends on the frequency. Therefore, the lower-right terms in the small-signal models Eqs. (9.4), (9.11), (9.44) are approximations, and so are the Eigenvalues calculated from these models.

Taking into consideration the above frequency-dependency, the analytical expressions of the Eigenvalues for the lossless case, as opposed to $\lambda_{1,2} = -\gamma \pm \sqrt{\gamma^2 - \frac{EV}{MX}\cos\delta}$ from Eq. (9.5), become

$$\lambda'_{1,2} = -\gamma' \pm \sqrt{\gamma'^2 - \frac{EV}{M\,\omega L}\cos\delta}, \text{with } \gamma' = \frac{D'}{2M} \text{ and } D' = D - \frac{P}{\omega} \qquad (9.56)$$

The values of this corrected expression for each transient can be observed in Table 9.3.

TABLE 9.3 Corrected analytic Eigenvalues for the "No LSD" case at various operation points.

λ'	(3..5) s	(7..9) s	(11..13) s
No LSD	−3.7443+11.4085i	−3.6814+11.1091i	−3.6134+10.0432i
λ'	(5..7) s	(9..11) s	(13..15) s
No LSD	−3.6811+11.1355i	−3.6128+10.0769i	−3.5451+6.8942i

It can be concluded, that the above frequency-dependency partially explains why the Eigenvalues in Tables 9.1 and 9.2 are not equal to the ones calculated from Eq. (9.51).

Secondly, Eq. (9.3) is only valid in steady-state, at constant frequency. However, during transients, when the frequency changes, Eq. (9.3) is only an approximation, and so are the Eigenvalues in cases where the power is used to govern the dynamics. Only the Eigenvalues using the delta-based SE are not affected.

And finally, the third reason for the deviations is that because of the first two reasons mentioned above, Eq. (9.55) is only approximating the transients, so the estimation procedure is inaccurate—except for the delta-based case, where the estimated Eigenvalue is within 0.05% of the analytically computed one (both for its real and imaginary values).

As we have seen in Tables 9.1 and 9.2, though the estimated Eigenvalues differ from the analytically calculated ones for the Voltage-control and Inertia based LSD, the dependency on the operating point is very small, at least for the Inertia-based LSD. We can calculate the spread of the real and imaginary values of the estimated Eigenvalues as

$$\Delta \mathrm{Re}_{type} = \frac{\mathrm{MAX}_{tr}\left\{real\left(\lambda_{type,tr}\right)\right\} - \mathrm{MIN}_{tr}\left\{real\left(\lambda_{type,tr}\right)\right\}}{\frac{1}{6}\sum_{tr=1}^{6} real\left(\lambda_{type,tr}\right)} \tag{9.57}$$

and

$$\Delta \mathcal{J}\Updownarrow_{type} = \frac{\mathrm{MAX}_{tr}\left\{imag\left(\lambda_{type,tr}\right)\right\} - \mathrm{MIN}_{tr}\left\{imag\left(\lambda_{type,tr}\right)\right\}}{\frac{1}{6}\sum_{tr=1}^{6} imag\left(\lambda_{type,tr}\right)} \tag{9.58}$$

where *tr* is the serial number of the transient and *type* is the LSD method as in Tables 9.1 and 9.2. The spreads can be observed in Table 9.4.

In summary, the Delta-based LSD provides exactly constant Eigenvalues. All other LSD methods provide approximately constant Eigenvalues, the

TABLE 9.4 Spread of the real and imaginary value of the estimated Eigenvalues.

Type	$\Delta \mathrm{Re}_{type}$ (%)	$\Delta \mathcal{J}m_{type}$ (%)
No LSD	−23.08	37.59
Voltage	−10.53	1.84
Inertia	−0.75	0.70
Delta	−0.04	0.01

second best method is the (Adaptive) Inertia based LSD, the least accurate is the Voltage control-based approach.

9.2.7 LSD control embedded in existing inertia emulation schemes

The developed LSD control methods can be added to several existing converter control schemes. The control parameters of the VSM and LSD loops have to be designed so that a time-scale separation between these and other dynamics (such as PLL or voltage-, current- or DC link control loops, as well as power line dynamics) can be obtained. This way the swing-related Eigenvalues will be independent of other dynamics.

9.2.7.1 VSM with cascaded control

An established VSM control scheme is based on the cascaded control loop concept [20] (Fig. 9.10):

The Voltage control-based LSD can replace the reactive power control loop and the virtual impedance loop as follows (as discussed in more detail in [12]) (Fig. 9.11):

The Inertia-based LSD can be implemented as an additional input loop to the virtual inertia loop (Fig. 9.12):

The Delta-based LSD can be implemented in a loop that replaces the virtual inertia and power control (a separate power droop may be necessary in this case) (Fig. 9.13):

9.2.7.2 Synchronverter

Another popular implementation of virtual inertia is the synchronverter concept [4] (Fig. 9.14).

The Voltage control-based LSD can replace the reactive power control loop (as discussed in more detail in [15]) (Fig. 9.15):

Inertia-based LSD can be added to alter the (Fig. 9.16) constants J and D:

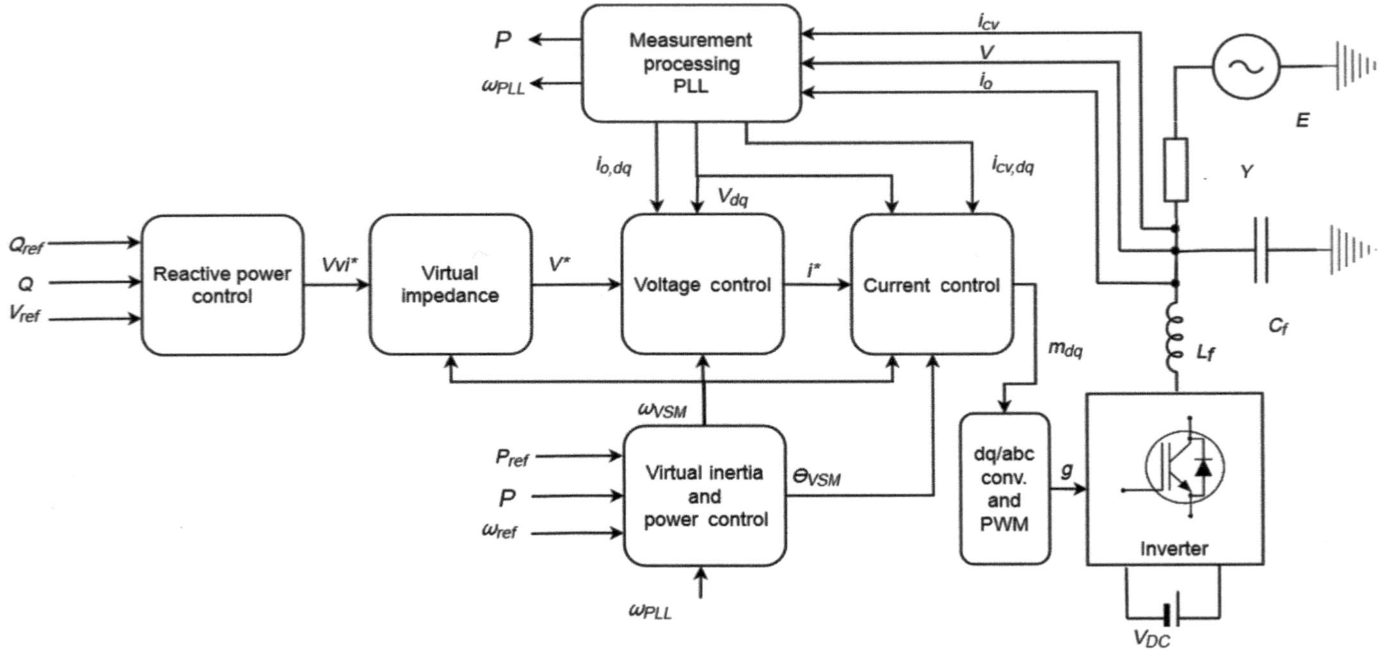

FIG. 9.10 VSM with cascaded control loops.

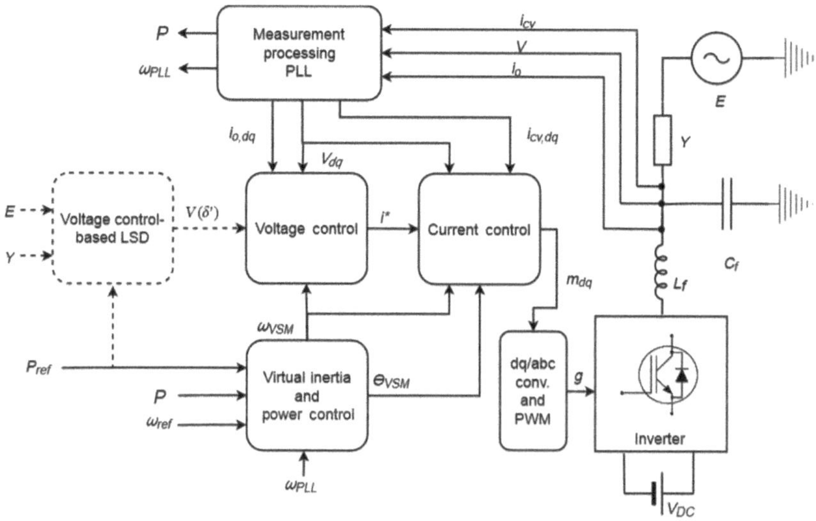

FIG. 9.11 Voltage control-based LSD embedded in a VSM with cascaded control loops.

The implementation of the delta-based LSD would completely replace the synchronverter equations.

9.2.7.3 HVDC converter

HVDC systems can also be used to participate in system frequency control. Fig. 9.17 shows the controller structure of the LSD-VSG in the HVDC converter that is responsible for controlling the DC link voltage based on the active power-voltage droop principle. The LSD behavior is achieved while the AC voltage is kept around the nominal value, and frequency support can be provided in case of disturbances in the AC grid [13].

Either voltage control-based or Inertia-based LSD approaches can be implemented as shown in Fig. 9.17.

9.3 LSD concepts for multimachine systems

The transition from single-machine to multimachine systems in the development of LSD solutions can be done using various approaches.

A *centralized method* assumes that all quantities that are necessary to carry out LSD control are measured at the network nodes and transmitted in real-time to a central computing engine. This engine determines the control actions and transmits them back to the converters in real-time.

Such an approach is practically not viable considering state-of-the-art measurement, communication, and control techniques. However, for theoretical analysis or benchmarking, such a centralized method could be of interest.

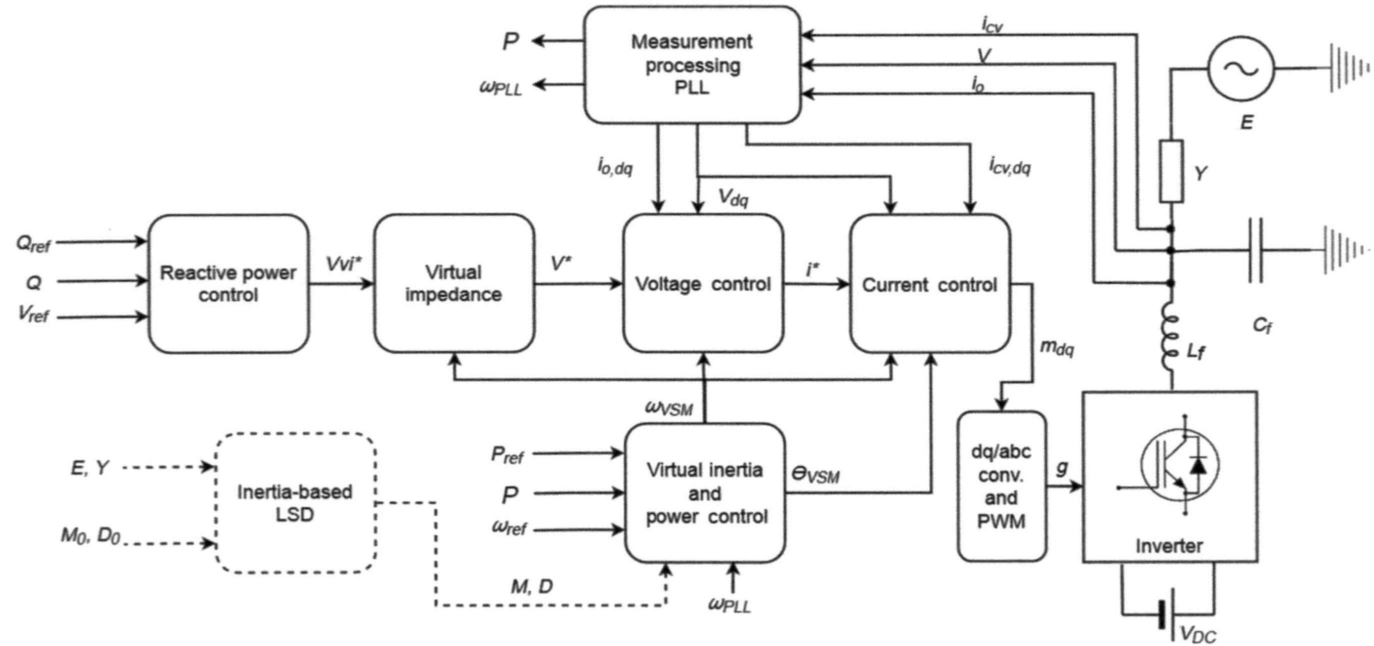

FIG. 9.12 Inertia-based LSD embedded in a VSM with cascaded control loops.

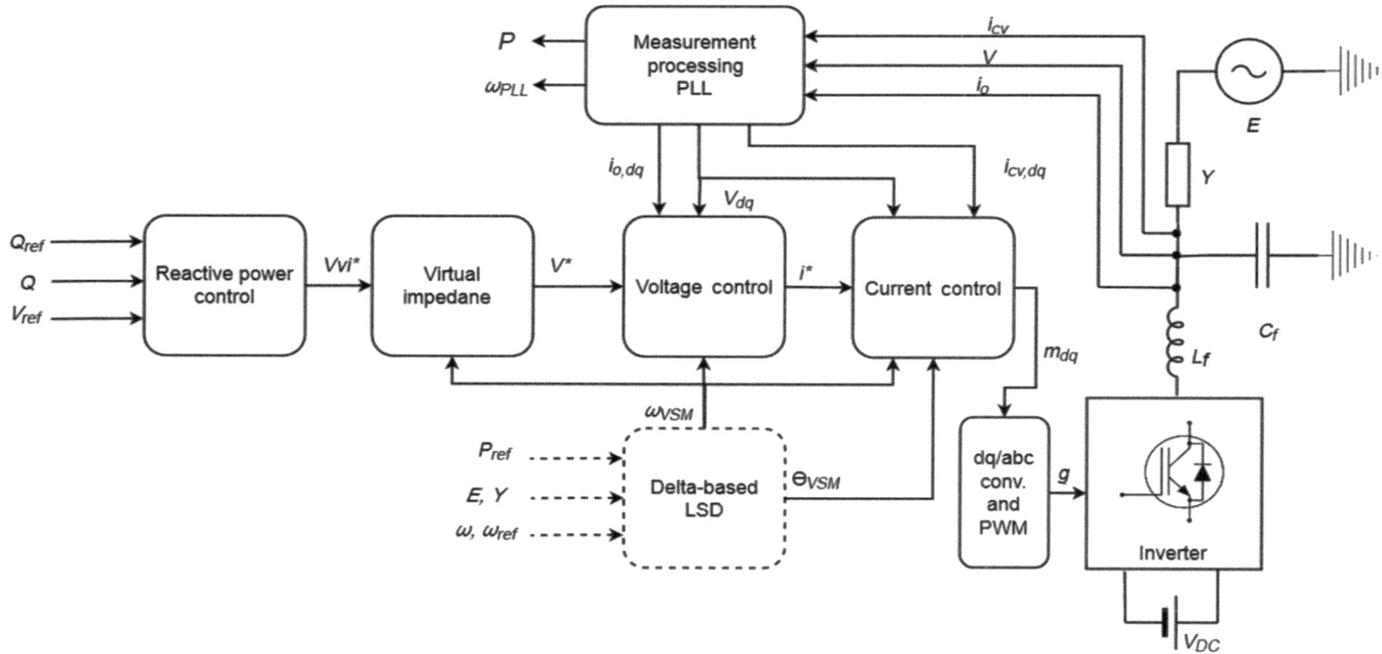

FIG. 9.13 Delta-based LSD embedded in a VSM with cascaded control loops.

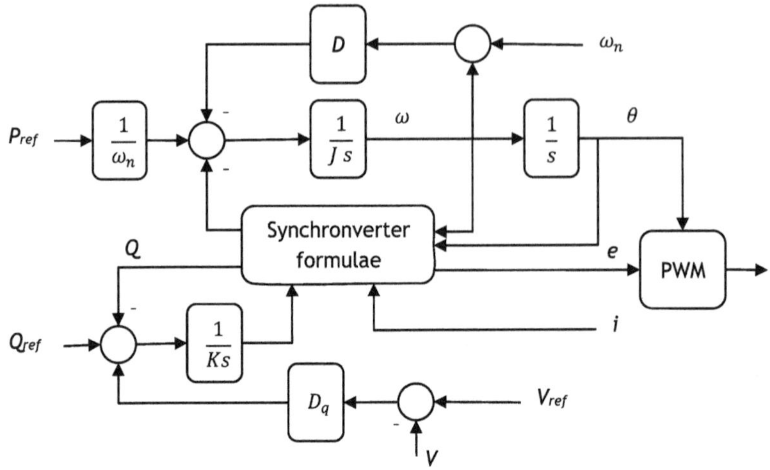

FIG. 9.14 Synchronverter controller.

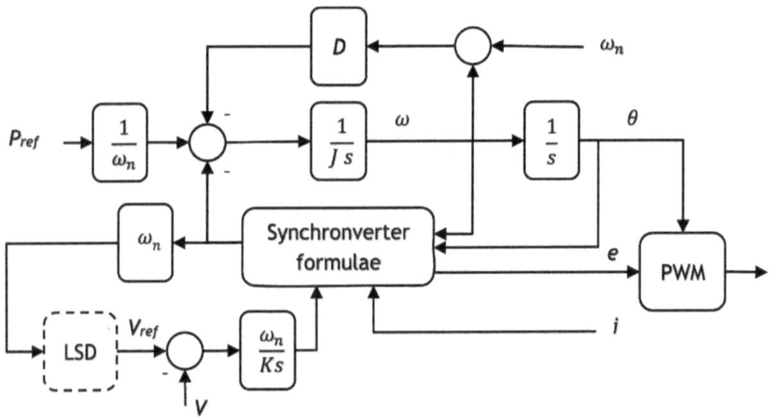

FIG. 9.15 Voltage control-based LSD with a Synchronverter.

In a *decentralized method* each converter determines its control actions locally, but by doing that the local controllers rely on a small number of centrally available information, sent to the converters periodically or in an event-based manner. This method puts less stringent requirements on the SCADA infrastructure, but still needs locally unavailable data.

A *distributed approach* is characterized by exclusively local measurement and control, which does not rely on any centrally provided data. This is the most robust and reliable approach.

In the following chapters we show examples for all three approaches and demonstrate that the distributed approach is practically feasible.

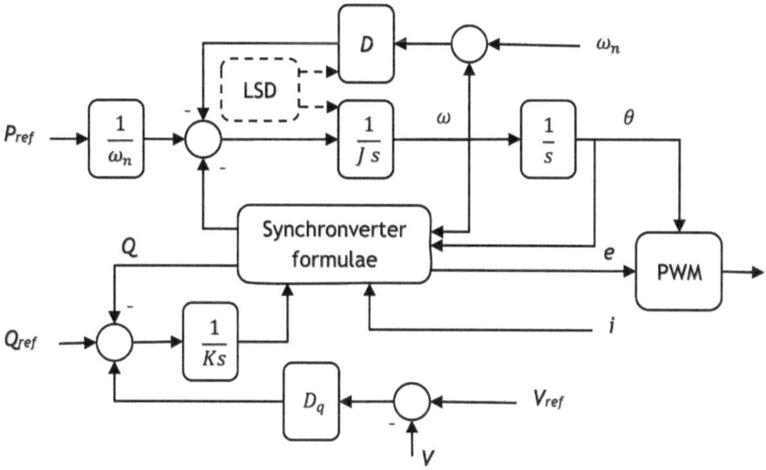

FIG. 9.16 Inertia-based LSD with a Synchronverter.

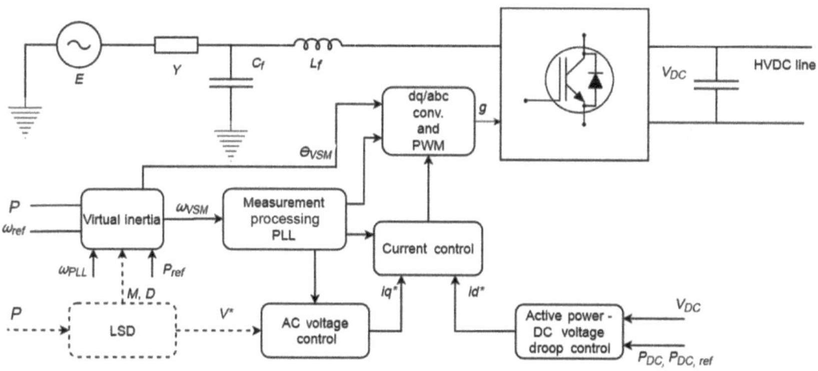

FIG. 9.17 Voltage control-based LSD-VSG in an HVDC converter.

9.3.1 Centralized approach—Voltage control based example

In this chapter we present the extension of the voltage control based approach to multimachine systems. First, we assume that branch conductances (as well as nodal conductances) can be neglected.

The active power injections at node i can be expressed as

$$P_i = \sum_{j \neq i} V_i V_j B_{ij} \sin\left(\delta_i - \delta_j\right) \tag{9.59}$$

with B_{ij} being the element of the susceptance matrix $\boldsymbol{B}$ in the i-th row and j-th column. Analogously to Eq. (9.10), the linear relation we wish to achieve can be expressed as

$$P_i = E^2(1-\varepsilon)\sum_{j\neq i} B_{ij}\left(\delta_i - \delta_j\right) \tag{9.60}$$

Recall that $B_{ii} = -\sum_{j\neq i} B_{ij}$, and therefore, if B_i denotes the ith row of $\boldsymbol{B}$:

$$-B_i\boldsymbol{\delta} = B_i \begin{bmatrix} \vdots \\ \delta_j \\ \vdots \\ \delta_i \\ \vdots \end{bmatrix} = -\left(\cdots + B_{ij}\delta_j + \ldots + B_{ii}\delta_i + \cdots\right)$$
$$= -\left(\ldots + B_{ij}\delta_j + \ldots - \delta_i\sum_{j\neq i} B_{ij} + \ldots\right) = \sum_{j\neq i} B_{ij}\left(\delta_i - \delta_j\right) \tag{9.61}$$

Therefore, Eq. (9.60) can be written as

$$\boldsymbol{P} = -E^2(1-\varepsilon)\boldsymbol{B\delta} \tag{9.62}$$

where $\boldsymbol{P}$ denotes the column vector of the node injection powers P_i and $\boldsymbol{\delta}$ represents the column vector of the nodal voltage angles δ_i. The angles can now be calculated by solving the linear Eq. (9.62) for $\boldsymbol{\delta}$. Then the voltage references can be calculated by solving Eq. (9.59) for all voltages. Eq. (9.59) is nonlinear, and a closed-form solution does not exist for node numbers $N>3$. In such cases, Eq. (9.59) can be solved numerically.

To obtain the Eigenvalues of the multimachine LSD-controlled system, we have to derive the expressions $\frac{\partial P_i}{\partial \delta_j}$ (following the approach in Eqs. (12.102)–(12.108) in Ref. [8]). From Eq. (9.60) it can be seen, that

$$\frac{\partial P_i}{\partial \delta_j} = -E^2(1-\varepsilon)B_{ij} \tag{9.63}$$

Therefore, the elements of the state-space matrix are constant:

$$\begin{bmatrix} \Delta\dot{\delta} \\ \Delta\dot{\omega} \end{bmatrix} = \begin{bmatrix} \boldsymbol{0} & \boldsymbol{I} \\ \left[-E^2(1-\varepsilon)\frac{B_{ij}}{M_j}\right] & \mathrm{diag}\left(-\frac{D_i}{M_i}\right) \end{bmatrix} \begin{bmatrix} \Delta\delta \\ \Delta\omega \end{bmatrix} \tag{9.64}$$

where [.] indicates a matrix structure of appropriate size, composed of the shown elements, $\boldsymbol{I}$ and $\boldsymbol{0}$ are the identity matrix and the zero of appropriate size, and *diag*(.) indicates a diagonal matrix, composed of the shown elements.

From Eq. (9.64) it follows that the Eigenvalues will be independent of the actual power flows or voltage angles.

In Ref. [12] we presented simulations and analyzed this centralized solution on a 4-bus network with purely reactive grid impedances.

However, extending the method to a multimachine system where the grid impedances have also a resistive component, poses some difficulties. We have shown in Section 9.2.1.2 that in such cases the Eigenvalues are not constant, but can take a finite number of possible values, depending on the actual sign of the angles.

Therefore, the voltage control-based method is not the best candidate for a multimachine system, if the grid impedances have resistive components.

9.3.2 Decentralized approach—Adaptive voltage control based example

This method is again based on the Voltage Control-based approach. Each converter is controlled in a way as if it was connected to a SMIB system, where the rest of the system, as seen by each converter, is represented by its Thévenin equivalent circuit.

The control rules are formally the same as described in Section 9.2.2, with the following differences:

- Instead of E the Thévenin equivalent voltage at each node $E_{Th,\ i}$ has to be used.
- Instead of B and G the Thévenin equivalent admittance components $B_{Th,\ i}$ and $G_{Th,\ i}$ have to be used.
- Instead of a constant voltage tolerance value ε an ε_i value has to be set individually for each node.

  The power flow from node i will be—analogously to Eq. (9.8) in the purely reactive case

$$P_i = -E_{Th,i}B_{Th,i}(1-\varepsilon_i)E_n\delta_i \tag{9.65}$$

  Where E_n is the nominal voltage. The role of ε_i is to ensure that a centrally calculated power flow is fulfilled, that is, the voltage $(1-\varepsilon_i)E_n$ at node i equals to the voltage calculated for the planned operation.

The dynamic model of machine i can be expressed as

$$\begin{bmatrix} \Delta\dot{\delta}_i \\ \Delta\dot{\omega}_i \end{bmatrix} = \begin{bmatrix} 0 & 1 \\ -\dfrac{(1-\varepsilon_i)E_nE_{Th,i}B_{Th,i}}{M_i} & -\dfrac{D_i}{M_i} \end{bmatrix} \begin{bmatrix} \Delta\delta_i \\ \Delta\omega_i \end{bmatrix} \tag{9.66}$$

It can be seen that the components of the system matrix are independent of the operating point.

However, during operation the parameters of the Thévenin equivalent circuit can change over time, due to less frequent events like network reconfiguration or fault clearance, etc. To compensate for these variations, an adaptation of the inertia and damping parameters can be performed, similarly to the adaptive voltage control-based approach:

$$M_i = M_{i,0} \frac{E_{Th,i} B_{Th,i}}{E_{Th,i,0} B_{Th,i,0}} \tag{9.67}$$

and

$$D_i = D_{i,0} \frac{E_{Th,i} B_{Th,i}}{E_{Th,i,0} B_{Th,i,0}} \tag{9.68}$$

where $M_{i,\,0}$ and $D_{i,\,0}$ are the nominal values of the emulated inertia and damping respectively, and $E_{Th,\,i,\,0}$ and $B_{Th,\,i,\,0}$ are the Thévenin equivalent values calculated for the planned operation.

In Ref. [14] simulations and analysis of this decentralized solution on a 3-bus network with purely reactive grid impedances was presented.

The application of the method assumes, that the centrally calculated voltage set point for each converter is available at the respective converter, together with the Thévenin equivalent circuit parameters seen from each converter. It has to be noted that the Thévenin equivalent voltage can be calculated once the active and reactive power flows from the converter are measured and if the Thévenin equivalent impedance is known. This impedance can either be provided from a central calculation or it can be estimated using local measurements, e.g., using a current injection method [21].

9.3.3 Distributed approach—Delta-based example with internal reactance method

The method presented in this Chapter has the advantage over previously presented multimachine LSD methods that it does not require the knowledge of any centrally available parameter, because this requirement could limit the practical applicability.

We have also seen that it is possible to establish an SE-like dynamic behavior that is defined by quantities other than those in the traditional SE: the delta-based (reverse) approach provides swing behavior based on the difference of angles rather than powers. We further exploit this approach here, because this makes it possible to separate the swing behavior from the nonlinear power flow that couples the individual swing responses in the traditional SE-based approach.

9.3.3.1 The delta-based LSD with internal reactance method

For the sake of the following analysis, we assume that the converter controller includes a voltage controller, and that this voltage control loop is much faster than the swing dynamics of the system. This means that we can keep assuming a constant converter voltage. We also assume that—like in the case of most DG converters—the filter includes a series reactance: we call it the *internal reactance* B_{int} for the sake of the underlying analysis. The filter can be e.g.,

(a) an LC filter (the controlled voltage is at the grid side of the internal reactance) or
(b) an LCL filter (the controlled voltage is at the inverter side of the internal reactance).

In both cases, the capacitor voltage can be considered as constant (V) when analyzing the swing dynamic behavior of the system. In case of the LCL filter, the grid-side reactance is used as the internal reactance. We further define the angle difference δ_{int} between the voltages at both ends of the internal reactance as the state variable that governs the swing dynamics.

Since the modified SE indirectly controls the active power of the converter (to track a reference value P_{ref}), the converter can be seen as a "PV" node from the point of view of a powerflow calculation. This means that the voltage angle and the reactive power of the converter will be the result of the interaction with the external network.

Other than in the standard modeling approach for small-signal stability analysis, where the system is reduced so that all nodes are generator nodes, the presented approach does not use model reduction. Therefore, instead of pure differential equations a *system of differential-algebraic equations (DAE)* has to be solved.

The *differential equations* can now be written as

$$\begin{aligned} \dot{\omega} &= (\gamma^2 + \Omega)\left(\delta_{\text{int},ref} - \delta_{int}\right) - 2\gamma\dot{\delta}_{int} \\ \dot{\delta}_{int} &= \omega - \omega_n \end{aligned} \tag{9.69}$$

In the following analysis we focus on a converter with an LC-filter. In Ref. [16] the equations for the LCL filter case are also presented.

At the marked point in Fig. 9.18 the following *algebraic equations* hold:

$$0 = P_{int} - P \text{ and } 0 = Q_{int} - Q \tag{9.70}$$

where

$$P = V^2 G - EV(B\sin\delta + G\cos\delta) \tag{9.71}$$

and

$$Q = -V^2(B - B_c) - EV(-B\cos\delta + G\sin\delta) \tag{9.72}$$

with B_c being the susceptance of the filter capacitance, and

$$P_{int} = -VV_{int}B_{int}\sin\delta_{int} \tag{9.73}$$

and

$$Q_{int} = V^2 B_{int} - VV_{int}B_{int}\cos\delta_{int} \tag{9.74}$$

The differential and algebraic equations (9.69)–(9.74) constitute the DAE. Let $\boldsymbol{x} = (\delta_{int}, \omega)$ denote the vector of state variables, $\boldsymbol{y} = (V_{int}, \delta)$ denote the

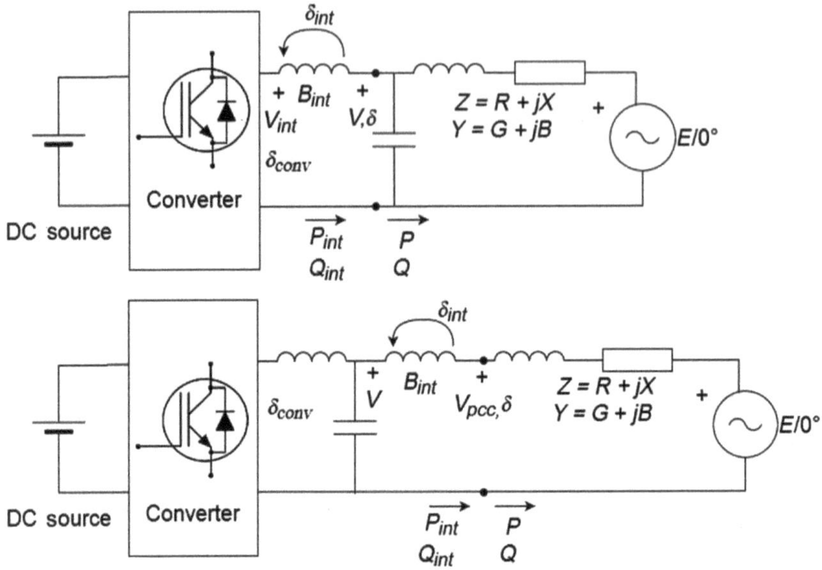

FIG. 9.18 Internal reactance based approach in case of (A) LC (B) LCL filter.

vector of algebraic variables. Let f denote the vector of differential equations (9.69) and g the vector of algebraic equations (9.70)–(9.74). We can write the DAE as

$$\begin{aligned} \dot{x} &= f(x, y) \text{ and} \\ 0 &= g(x, y) \end{aligned} \tag{9.75}$$

For small-signal stability analysis, Eq. (9.75) is linearized at an equilibrium point to obtain the state matrix

$$J = f_x - f_y g_y^{-1} g_x \tag{9.76}$$

where $f_x = \partial f / \partial x$, $f_y = \partial f / \partial y$, $g_x = \partial g / \partial x$, and $g_y = \partial g / \partial y$ [22].

The dynamic behavior of the DAE is governed by the Eigenvalues of the state matrix J. In the special case when f_y or g_x is a zero matrix, then $J = f_x$, i.e. the dynamic behavior is governed by the Eigenvalues of Eq. (9.69).

We will show that it is possible to design the parameters of Eq. (9.69) so that $f_y = 0$.

Regarding the reference angle, we can use

$$\delta_{\text{int},ref} = \text{asin}\left(-\frac{P_{ref}}{B_{int} V V_{\text{int},ref}} \right) \tag{9.77}$$

The values of P_{ref}, B_{int} and V can be considered constant and are known locally.

Further we have to consider, that the converter is not directly controlling δ_{int}, but it is controlling the angle δ_{conv}, so that the reference angle for the converter itself has to be

$$\delta_{conv,\,ref} = \delta_{ref} + \delta_{int,\,ref} \tag{9.78}$$

From Eqs. (9.77), (9.78) it follows that two parameters, namely $V_{int,ref}$ and δ_{ref} have to be determined, and they have to be determined in such a way that they are independent on $\boldsymbol{y}$. (For this reason we cannot use a PLL-measured δ instead of δ_{ref}.)

For the determination of $V_{int,ref}$ and δ_{ref} we offer two possible scenarios below.

(i) On the one hand, these values can be obtained as a result of a central power flow calculation performed by a system operator. In case of an LC-filter, considering that the inverter is a "PV" node at its PCC, the result of the power flow calculation is the reactive power Q_{ref} at the PCC and δ_{ref} itself. From Q_{ref} we have

$$V_{int,\,ref} = \frac{\sqrt{P_{ref}^2 + \left(Q_{ref} - VB_{int}\right)^2}}{B_{int}V} \tag{9.79}$$

This choice is necessary and sufficient to ensure, that in steady state the active power equals its reference. The disadvantage of this method is that it still relies on two pieces of centrally available information (namely, either a voltage or a reactive power, and an angle).

(ii) On the other hand, it is also possible to use a constant value for $V_{int,\,ref}$ and δ_{ref} other than the above powerflow results. In this case all the control parameters are available locally, no centrally calculated value is necessary—which is a great advantage for practical realization. However, the steady-state active power will not equal P_{ref}. This disadvantage can be overcome by using a PI controller, as shown in Fig. 9.19. The parameters of the PI controller shall be chosen so that this secondary-type controller is significantly slower than the swing behavior in Eq. (9.69). This ensures that the dynamics of the SE are not influenced by the PI controller.

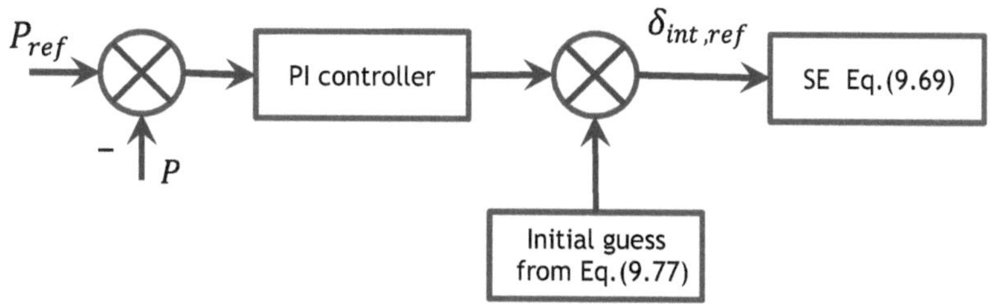

FIG. 9.19 PI controller to set angle reference.

The (fast) voltage controller and this power control loop can together ensure, that in steady-state $P = P_{ref}$ and $V = V_{ref}$.

Choosing $V_{int,\ ref}$ and $\delta_{int,\ ref}$ according to the above considerations, Eq. (9.69) is independent on $\boldsymbol{y}$. This means, that the Eigenvalues of the system are constant, the same as in Eq. (9.51).

In case of a multimachine system with N converters, the DAE that governs the swing dynamics consists of multiple instances of Eq. (9.69) and the power flow equations. Assuming that all converters use the same values for the parameters γ and Ω, the system matrix will then have the form:

$$\boldsymbol{J} = \boldsymbol{f_x} = \begin{bmatrix} \boldsymbol{0}_{N\times N} & \boldsymbol{I}_{N\times N} \\ -\mathrm{diag}(\gamma^2 + \Omega)_{N\times N} & -\mathrm{diag}(2\gamma)_{N\times N} \end{bmatrix} \tag{9.80}$$

where I is the identity matrix and $diag(.)$ indicates a diagonal matrix. The Eigenvalues will be as in Eq. (9.51), N times.

9.3.3.2 *Simulation of a multimachine system*

In this chapter we show time domain simulation results of a 4-bus system in which a converter is connected to each bus via an internal reactance, and each converter is modeled as a controllable voltage source that applies the delta-based LSD approach. The network data can be seen in Table 9.5.

The nominal voltage is $E = 231\ V_{rms}$. For the sake of comparison, we choose the damping and inertia constants $D = 15{,}000\ Ws$ and $M = 2000\ Ws^2$ for each converter. For the delta-based LSD we need parameters γ and Ω. From D and M we have $\gamma = 3.75$. We calculate an average

TABLE 9.5 Network impedances.

Nodes	*X* (Ohm)
1–2	0.1
1–3	0.15
2–3	0.1
1–4	0.2
3–4	0.25
$1_{internal}$	0.1
$2_{internal}$	0.05
$3_{internal}$	0.02
$4_{internal}$	0.12

$$\Omega = -\frac{1}{4}\sum_{i=1}^{4}\left(\gamma^2 - \frac{E^2}{X_{i,internal}M}\right) = 575\,\frac{1}{s^2} \tag{9.81}$$

for all converters. This means, we expect Eigenvalues

$$-\gamma \pm j\sqrt{\Omega} = -3.75 \pm j23.98\,\frac{1}{s}$$

We define the following reference values during the simulation (Tables 9.6 and 9.7):

From these values the references of the active and reactive powers are determined using powerflow calculations (Tables 9.8 and 9.9).

The simulation results can be observed in Fig. 9.20. In this figure, the active power reference values and the active power outputs are shown, for the two scenarios described above:

(i) using the correct reference values for both $V_{int,\ ref}$ and $\delta_{int,\ ref}$ from a central powerflow calculation;
(ii) using $V_{int,\ ref} = E$ and $\delta_{int,\ ref} = 0$.

The analysis of the results shows that (unlike in scenario (ii)) in scenario (i) the reference powers are tracked without a steady-state error.

TABLE 9.6 Simulation scenarios (V_{ref} values relative to E, V).

Node	1..4 s	4..8 s	8..12 s	12..16 s
1	1	1	1	0.95
2	1	1	0.98	0.90
3	1	1	1	1
4	1	1	1.02	1.10

TABLE 9.7 Simulation scenarios (δ_{ref} values, degrees).

Node	1..4 s	4..8 s	8..12 s	12..16 s
1	0°	0°	0°	0°
2	0.5°	−0.5°	1°	−3°
3	−0.5°	0.5°	−0.5°	3°
4	0.2°	−0.2°	0.5°	1°

TABLE 9.8 Simulation scenarios (P_{ref} values, W).

Node	1..4 s	4..8 s	8..12 s	12..16 s
1	−2483.5	2483.5	−8397.0	1324.6
2	13,969.3	−13,969.3	22,815.4	−74,077.3
3	−15,024.8	15,024.8	−20,592.9	76,080.8
4	3539.0	−3539.0	6174.5	−3328.1

TABLE 9.9 Simulation scenarios (Q_{ref} values, var).

Node	1..4 s	4..8 s	8..12 s	12..16 s
1	35.5	35.5	5439.7	−28,440.0
2	101.6	101.6	−20,658.7	−68,781.2
3	110.7	110.7	6629.2	53,040.6
4	17.6	17.6	9840.6	67,687.2

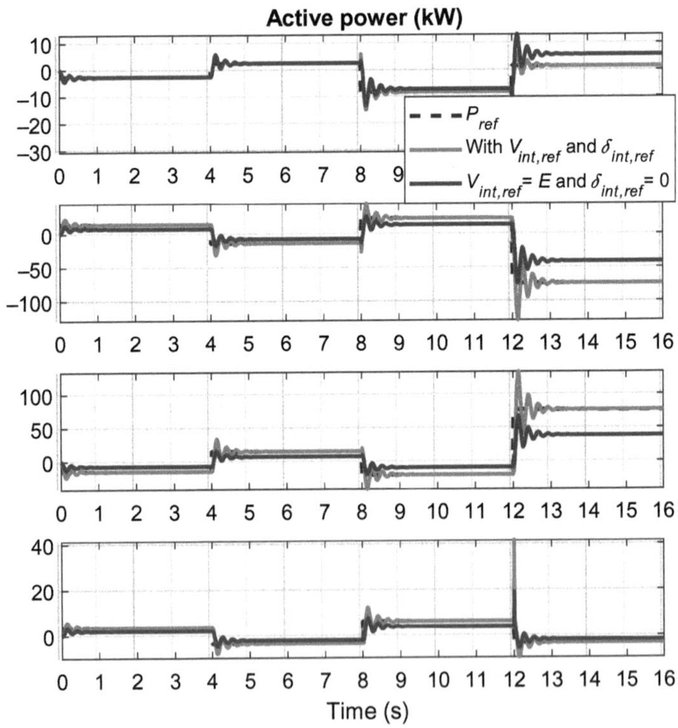

FIG. 9.20 Four-machine system with delta-based LSD using the internal reactance method—active power of the four converters.

The analysis of the power transients using the curve fitting method as introduced in Section 9.2.6 reveals that the **Eigenvalues (calculated from every transient at each converter) are exactly $-3.75 \pm j23.98$** (within 0.1% accuracy for both the real and imaginary components), which is the expected eigenvalue pair.

Note that in these simulations we did not implement either the voltage controller or the power control loop. These should be designed so that a time-scale separation is achieved. This way

- the dynamics of the voltage control, the swing behavior and the secondary power control can be separated
- the Eigenvalues of the swing dynamics is independent of the operating point, of the grid impedances and voltages and can be kept at a constant design value valid for all converters
- the voltage and power references can be tracked without any centrally calculated information.

We compare the above results with a simulation where no LSD is applied, only the traditional SE in Eqs. (9.1), (9.2). The results can be seen in Fig. 9.21.

From the transients it is obvious that the dynamics is not uniform either across machines or at various operating points. The identified Eigenvalues are given in Table 9.10.

This confirms that without LSD the system dynamics depends on the operating point and is not uniform across the system.

9.4 Conclusion

This chapter presented a systematic analysis of various LSD concepts for SMIB systems, with resistive-inductive network impedance. We concluded that the Delta-based LSD provides exactly constant Eigenvalues, independent of the operating point of the converter. All other LSD methods provide approximately constant Eigenvalues, the second best method is the Adaptive inertia based LSD, followed by the Inertia based method, and the least accurate is the Voltage control-based approach. Nonetheless, all approximations except the Voltage-control based LSD can be considered very good. (The error is smaller than 0.7% for the Inertia-based and smaller than 0.04% for the Delta-based approaches.) We have also identified for all LSD methods how the dynamics depend on various systems parameters, as well as the required input parameters for the implementation of the algorithms.

We analyzed the possible approaches for the extension of the concept to multimachine systems. Our goal was to find a method, that

- can be implemented in a multimachine system based exclusively on locally available parameters
- is truly independent on the operating point of the system.

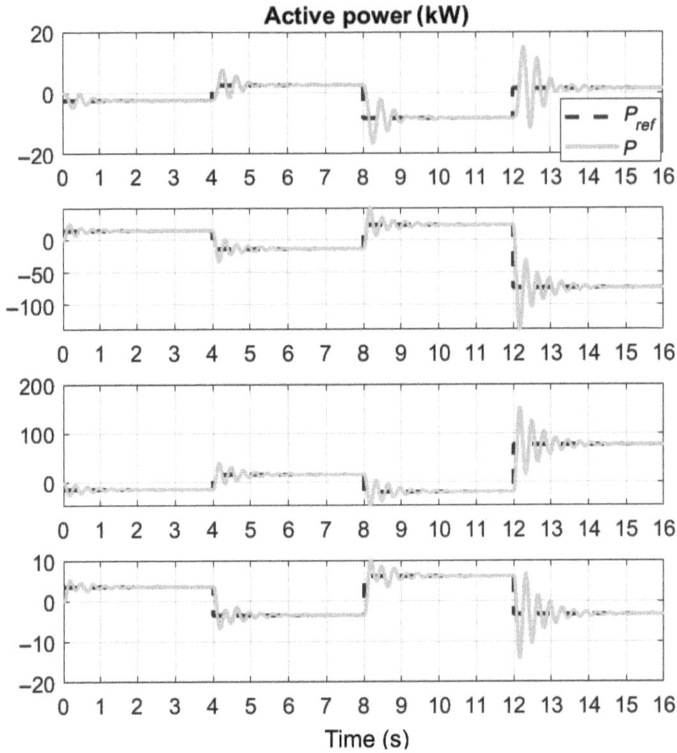

FIG. 9.21 Four-machine system without LSD (only SE)—active power of the four converters.

TABLE 9.10 Estimated Eigenvalues for each converter at various operation points without LSD.

Node	4..8 s	8..12 s	12..16 s
1	−1.4823 ± 20.2089i	−1.4190 ± 20.2982i	−1.7359 ± 19.3649i
2	−1.6310 ± 19.7795i	−1.5977 ± 19.7441i	−1.6747 ± 19.6101i
3	−2.001 ± 19.9581i	−2.1635 ± 19.9840i	−2.0504 ± 19.8578i
4	−2.3475 ± 16.9296i	−2.3837 ± 16.1822i	−2.9149 ± 17.4476i

We have proved that the delta-based LSD concept fulfills the above requirements. With proper tuning of the voltage controller and a power control PI loop,

- the dynamics of the voltage control, the swing behavior and the secondary power control can be separated;

– the Eigenvalues of the swing dynamics is independent of the operating point, of the grid impedances and voltages and can be kept at a constant design value valid for all converters;
– the voltage and power references can be tracked without any centrally calculated information.

References

[1] H. Bevrani, Robust Power System Frequency Control, second ed., Springer International Publishing, Switzerland, 2014.

[2] A. Benchaib, Advanced Control of AC/DC Power Networks, System of Systems Approach Based on Spatio-Temporal Scales, Wiley, USA, 2015.

[3] J. Rocabert, A. Luna, F. Blaabjerg, P. Rodríguez, Control of power converters in AC microgrids, IEEE Trans. Power Electron. 27 (11) (2012) 4734–4749.

[4] Q.C. Zhong, P.L. Nguyen, Z. Ma, W. Sheng, Self-synchronized synchronverters: inverters without a dedicated synchronization unit. IEEE Trans. Power Electron. 29 (2) (2014) 617–630, https://doi.org/10.1109/TPEL.2013.2258684.

[5] S. D'Arco, J.A. Suul, Virtual synchronous machines—classification of implementations and analysis of equivalence to droop controllers for microgrids. in: 2013 IEEE Grenoble Conference, Grenoble, 2013, pp. 1–7, https://doi.org/10.1109/PTC.2013.6652456.

[6] M. Ashabani, F.D. Freijedo, S. Golestan, J.M. Guerrero, Inducverters: PLL-less converters with auto-synchronization and emulated inertia capability, IEEE Trans. Smart Grid 7 (3) (2016) 1660–1674.

[7] B. Johnson, M. Rodriguez, M. Sinha, S. Dhople, Comparison of virtual oscillator and droop control, in: 2017 IEEE 18th Workshop on Control and Modeling for Power Electronics (COMPEL), Stanford, CA, 2017, pp. 1–6.

[8] J. Machowski, J. Bialek, J. Bumby, Power System Dynamics: Stability and Control, Wiley, 2008. ISBN 978-0-470-72558-0.

[9] B.K. Poolla, D. Groß, F. Dörfler, Placement and implementation of grid-forming and grid-following virtual inertia and fast frequency response, IEEE Trans. Power Syst. 34 (4) (2019) 3035–3046.

[10] P. Kundur, et al., Definition and classification of power system stability IEEE/CIGRE joint task force on stability terms and definitions, IEEE Trans. Power Syst. 19 (3) (2004) 1387–1401.

[11] D. Raisz, A. Musa, F. Ponci, A. Monti, Linear and uniform system dynamics of future converter-based power systems, in: 2018 IEEE Power & Energy Society General Meeting (PESGM), Portland, OR, 2018, pp. 1–5.

[12] D. Raisz, A. Musa, F. Ponci, A. Monti, Linear and uniform swing dynamics, IEEE Trans. Sustain. Energy 10 (3) (2019) 1513–1522.

[13] A. Musa, A. Kaushal, S.K. Gurumurthy, D. Raisz, F. Ponci, A. Monti, Development and stability analysis of LSD-based virtual synchronous generator for HVDC systems, in: IECON 2018—44th Annual Conference of the IEEE Industrial Electronics Society, Washington, DC, 2018, pp. 3535–3542.

[14] D. Nouti, D. Raisz, F. da Ponci, A. Monti, Decentralized linear swing dynamics in a multi-machine system with inverters, in: Accepted for Publication at the 6th IEEE International Energy Conference, 13–16 April, 2020, Gammarth, Tunisia, 2020.

[15] D. Deepak, D. Raisz, A. Musa, F. Ponci, A. Monti, Inertial control applied to synchronverters to achieve linear swing dynamics. in: 2019 Electric Power Quality and Supply Reliability Conference (PQ) & 2019 Symposium on Electrical Engineering and Mechatronics (SEEM), Kärdla, Estonia, 2019, pp. 1–6, https://doi.org/10.1109/PQ.2019.8818273.
[16] D. Raisz, D. Deepak, F. da Ponci, A. Monti, Linear and Uniform Swing Dynamics in Multimachine Converter-Based Power Systems, submitted to the International Journal of Electrical Power & Energy Systems(March 2020).
[17] A. Kusko, M.T. Thompson, Power Quality in Electrical Systems, McGraw-Hill, New-York, 2007.
[18] M.L. Crow, et al., Identification of electromechanical modes in power systems, in: Technical Report of the IEEE TF on Identification of Electromechanical Modes, June 2012. ISBN 978-1-4799-1000-7.
[19] J.F. Hauer, C.J. Demeure, L.L. Scharf, Initial results in Prony analysis of power system response signals, IEEE Trans. Power Syst. 5 (1) (1990) 80–89.
[20] S. D'Arco, J.A. Suul, O.B. Fosso, Control system tuning and stability analysis of virtual synchronous machines, in: 2013 IEEE Energy Conversion Congress and Exposition, Denver, CO, 2013, pp. 2664–2671.
[21] A. Riccobono, M. Mirz, A. Monti, Noninvasive online parametric identification of three-phase AC power impedances to assess the stability of grid-tied power electronic inverters in LV networks, IEEE Trans. Emerg. Sel. Topics Power Electron. 6 (2) (2018) 629–647.
[22] Y.V. Makarov, Z.Y. Dong, D.J. Hill, A general method for small signal stability analysis, IEEE Trans. Power Syst. 13 (3) (1998) 979–985.

Chapter 10

Long-term voltage control

Mohammed Ahsan Adib Murad[a], Massimiliano Chiandone[b], Giorgio Sulligoi[b], and Federico Milano[a]
[a]*School of Electrical and Electronic Engineering, University College Dublin, Dublin, Ireland,*
[b]*Department of Engineering and Architecture, University of Trieste, Trieste, Italy*

10.1 Introduction

Long-term voltage control is aimed at maintaining the voltage magnitude of buses at the transmission and distribution levels within a given range and properly share the reactive power among available resources. Traditionally, long-term controllers are the voltage control of underload tap changer (ULTC) transformers that interface transmission and distribution networks and, in some countries, the secondary voltage regulation (SVR) that coordinates the reactive power supply of conventional synchronous machines. This chapter discusses these two controllers but with an unconventional focus. In fact, the interaction of ULTCs with and the coordination of an SVR of renewable energy sources (RESs) is presented and illustrated through several examples.

10.1.1 ULTC transformers

Most transformers in distribution networks and, in particular, those interfacing the transmission with the distribution systems have underload tap changing capability. The modeling of such transformers is crucial for voltage stability analysis [1] due to the presence of nonlinearity (dead band, time delay, discrete tap positions) in these transformers. Even though the circuit model of tap changing transformers is well known, the model of the controller of such devices differs depending on the applications and/or implementations [2].

As the integration of stochastic distributed renewable energy resources increases, undesirable voltage fluctuations are observed in different levels of power systems. While this behavior is expected, to properly reproduce the precise dynamic behavior of ULTC regulators through simulations is not a trivial task. Moreover, as thoroughly discussed in this chapter, the implementation of the ULTC controller significantly impacts on the overall system dynamic response. This chapter addresses this modeling issue from a dynamic point

Converter-Based Dynamics and Control of Modern Power Systems
https://doi.org/10.1016/B978-0-12-818491-2.00010-9

of view, considering stochastic variation of the load power consumption and wind power generation at the distribution system level.

The effect of stochastic distributed generation such as wind power and photovoltaic on the frequency of tap change and performance of the ULTCs have been studied in [3–5]. These studies are relevant from the economic point of view as 50% of maintenance cost of such transformers is related to the number of tap operations. However, the aforementioned studies are based on step-wise power flow solutions and do not consider the dynamic behavior of ULTC controllers. Studies based on time domain or quasisteady-state simulations are considered in [6, 7], respectively. These references do not consider stochastic modeling.

Most of the previous studies showed the behavior of ULTC transformers considering either steady-state power flow or quasisteady-state analyses. This is adequate enough for the appraisal of power system operation. However, when considering a short period, for example, within a time frame of 5–15 min, stochastic fluctuations due to loads and distributed generation can lead to variations of the tap changers that might not be captured using a steady-state or quasisteady-state approaches. That is why the focus of this chapter is on ULTC operations occurring in a time scale of 15 min.

10.1.2 Secondary voltage regulation

Similarly to the automatic generation control (AGC), the SVR is a regional control that coordinates several (conventional) generation units and allows uniformly sharing the reactive power generation among such units. Unlikely the AGC, however, the SVR is not particularly common among TSOs. While SVR is implemented manually by some TSOs, for example, PJM and MISO in the United States [8], RTE in France and ENEL in Italy were pioneers in the design and implementation of automatic SVR schemes. Recently, SVR has been adopted in other countries, for example, South Africa [9].

The SVR allows reducing the risk of shortage of reactive power in the system and thus of the occurrence of limit-induced bifurcations (LIBs). In an ideal scenario, in fact, the generators that participate to the SVR reach their reactive power limit at the same time, thus leading to a multiple LIB rather than to a sequence of LIBs. This, in general, allows a higher loading level of the system. It has been argued, for example, that if the SVR had actually been in place and properly working, the blackout that shut down the whole Italian peninsula in 2003 would not have happened [10].

The increasing penetration of CIG has led to the need to utilize such devices to support the regulation of the network. The first obvious regulators to be implemented were primary controllers of voltage and frequency. With regard to voltage support, it is now commonly accepted that RES power plants should not work at constant unitary power factor, regardless the fact that they are connected to the transmission network or to the distribution network. Moreover, as noticed in [11], existing SVR based on reactive power sources offered by traditional fossil fuels power plants shows limited controlling capability when new RES generators replace equivalent amounts of power from traditional thermal and hydroelectric large plants.

For these reasons and thanks to the development of efficient communication systems, SVR has also been considered for applications to distributed generation including wind and solar PV power plants, for example, [12–14] and micro-grids, for example, [15, 16]. It thus appears feasible and sensible, to implement secondary controllers also for CIG, whose installed capacity has considerably increased worldwide in recent years [17].

Several proposals on the utilization of solar photo-voltaic generation (SPVG) for voltage support are available. For example, in [18], the authors propose to modulate the reactive power for voltage control functionality and bound the active power injected ramp, whereas, in [19], the authors propose to operate PV power plants as STATCOM devices. In this vein, recent regulation by the Italian Grid Code impose for all PV plants to be able to follow a voltage reference signal remotely send by the TSO and to modulate the reactive power exchanged by the plant according to a signal from the TSO.

Since power plants based on RES, and specifically SPVG, are more often than not connected to LV and MV distribution networks, several countries have issued specific connection rules. For example, in Italy, Refs. [20, 21] concern connection rules of PV power plants to LV and MV, respectively. The standards discussed previously specify that a PV plant has to be equipped with proper control systems in order to provide:

- control of the active power injected;
- control of the reactive power injected or absorbed;
- low-voltage ride-through functionality; and
- remote disconnection functionality.

A considerable number of plants are connected directly to the HV national transmission grid. Only in Italy, 170 PV plants have a nominal power greater than 5 MW and a few dozen are connected with the HV grid (at least all of these with a power greater than 10 MW). For these power plants connected to the HV grid, other than the TSO grid code, usually other regulations have been published. In [22], a general exposition of rules to be applied in all Europe can be found. The possibility to use the connected PV units in the provision of the reactive power compensation ancillary service is shown in [23], which discusses the well-known dynamic reactive power compensation obtained through an appropriate control scheme of power electronic devices [24].

10.1.3 Organization

The remainder of this chapter is organized as follows. Section 10.2 presents the discrete and continuous models of ULTCs and their controllers for dynamic studies and discusses the dynamic response of such devices considering both deterministic and stochastic scenarios. The latter includes load and wind speed models formulated as stochastic differential equations that properly capture the probability distribution and autocorrelation of the stochastic processes. Section 10.3 presents an SVR scheme for large PV power plants with one HV connection point of a hierarchical control strategy similar to that in use in high-voltage transmission networks. This control unit basically implements the reactive power regulators (RPRs) of the SVR that is discussed in the Italian Grid Code.

10.2 Underload tap changer

Transformers are ubiquitous in transmission and distribution systems. They connect sections of the network at different voltage levels. Depending on the system structure, these transformers are:

- step-up transformers at generator terminals;
- transformers connecting different transmission voltage levels; and
- transformers feeding a distribution system.

Voltage control through changing transformer ratios using "taps" at different voltage level is a common strategy. For example, transformers feeding to a distribution system compensate changes in the voltage due to changes in the load without interruption employing an automatic tap changing mechanism. Such transformers are often known as load tap changing (LTC) or ULTC or on load tap changing (OLTC) transformer. Details of fundamentals on transformers can be found in [25]. We describe next the modeling of the circuit of ULTC transformers as well as a variety of discrete and continuous voltage control schemes.

10.2.1 Modeling

The general scheme of an ULTC transformer with automatic voltage control is shown in Fig. 10.1. Three main physical components compose such transformers: (i) an automatic voltage regulator; (ii) a tap changer with a switching mechanism; and (iii) the main transformer.

The voltage regulator includes: (i) a measuring element to measure the voltage (or reactive power) on the connected bus; (ii) a unit to compare the difference between measured and reference quantity; (iii) a deadband element that reduces the sensitivity of the controller; and (iv) a time-delay element that limits the number of variations of the tap position.

The tap changer selects a tap to change from the previous position using a switching principal, and there exists several switching mechanism. Few commonly found mechanisms are: the high-speed resistor type, the reactor type, and the vacuum type [26]. The whole process of the tap changer is achieved through a driving mechanism.

An illustration of the switching sequence of a resistor-oil type ULTC comprising a diverter switch and a tap selector in Fig. 10.2. The tap selector first selects the tap at no load (see A–C in Fig. 10.2). Next, the diverter switch

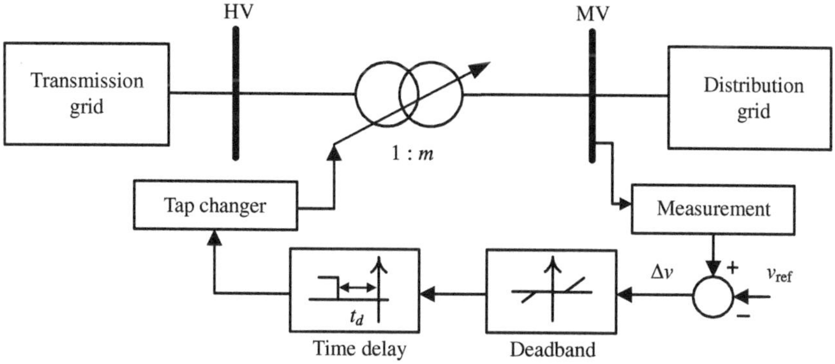

FIG. 10.1 A tap-changer underload transformer with a voltage controller.

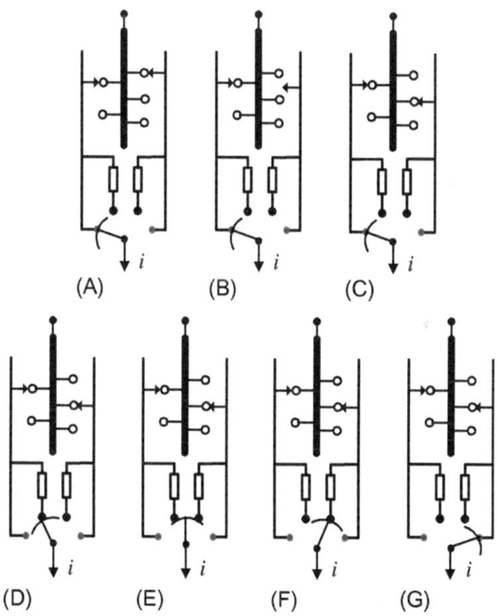

FIG. 10.2 Sequence of switching of tap changer: tap selector (A–C) and diverter switch (D–G) [26]. *(From W. Breuer, K. Stenzel, The tap-changer for power- and industrial transformers, Maschinenfabrik Reinhausen GmbH, Germany, 1982. Used with permission from Maschinenfabrik Reinhausen GmbH.)*

transfers the load current selected tap from the tap in operation (see D–G in Fig. 10.2). A driving mechanism completes the switching tasks.

10.2.1.1 ULTC circuit

Fig. 10.3 shows the equivalent circuit of a two-winding transformer assuming the tap is on the primary [27]. As $\overline{v}_h' = \overline{v}_h/m$, the current injections at buses h' and k are

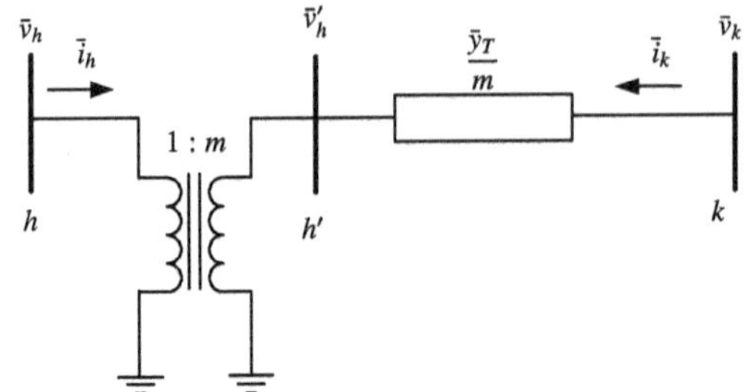

FIG. 10.3 Equivalent circuit of the transformer with tap ratio module and series impedance. *(From F. Milano, Transmission devices, in: Power System Modelling and Scripting. Power Systems, vol. 0, Springer, Berlin, Heidelberg, 2010.)*

$$\begin{bmatrix} \bar{i}_h \\ \bar{i}_k \end{bmatrix} = \bar{y}_T \begin{bmatrix} \frac{1}{m^2} & -\frac{1}{m} \\ -\frac{1}{m} & 1 \end{bmatrix} \begin{bmatrix} \bar{v}_h \\ \bar{v}_k \end{bmatrix}. \tag{10.1}$$

Considering the physical buses h and k and including magnetization and iron losses on the primary winding (see Fig. 10.4), one obtains

$$\begin{bmatrix} \bar{i}_h \\ \bar{i}_k \end{bmatrix} = \begin{bmatrix} g_{\mathrm{Fe}} + jb_\mu + \bar{y}_T \frac{1}{m^2} & -\bar{y}_T \frac{1}{m} \\ -\bar{y}_T \frac{1}{m} & \bar{y}_T \end{bmatrix} \begin{bmatrix} \bar{v}_h \\ \bar{v}_k \end{bmatrix}, \tag{10.2}$$

where $\bar{y}_T = (r_T + jx_T)^{-1}$; g_{Fe}, b_μ, r_T, and x_T are transformer iron loss, magnetizing susceptance, resistance, and reactance, respectively. Finally, the power injections at buses h and k are

$$\begin{aligned} p_h &= v_h^2(g_{\mathrm{Fe}} + g_T/m^2) - v_h v_k (g_T \cos\theta_{hk} + b_T \sin\theta_{hk})/m, \\ q_h &= -v_h^2(b_\mu + b_T/m^2) - v_h v_k (g_T \sin\theta_{hk} - b_T \cos\theta_{hk})/m, \\ p_k &= v_k^2 g_T - v_h v_k (g_T \cos\theta_{hk} - b_T \sin\theta_{hk})/m, \\ q_k &= -v_k^2 b_T + v_h v_k (g_T \sin\theta_{hk} + b_T \cos\theta_{hk})/m. \end{aligned} \tag{10.3}$$

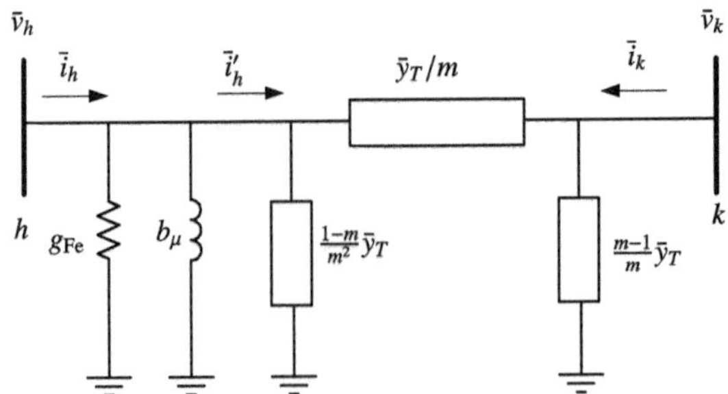

FIG. 10.4 Equivalent circuit of a transformer. *(From F. Milano, Transmission devices, in: Power System Modelling and Scripting. Power Systems, vol. 0, Springer, Berlin, Heidelberg, 2010.)*

10.2.1.2 ULTC control

In principle, all the elements that compose the ULTC voltage controller should be properly modeled. In the literature, however, mostly only simplified models are considered [1]. In this section, two commonly used control models are considered based on the references [1, 2, 28, 29], namely, (i) the discrete model and (ii) the continuous model.

Discrete model

In this model, the tap ratio m is a discrete variable that can take only fixed values in the range of $m^{\max}$ and $m^{\min}$ by a fixed step Δm. The tap ratio can move up or down by one step Δm if the controlled voltage v_k deviates more than a given dead band db with respect to the reference voltage v^{ref} for longer than a given period Δt. The switching logic of the tap ratio is

$$m(t) = m(t-\Delta t) + f(e(t), c(t), \tau(t))\,\Delta m, \tag{10.4}$$

where e models the dead band, f the time delay, c is a memory function that stores the time elapsed since the tap change, and $\tau = \tau_d + \tau_m$ is the time delay. τ_d is the adjustable time delay of the controller and τ_m mechanical switching time delay. The e and c are expressed as

$$e(\Delta v(t), m(t-\Delta t), db, m^{\max}, m^{\min}) = \begin{cases} 1, & \text{if } \Delta v(t) > db \text{ and } m(t-\Delta t) < m^{\max} \\ -1, & \text{if } \Delta v(t) < -db \text{ and } m(t-\Delta t) > m^{\min} \\ 0, & \text{otherwise,} \end{cases} \tag{10.5}$$

$$c(e(t), c(t-\Delta t)) = \begin{cases} c(t-\Delta t) + \Delta t, & \text{if } e(t) = 1 \text{ and } c(t-\Delta t) \geq 0 \\ c(t-\Delta t) - \Delta t, & \text{if } e(t) = -1 \text{ and } c(t-\Delta t) \leq 0 \\ 0, & \text{otherwise,} \end{cases} \tag{10.6}$$

where t is the current simulation time and $t - \Delta t$ is the previous simulation step.

The function f depends upon the mode of operation: sequential or nonsequential mode. The timer is reset after each tap change in nonsequential mode, whereas in the sequential mode, the timer is reset only after the voltage is back within the dead band range. The function f is as follows:

- for nonsequential mode,

$$f(e(t), c(t), \tau(t)) = \begin{cases} 1, & \text{if } e(t) = 1 \text{ and } c(t) > \tau(t) \\ -1, & \text{if } e(t) = -1 \text{ and } c(t) < \tau(t) \\ 0, & \text{otherwise.} \end{cases} \tag{10.7}$$

- For sequential mode,

$$f(e(t),c(t),\tau(t)) = \begin{cases} 1, & \text{if } e(t)=1 \text{ and } c(t)>\tau(t), \\ & \text{for subsequent taps if } e(t)=1 \text{ and } c(t)>\tau_m(t) \\ -1, & \text{if } e(t)=-1 \text{ and } c(t)>\tau(t), \\ & \text{for subsequent taps if } e(t)=-1 \text{ and } c(t)>\tau_m(t) \\ 0, & \text{otherwise.} \end{cases} \tag{10.8}$$

The time delay τ can be fixed or variable. In case of variable time delay, the higher the voltage error, the faster the tap change. Depending on different time-delay settings, four different discrete models are considered for both sequential and nonsequential modes and summarized in the following:

- D1: both delays are fixed, $\tau_d = \tau_{d,\,0}$ and $\tau_m = \tau_{m,\,0}$;
- D2: τ_d is a combination of fixed and variable time, $\tau_d = \tau_{d,0} \cdot \frac{db}{|\Delta v|} + \tau_{d,1}$, and τ_m is fixed, $\tau_m = \tau_{m,\,0}$;
- D3: τ_d is variable, $\tau_d = \tau_{d,0} \cdot \frac{db}{|\Delta v|}$ and τ_m is fixed, $\tau_m = \tau_{m,\,0}$;
- D4: both delays are variable, $\tau_d = \tau_{d,0} \cdot \frac{db}{|\Delta v|}$ and $\tau_m = \tau_{m,0} \cdot \frac{db}{|\Delta v|}$.

Continuous model

The continuous control model approximates the tap ratio step Δm to be small so that tap ratio m can vary continuously. The time delays are approximated as a lag transfer function, and the tap ratio differential equation is given by

$$\dot{m}(t) = -Hm(t) + K(v_k(t) - v_{\text{ref}}), \quad m^{\min} \le m(t) \le m^{\max} \tag{10.9}$$

where H, K, v_k, and v_{ref} are the integral deviation, inverse time constant, secondary bus voltage, and controlled reference voltage, respectively. The dead band is not included in this model. Note that it is possible to create an equivalent continuous model of each discrete model [2].

10.2.2 Examples

The impact of the voltage control action of different ULTC controller implementations to the long-term voltage stability is studied in this section using deterministic case studies.

10.2.2.1 Case study 1

The test system consists of four buses, one slack bus, one tap changing transformer, three transmission lines, and one dynamic load. The scheme of this test system, which is based on [30], is shown in Fig. 10.5. Data are given in Table 10.1.

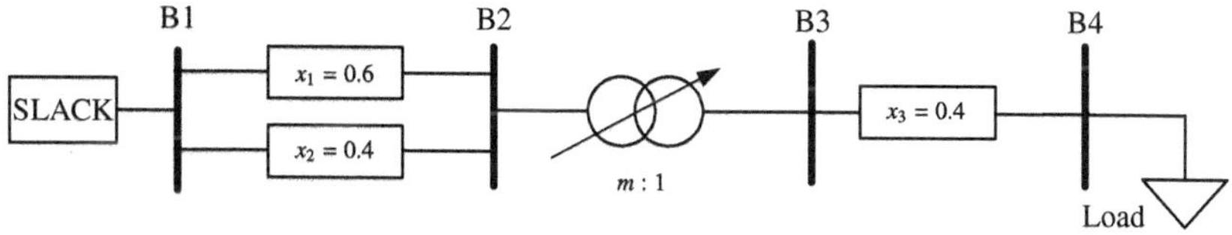

FIG. 10.5 A single generator connected to an infinite bus [30].

TABLE 10.1 Base case parameters of the test system of Fig. 10.5.

Component	Parameters
Discrete model	$db = 4\%$, $\Delta m = 0.0125$, $\tau_{d,0} = 15$, $\tau_{d,1} = 5$, $\tau_{m,0} = 8$
Continuous model	$H = 0.001$, $K = 0.043$
All ULTCs	$m^{max} = 1.1$, $m^{min} = 0.8$, $r_t = 0$, $x_t = 0.4$
Load	$p_o = 0.4$ pu(MW), $T_p = 5$ s, $\alpha_t = 2$, $\alpha_s = 0$

The continuous dynamics of the real-power load in bus 4 considered as a first-order dynamic exponential recovery load given by

$$\begin{aligned} T_p\dot{x}_p(t) &= -x_p(t) + p_o(v_t(t)^{\alpha_s} - v_t(t)^{\alpha_t}), \\ p_L(t) &= x_p(t) + p_o v_t(t)^{\alpha_t}, \end{aligned} \tag{10.10}$$

where x_p is the load state driving the actual load demand p_L, v_t is the voltage at the load bus, and p_o is the nominal active power load. The load undergoes an initial transient given by the term $p_o v_t^{\alpha_t}$ during a voltage disturbance and recovery of the load dictated by the time constant T_p.

The contingency consists of a line outage ($x = 0.4$) between B1 and B2 at $t = 5$ s. Figs. 10.6 and 10.7 show the trajectories of voltage at bus 3 for nonsequential and sequential discrete models, respectively.

Following the disturbance, the voltage response is faster using the sequential models compared to the nonsequential models. This is due to the reset logic of the time delay. Comparing all nonsequential models (NS D1–NS D4) and sequential models (S D1–S D4), the voltage response is comparatively faster when moving from delay settings D1–D4 because the time-delay changes dynamically depend on the voltage deviation. Moreover, all ULTCs restore the voltage at the same voltage so the number of tap operations is the same for all ULTCs. These results indicate that the sequential model with variable time delay (D4) restores the voltage within the dead band compared to all other discrete tap changers.

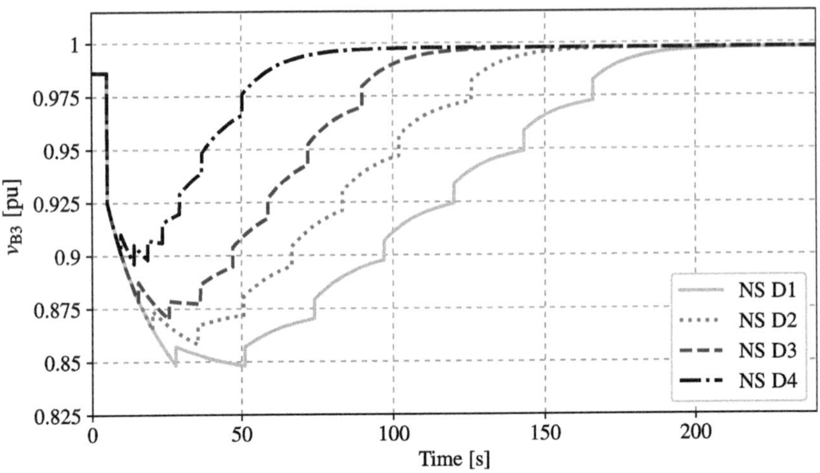

FIG. 10.6 Voltage magnitude at bus 3 using nonsequential discrete ULTC models.

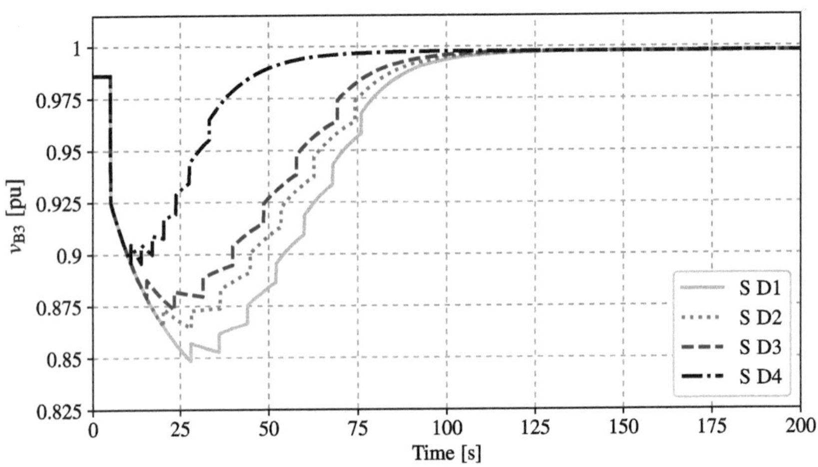

FIG. 10.7 Voltage magnitude at bus 3 using sequential discrete ULTC models.

10.2.2.2 Case study 2

The Nordic test system presented in [31] is used for the second deterministic case study. The system includes 74 buses; 102 branches, of which 20 step-up and 22 distribution transformers with ULTCs; 20 generators, of which 7 are round rotor and 13 are salient pole types, with turbine governors (TGs), automatic voltage regulations (AVRs), power system stabilizers (PSSs), and overexcitation limiters (OELs). All models used in this case study match those reported in [31], except for the OELs, which are modeled as in [27]. The parameters of all devices are also given in [31]. All distribution transformers are

TABLE 10.2 Delays of ULTCs.

Transformer	Delays			Transformer	Delays		
	τ_{d0}	τ_{d1}	τ_m		τ_{d0}	τ_{d1}	τ_m
11–1011	30	5	8	41–4041	31	5	9
12–1012	30	5	9	42–4042	31	5	10
13–1013	30	5	10	43–4043	31	5	11
22–1022	30	5	11	46–4046	31	5	12
1–1041	29	5	12	47–4047	30	5	8
2–1042	29	5	8	51–4051	30	5	9
3–1043	29	5	9	61–4061	30	5	10
4–1044	29	5	10	62–4062	30	5	11
5–1045	29	5	11	63–4063	30	5	12
31–2031	29	5	12	71–4071	31	5	9
32–2032	31	5	8	72–4072	31	5	11

equipped with a ULTC controller with dead band $db = 2\%$, maximum and minimum tap position are $m^{\max} = 1.2$ and $m^{\min} = 0.8$ with a step size $\Delta m = 0.01$. The values for the time delays are given in Table 10.2.

The system consists of four areas: North, Central, Equivalent, and South. The base case of the system is heavily loaded with large power transfers from North to Central areas.

The comparison discussed in this section considers the dynamic response following a three-phase fault at bus 4032, occurring at $t = 1$ s and cleared at $t = 1.06$ s by opening the line between buses 4032–4044. Transient response of voltage of a distribution bus at the Central area is shown in Figs. 10.8 and 10.9 for nonsequential and sequential discrete tap changers, respectively.

The line trip following the fault forces power to flow North-Central corridor over the remaining lines. However, the reactive power capabilities of the Central and Northern generators impacts the maximum power delivered to the central loads. On the other hand, the ULTCs try to restore the voltages of the distribution buses and load powers. In this case, however, the amount of power that the ULTCs have to restore is greater than the maximum power that can be delivered by the generation and the transmission system and, hence, a voltage instability occurs, which eventually leads to a voltage collapse. The collapse occurs in the time scale of minutes, hence the notation *long-term voltage instability*.

Similar to case study 1 in Section 10.2.2.1, due to the dependency of the time delay of the tap switching on the voltage error, nonsequential and sequential

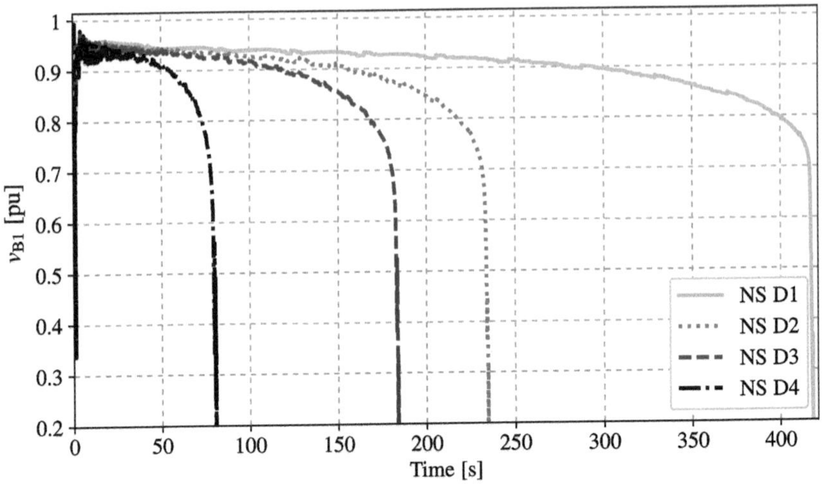

FIG. 10.8 Voltage magnitude at bus 1 using nonsequential discrete ULTC models.

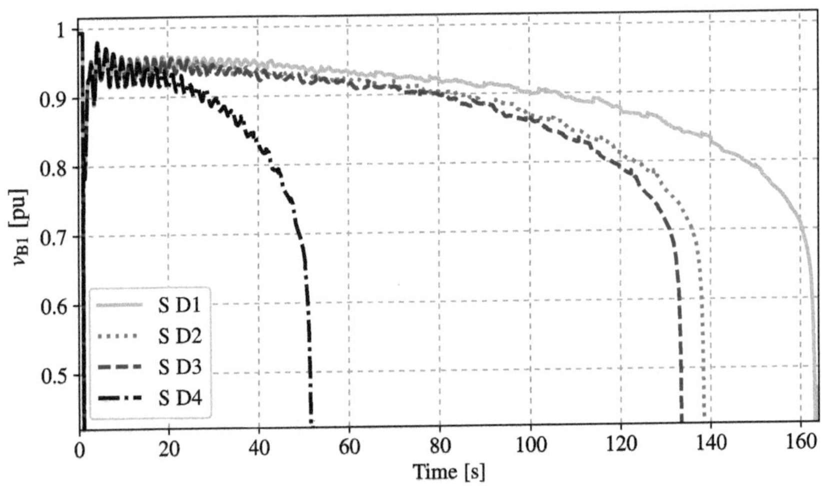

FIG. 10.9 Voltage magnitude at bus 1 using sequential discrete ULTC models.

ULTC controllers show different transient responses. Of course, due to its slow response, the nonsequential ULTC controllers take a longer time than sequential ULTC controllers to drive the system to collapse.

10.2.3 Stochastic modeling

This section outlines the load consumption and wind speed models considered in the stochastic case study discussed in the example of Section 10.2.4. These models include stochastic perturbations modeled by means of the following Itô-type differential equation:

$$\dot{\eta}(t) = a(\eta(t),t) + b(\eta(t),t)\,\xi(t), \tag{10.11}$$

where η is the state variable that describes the stochastic process, a and b are the drift and the diffusion terms, respectively, and ξ is the white noise, that is, the formal time derivative of the Wiener process [32]. Eq. (10.11) is a general expression that can take into account both Gaussian and non-Gaussian processes and is thus appropriate to model load power variations [33] and wind speed fluctuations [34].

10.2.3.1 Voltage-dependent load

The well-known voltage-dependent load (VDL) model is given by [27]

$$\begin{aligned} p_L(t) &= -p_{L,o}(v(t)/v_o)^\gamma, \\ q_L(t) &= -q_{L,o}(v(t)/v_o)^\gamma, \end{aligned} \tag{10.12}$$

where $p_{L,\,o}$ and $q_{L,\,o}$ are the active and reactive powers at the nominal voltage v_o; v is the voltage magnitude of the bus where the load is connected; and γ is the power exponent.

Merging together the stochastic equation (10.11) and the load equations (10.12) lead to a stochastic VDL (SVDL) model. Since load variations are approximately Gaussian and show a constant standard deviation, we define the diffusion terms a and b in Eq. (10.11) to resemble an Ornstein-Uhlenbeck process [33]. The resulting SVDL model is

$$\begin{aligned} p_L(t) &= (-p_{L,o} + \eta_p(t))(v(t)/v_o)^\gamma, \\ q_L(t) &= (-q_{L,o} + \eta_q(t))(v(t)/v_o)^\gamma, \\ \dot{\eta}_p(t) &= \alpha_p(\mu_p - \eta_p(t)) + b_p\xi_p(t), \\ \dot{\eta}_q(t) &= \alpha_q(\mu_q - \eta_q(t)) + b_q\xi_q(t), \end{aligned} \tag{10.13}$$

where the α terms are the speed at which the stochastic variables η are "attracted" toward the mean values μ, and the b terms represent the volatility of the processes.

10.2.3.2 Wind speed

To emulate the wind speed, a and b in Eq. (10.11) must be defined so that the probability distribution of η is a Weibull process. It is also important to reproduce the autocorrelation of the wind speed, which is assumed to be exponentially decaying. This can be achieved through the regression theorem and the stationary Fokker-Planck equation, as thoroughly discussed in [34]. The resulting drift and diffusion terms are

$$
\begin{aligned}
a(\eta(t)) &= -\alpha \cdot (\eta(t) - \mu_W), \\
b(\eta(t)) &= \sqrt{b_1(\eta(t)) \cdot b_2(\eta(t))},
\end{aligned}
\tag{10.14}
$$

where α is the autocorrelation coefficient; μ_W is the mean of the Weibull distribution; and

$$
\begin{aligned}
b_1(\eta(t)) &= \frac{2\alpha}{p_W(\eta(t))}, \\
b_2(\eta(t)) &= \lambda \cdot \Gamma\left(1 + \frac{1}{k}, \left(\frac{\eta(t)}{\lambda}\right)^k\right) - \mu_W \cdot e^{-(\eta(t)/\lambda)^k},
\end{aligned}
$$

where p_W is the Weibull probability density function (PDF); Γ is the incomplete Gamma function; k and λ are the shape and scale parameters of the Weibull distribution, respectively.

10.2.4 Example

The test network considered in this example is a small Irish distribution system with both radial and meshed configurations [35, 36]. The network includes eight buses, six loads, two wind generations, one slack bus, and eight transmission lines. The operating nominal voltage of B1–B8 is 38 kV, and the buses are fed by an ULTC type step down transformer from a 110 kV network. The network topology is shown in Fig. 10.10. Network parameters can be found in [35].

The active and reactive power loading of the system are 15.02 MW and 8.29 MVAr, respectively, and the nominal wind farm capacities are 12 and 20 MW at B5 and B3, respectively. In total, 5% of the loads is a SVDL

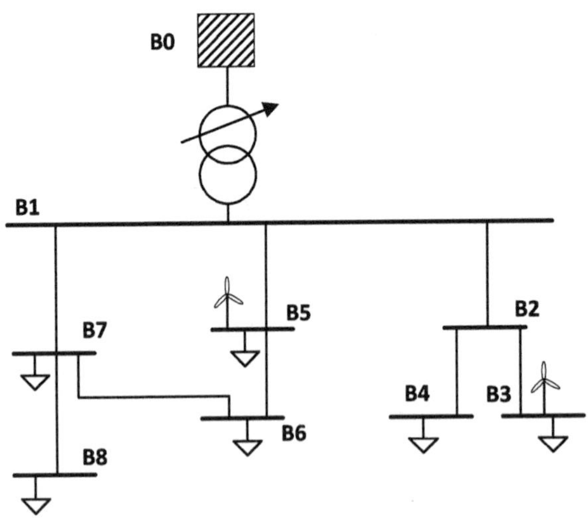

FIG. 10.10 Topology of the test distribution network [35, 36].

TABLE 10.3 Average number of tap operations using sequential discrete models.

Cases	Parameters			D1	D2	D3	D4
	db	τ_{d0}	τ_{m0}				
C1	2.2	15	8	4.40	3.46	4.72	7.94
C2	2.4	15	8	2.43	2.23	2.56	2.60
C3	2.5	15	8	1.79	1.68	1.83	1.85
C4	2.5	20	10	1.60	1.43	1.64	1.66
C5	3	15	8	0.97	0.95	0.97	0.97

modeled as in Eq. (10.13), and 95% is modeled as constant PQ ($\gamma = 0$ in Eq. 10.13). The wind generator model is a fifth-order doubly fed induction generator with a variable-speed wind turbine having discrete pitch control, first-order AVR, turbine governor, and maximum power point tracking (MPPT). The input to wind turbine is a stochastic wind modeled as in Eq. (10.14).

Five different case studies are considered for different delay and dead band settings of the discrete ULTC models. A total of 500 15-min Monte Carlo simulations are considered for each model and parameter set. The average number of tap operations using sequential type ULTCs for different dead band and time delay settings are given in Table 10.3. As expected, the higher the time delay and dead band, the lower the number of tap changes. This case study only considers sequential type ULTCs. Due to the stochastic variation of the load and wind speed, nonsequential ULTC controller models also yield similar results as sequential types because of similar delay logic of first tap switch.

Figs. 10.11 and 10.12 show 500 trajectories of the voltage at bus 1 for cases C1 and C5 (see Table 10.3) with D1 type discrete model. These figures also include the mean, μ, and $\mu \pm 3\sigma$, where σ is the standard deviation. The mean and the standard deviation of the 500 trajectories are calculated at every time instant of the simulation interval, which is 0.01 s. The average voltage trajectory is similar in both figures; however, C1 shows more than four times the tap operations of C5. Thus, stochastic variations are important to take account for preventing unnecessary tap changes.

10.2.5 Remarks

The deterministic and stochastic case studies discussed previously show that dead band and delay settings of ULTCs play a crucial role for the voltage restoration or collapse and number of tap operation. When considering stochastic processes, the time-domain analysis is necessary to properly account for ULTC

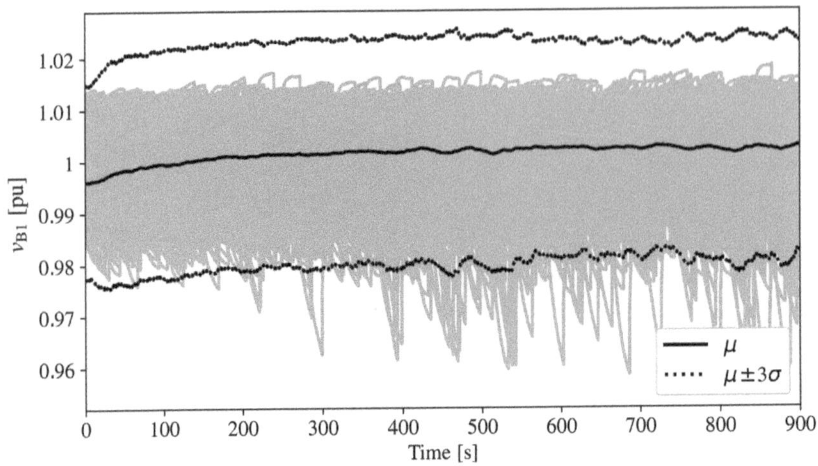

FIG. 10.11 Five-hundred stochastic trajectories and statistical properties of the bus 1 voltage using a sequential discrete model (D1) for case C1.

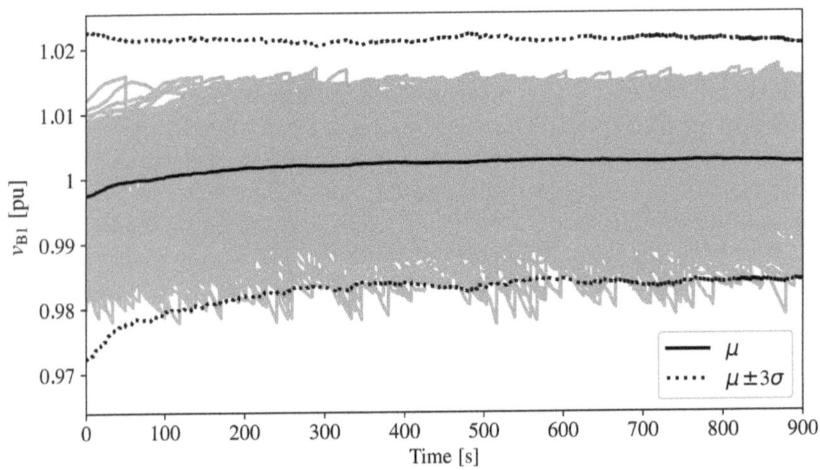

FIG. 10.12 Five-hundred stochastic trajectories and statistical properties of the bus 1 voltage using a sequential discrete model (D1) for case C5.

tap variations as these cannot be rightly captured considering steady-state analysis. Simulation results also allow concluding that, depending on the ULTC control model, the number of tap operation can be significantly different, so it is important to accurately implement the right control logic. This is another aspect that cannot be captured by conventional steady-state analyses.

10.3 Secondary voltage regulation

This section focuses on voltage regulation for large PV power plants and proposes the application to large PV power plants of a hierarchical control strategy similar to that in use in HV transmission networks [37]. A central control unit coordinates the reactive power of each converter to regulate the voltage of the point of connection. Commercially available converters usually accept as references the following signals: reactive power, power factor, or grid voltage. This control unit basically implements the RPRs of the SVR that is discussed in the Italian Grid Code. The interested reader can find a comprehensive description of the Italian SVR scheme in [38].

While we discuss only applications to PV power plants, the hierarchical voltage control scheme discussed in this section can be applied to different kinds of power plants connected to HV networks. For example, references [39, 40] show that such a control can be successfully used for the cluster of hydro power plants and wind farms, respectively, participating to TSO voltage regulation.

The application of a unique control system, for both traditional power plants and RES, allows standardizing the system dynamic response and increases the overall stability of the network.

10.3.1 Control strategy

Fig. 10.13 shows a synoptic scheme of the SVR. It is assumed that, at the point of connection, the PV has to provide the voltage control capability as discussed in [11]. The SVR controller receives the reference voltage from the TSO. The pilot bus is the point of connection of the PV power plant. Then, an external voltage control loop, implemented by the busbar voltage regulator (BVR), computes a reference of reactive power level $q_{\rm lev}$ ($q_{\rm lev} \in [-1, 1]$), which is then multiplied by the vector of the capability of each generator.

The resulting vector $q_{\rm ref}$ of reactive power limits is then compared with the actual reactive power of each generator. The error is then processed by the dynamic decoupling (DD) block and sent to the generator reactive power regulators (GRPRs) [42]. The DD block is a key for the dynamic response of the system and is further discussed in the following section. The resulting GRPRs output control signals are used as reference signals for the conventional voltage control of the converter. It is also possible to exclude the external voltage control loop and send a $q_{\rm lev}$ reference signal directly to the reactive power control loop.

Several studies exist on the possibility to control reactive power of a VSC independently from the active power. In [43], the transient response of the reactive power to the change of reactive power reference is shown. The results show also the decoupling of the p and q injections. The fast response for the reactive power control can also be found in [44]. Therefore, in the remainder of this chapter, the converters are simplified with a first-order model.

The control system shown in Fig. 10.13 presents specific features, such as speed of response and accuracy of the steady-state response. Moreover, it ensures the stability of the system, even in case of asymmetrical disturbances (thanks to the dynamic decoupling array), and the ability to allocate quotas of reactive power among the different generators according to the actual needs and contingencies. The control system shown in Fig. 10.13 is also simple to

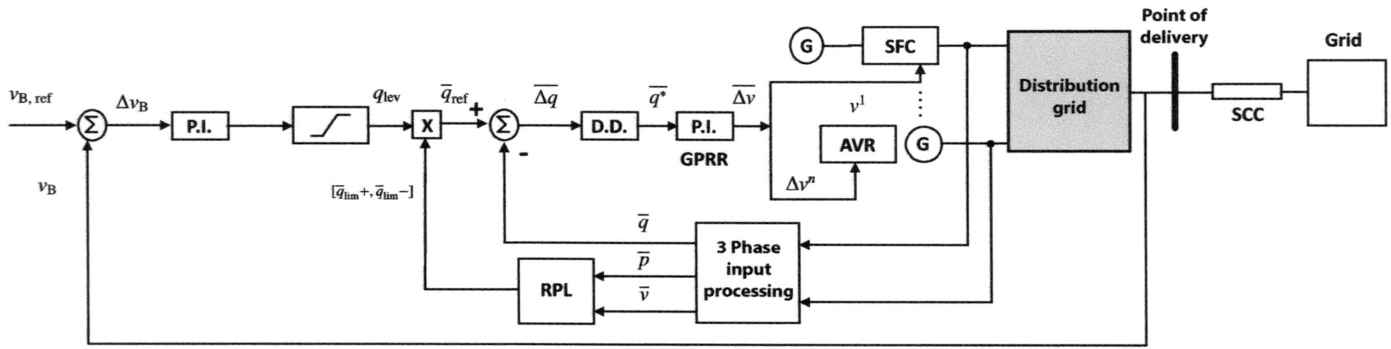

FIG. 10.13 Synoptic scheme of the secondary voltage regulation [41].

implement (as it is based only on one common signal q_{ref}) and is well known by the national TSO as it has been applied to conventional power plants that participate to the secondary voltage control.

10.3.2 Coupling of large RES power plants

Large RES power plants that are connected to the HV transmission system are typically composed of a set of generators connected through an MV distribution network and having a single point of connection (PoC) at the HV level. For the sake of example, Fig. 10.17 shows the topology of a real-world cluster of hydroelectric power plant. The MV distribution network can be described by the well-known power flow equations, as follows:

$$
\begin{aligned}
p_k &= \sum_{i=1}^{n} v_k v_i Y_{ki} \cos(\theta_k - \theta_i - \gamma_{ki}) - p_{o,k}, \quad k = 1,2,\ldots,n, \\
q_k &= \sum_{i=1}^{n} v_k v_i Y_{ki} \sin(\theta_k - \theta_i - \gamma_{ki}) - q_{o,k}, \quad k = 1,2,\ldots,n,
\end{aligned}
\tag{10.15}
$$

where p, q, v, Y, θ, and γ are the generator active power at node, the generator reactive power at node, the node voltage, the module of admittance coefficients, the phase angle of the node voltage, and the phase angle of the admittance coefficients, respectively. Finally, p_o and q_o are the active and reactive load power consumptions, respectively, at the network buses. Linearizing Eq. (10.15) at a given operating point leads to the following matrix form:

$$
\begin{bmatrix} \Delta \boldsymbol{p} \\ \Delta \boldsymbol{q} \end{bmatrix} = \begin{bmatrix} \boldsymbol{p}_v & \boldsymbol{p}_\theta \\ \boldsymbol{q}_v & \boldsymbol{q}_\theta \end{bmatrix} \times \begin{bmatrix} \Delta \boldsymbol{v} \\ \Delta \boldsymbol{\theta} \end{bmatrix} \tag{10.16}
$$

where $\Delta \boldsymbol{p}$, $\Delta \boldsymbol{q}$, $\Delta \boldsymbol{v}$, and $\Delta \boldsymbol{\theta}$ are the vectors of the power and voltage variations; and the Jacobian matrices $\boldsymbol{p}_v = \frac{\partial p}{\partial v}$, $\boldsymbol{p}_\theta = \frac{\partial p}{\partial \theta}$, $\boldsymbol{q}_v = \frac{\partial q}{\partial v}$, and $\boldsymbol{q}_\theta = \frac{\partial q}{\partial \theta}$ represent the link between the active and reactive power with the module and bus voltage phasors. The elements of the Jacobian matrices embed the information on the characteristic parameters of the transmission lines of the network. For the purposes of the voltage control, active power variations are neglected, hence leading to:

$$
\Delta \boldsymbol{q} = \boldsymbol{q}_v \, \Delta \boldsymbol{v}, \tag{10.17}
$$

from where it appears that $\boldsymbol{q}_v$ defines the electric coupling between generator reactive powers and voltage magnitudes. We assume that $\boldsymbol{q}_v$ is full rank, as it is in most practical applications. The discussion of idiosyncratic cases where $\boldsymbol{q}_v$ is singular is out of the scope of this chapter.

Starting from the inverse of the electric coupling matrix, it is possible to calculate the dynamic decoupling matrix, as follows:

$$
\mathbf{D} = \boldsymbol{q}_v^{-1}, \tag{10.18}
$$

where **D** is formally defined as $\frac{\partial v}{\partial q}$ and is thus defined as

$$\Delta v = \mathbf{D}\,\Delta q. \tag{10.19}$$

Note that the coefficients of matrix **D** can also be calculated through a numerical sensitivity analysis, as discussed in [42].

10.3.3 Examples

10.3.3.1 Case study 1

The proposed SVR scheme is applied to a real PV plant of a nominal power of 48 MVA connected to the high-voltage transmission grid. Each photovoltaic field (made of series and parallel connected photovoltaic modules) is connected to two centralized, that is, equipped with a unique MPPT, converters. The low-voltage converter outputs are raised to 20 kV via a double winding transformer in 45 "converter stations." The output of each converter station is connected with the output of the nearby station forming four groups that are connected to a central station from where four medium voltage circuits are sent to a transforming station. In the transforming station, the four circuits are connected to the medium voltage side of a transformer that raises the voltage up to 132 kV that is the nominal voltage of the transmission line, where the PV plant is connected. The single-line diagram of the plant topology is shown in [41].

Simulations have been carried out assuming an initial $v_{B,\mathrm{ref}} = 1.005$ pu and the same reactive power injections from all generators. At $t = 500$ s, the reference voltage $v_{B,\,\mathrm{ref}}$ step to 1.03 pu; consequently, the regulator imposes a new value of reactive power to be supplied by the generators. Fig. 10.14 shows the voltage profile at the point of delivery as well as the trajectory of the reactive power delivered by the generator expressed in pu with respect to its nominal power.

As expected, the proposed control allows distributing the reactive power equally among the generators of the power plant; furthermore, there is no steady-state error, the dynamic of the system does not show oscillations. The dynamic response of the proposed control scheme is thus very similar to that of an SVR installed in conventional power plants.

10.3.3.2 Case study 2

Three network topologies are considered. These are depicted in Figs. 10.15–10.17. Based on the method described in the previous paragraph, it is also possible to calculate the sensitivity coefficients **D** that combine the PoC N0 to all other nodes of the grid.

Table 10.4 shows, for all the three example plants, the influence of each generator on the voltage variations at the point of delivery. If all values have similar magnitude, the generators can effectively and similarly participate to the voltage regulation.

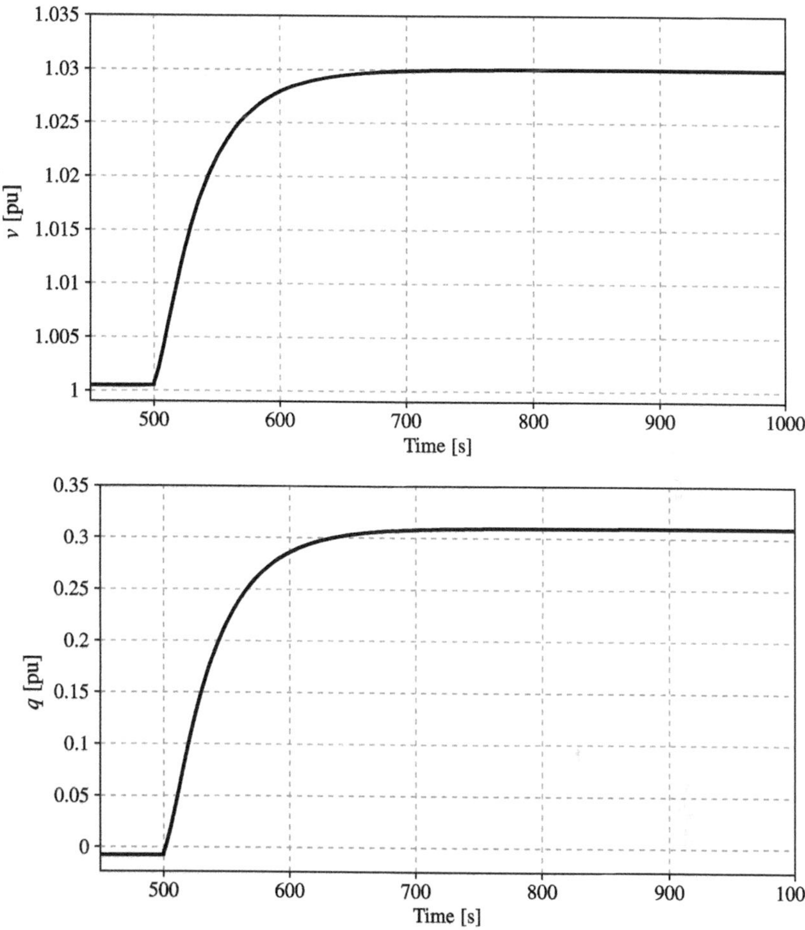

FIG. 10.14 Case study 1: Voltage profile at the point of connection and generator reactive power [41].

For all network topologies, regulator coefficients and time constants are chosen to ensure stability and a dynamic similar to those of traditional power plants. The internal voltage loop, for example, AVR, is characterized by fast timescale in the order of tenths of a second; the RPR has a time constant of few seconds and finally the external voltage loop (BVR) time constant is typically of the order of 10 s.

The first simulation has been solved considering topology A. Two generators connected to a common bus bar are working in asymmetrical states (the reactive power of the generators are different although their nominal power is equal). At time $t = 50$ s, the SVR is switched on trying to equate the reactive power of both generators. Without the decoupling matrix (see Fig. 10.18), reactive power

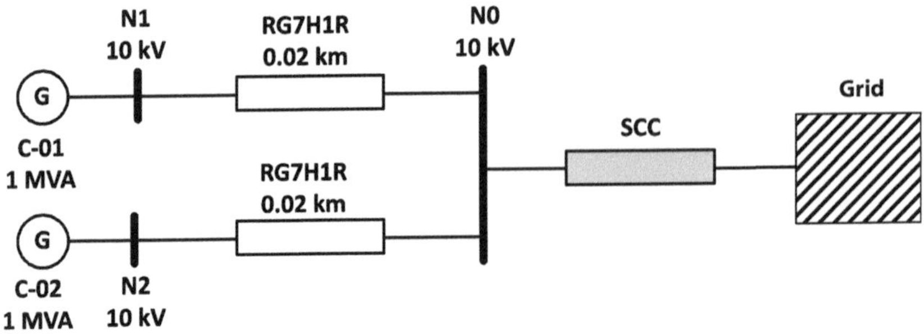

FIG. 10.15 Topology of plant A [45].

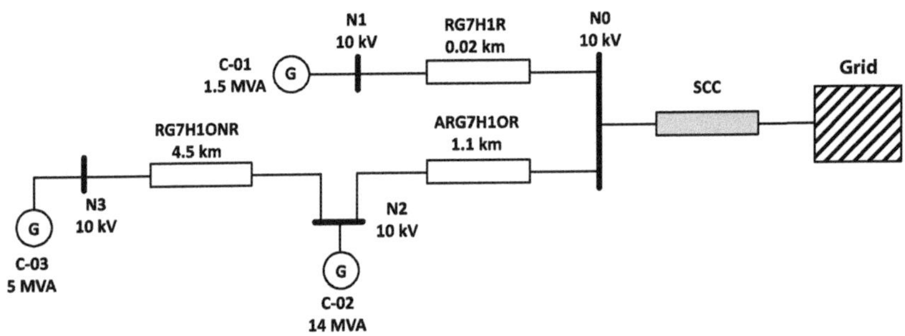

FIG. 10.16 Topology of plant B [45].

oscillations occur (all reactive power plots are in per unit of 100 MVA). The oscillations are in phase quadrature that implies a reactive power flow between the generators.

Although the topology of the network A and its initial operating conditions have been chosen to emphasize the coupling phenomenon, such asymmetric perturbations happens every time the system is started or a single generator is inserted into the cluster. Fig. 10.19 shows the effect of the same perturbation on the system when the decoupling matrix is included in the SVR. In this case, transients are exponentially decaying and do not oscillate. This kind of reactive power flow among generators has to be limited for several reasons: it represents a stress conditions for alternators, causes additional losses, can interfere with voltage protections, and represents a voltage disturbance for load directly fed by the generators.

Figs. 10.20 and 10.21 show the step response of the voltage reference for the network topology B, respectively, without and with the DD matrix. Figs. 10.22 and 10.23 show similar results for the topology C. In particular, reactive power responses in Figs. 10.8 and 10.10 demonstrate the positive effect of the decoupling action: no flow of reactive power occurs among generators. It is worth noticing that this phenomenon is not appreciable looking only at the voltage response.

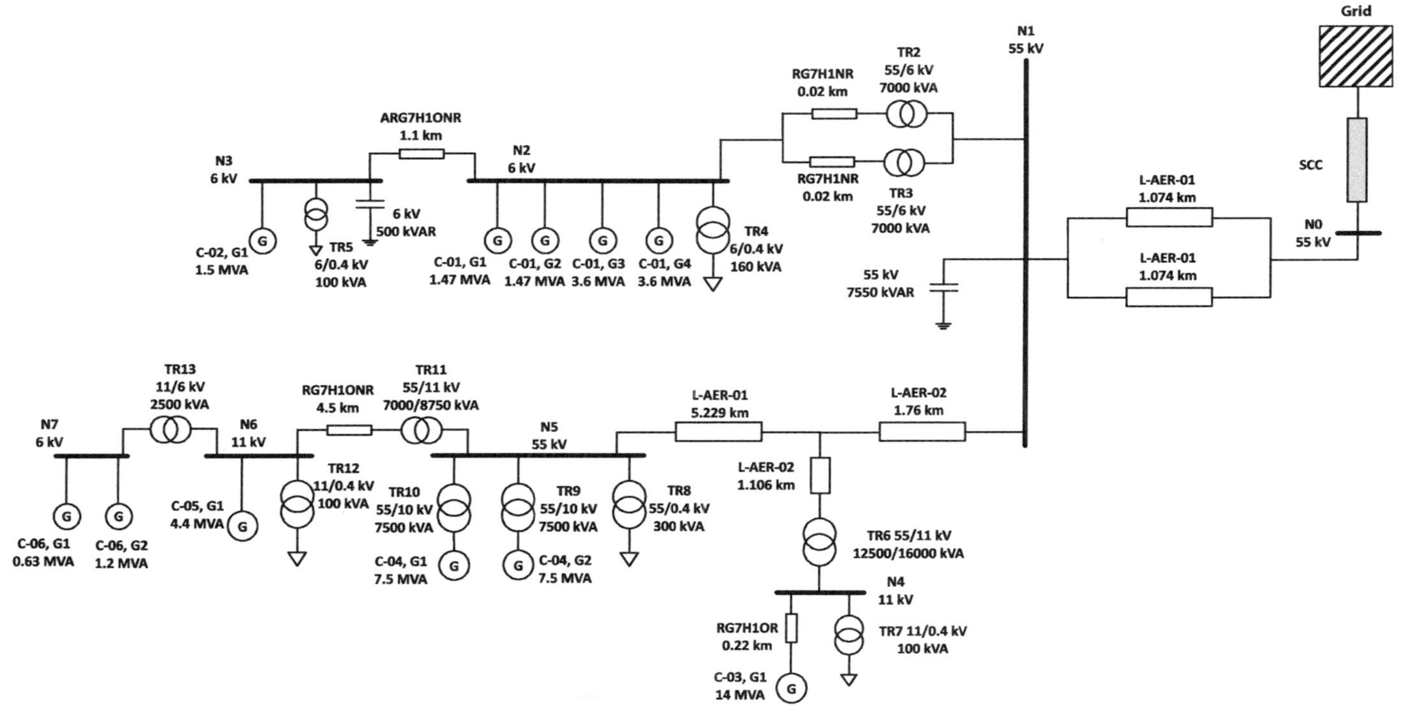

FIG. 10.17 Topology of plant C [45].

TABLE 10.4 Sensitivity coefficients [45].

$\frac{\partial v}{\partial q}$	N0	Plant	$\frac{\partial v}{\partial q}$	N0	Plant
G1	0.777	Plant A	C-02	0.0714	Plant C
G2	0.777		C-03	0.0675	
G1	0.634	Plant B	C-04.GR1	0.0680	
G2	0.610		C-04.GR2	0.0680	
G3	0.593		C-05	0.0689	
C-01	0.0717	Plant C	C-06	0.0657	

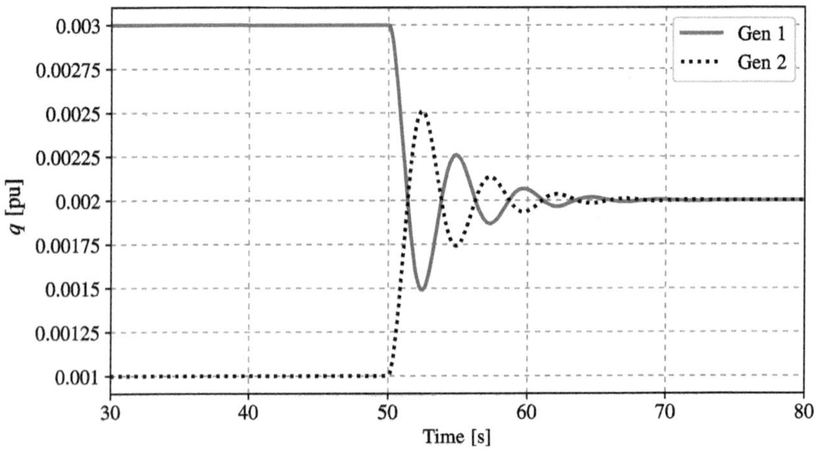

FIG. 10.18 Network A: Reactive power response without dynamic decoupling [45].

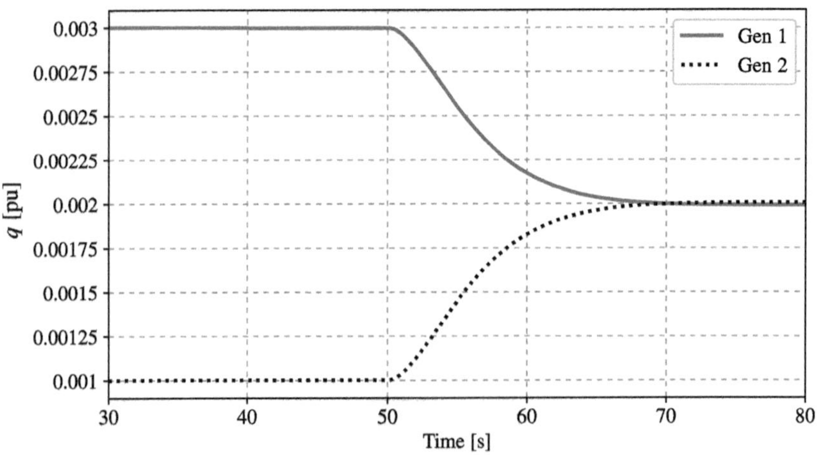

FIG. 10.19 Network A: Reactive power response with dynamic decoupling [45].

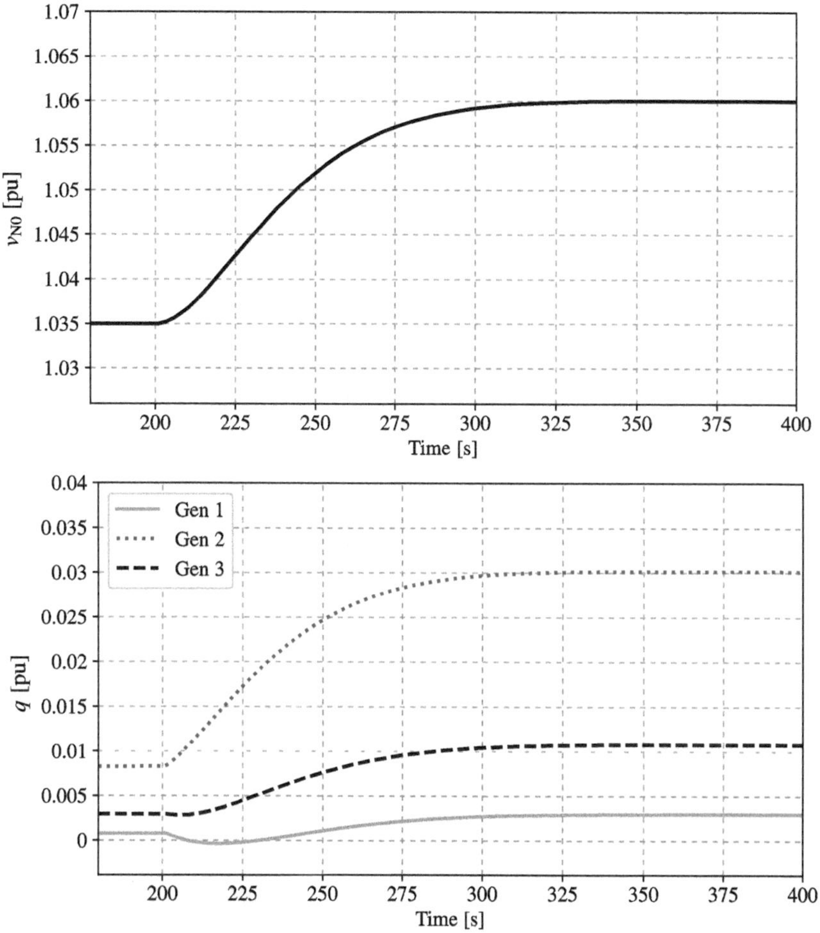

FIG. 10.20 Network B: Voltage, and reactive power response for a step in reference voltage without the decoupling matrix [45].

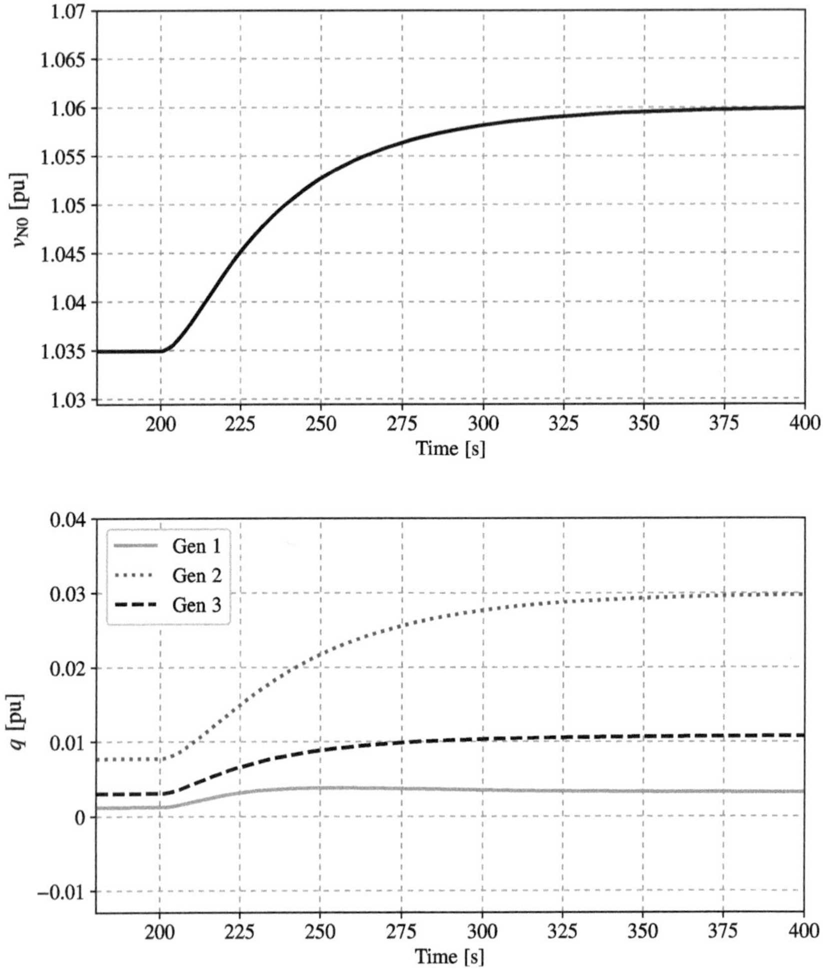

FIG. 10.21 Network B: Voltage and reactive power response for a step in reference voltage with the decoupling matrix [45].

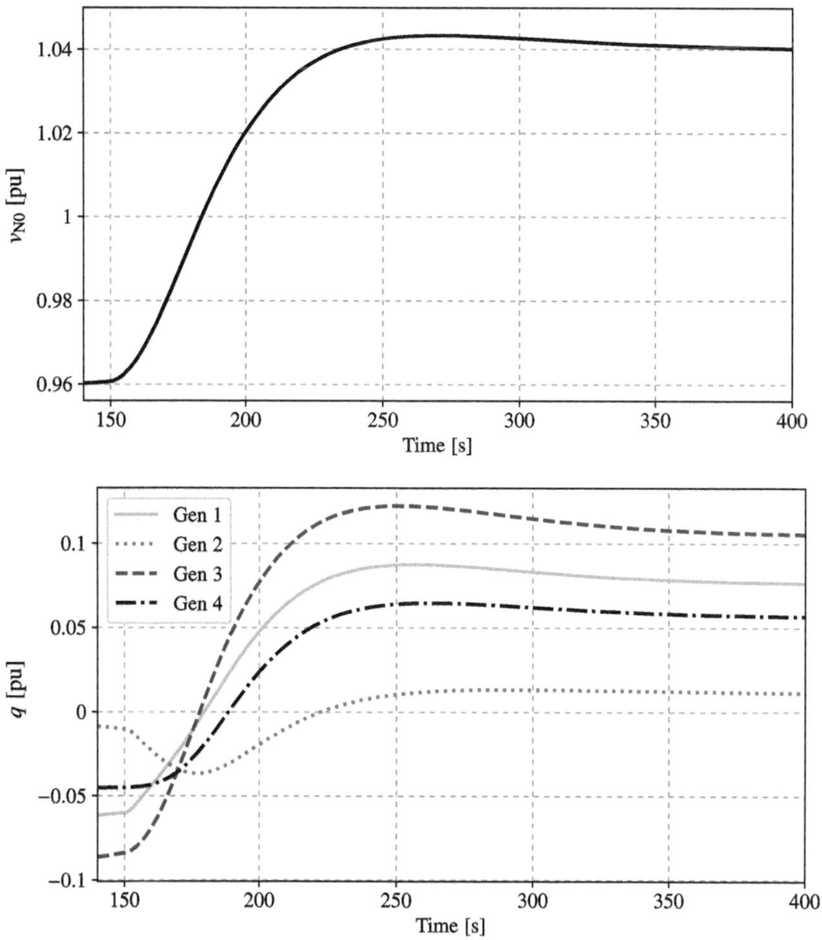

FIG. 10.22 Network C: Voltage and generators reactive power response without the decoupling matrix [45].

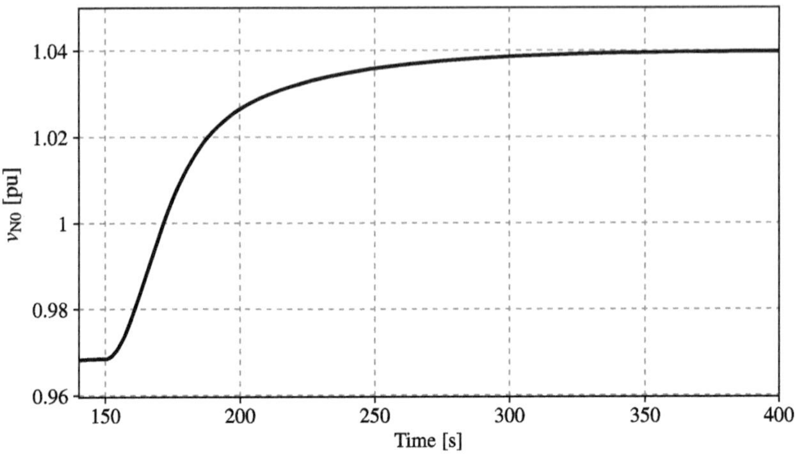

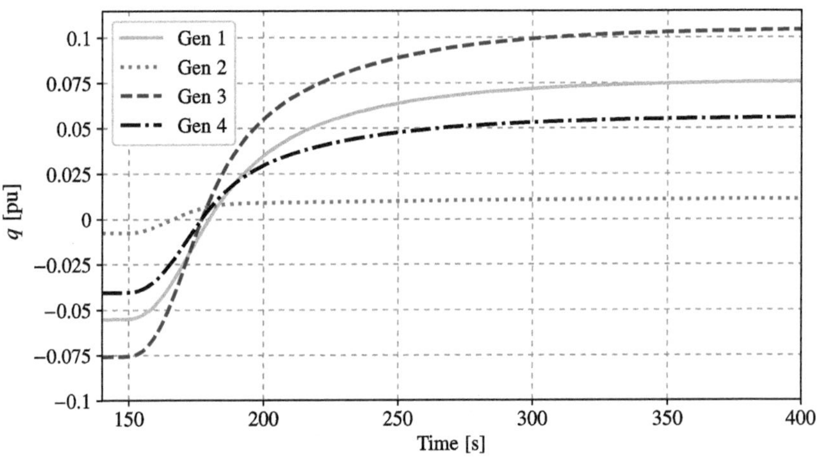

FIG. 10.23 Network C: Voltage and generators reactive power response with the decoupling matrix [45].

References

[1] T. Van Cutsem, C. Vournas, Voltage Stability of Electric Power Systems, Springer Science & Business Media, Germany, 2007.

[2] F. Milano, Hybrid control model of under load tap changers, IEEE Trans. Power Deliv. 26 (4) (2011) 2837–2844.

[3] S.N. Salih, P. Chen, O. Carlson, The effect of wind power integration on the frequency of tap changes of a substation transformer, IEEE Trans. Power Syst. 28 (4) (2013) 4320–4327, https://doi.org/10.1109/TPWRS.2013.2266260.

[4] S.S. Baghsorkhi, I.A. Hiskens, Impact of wind power variability on sub-transmission networks, IEEE PES General Meeting, July ISBN 1932-5517, 2012, pp. 1–7, https://doi.org/10.1109/PESGM.2012.6345683.

[5] C. Long, A.T. Procopiou, L.F. Ochoa, G. Bryson, D. Randles, Performance of OLTC-based control strategies for LV networks with photovoltaics, IEEE PES General Meeting, July ISBN 1932-5517, 2015, pp. 1–5, https://doi.org/10.1109/PESGM.2015.7285618.

[6] D. Ranamuka, A.P. Agalgaonkar, K.M. Muttaqi, Examining the interactions between DG units and voltage regulating devices for effective voltage control in distribution systems, IEEE Trans. Ind. Appl. 53 (2) (2017) 1485–1496, https://doi.org/10.1109/TIA.2016.2619664.

[7] G.K. Ari, Y. Baghzouz, Impact of high PV penetration on voltage regulation in electrical distribution systems, International Conference on Clean Electrical Power (ICCEP), June 2011, pp. 744–748.

[8] Y. Song, M. Begović, Secondary voltage and stability control, 15th International Power Electronics and Motion Control Conference (EPE/PEMC), September 2012, https://doi.org/10.1109/EPEPEMC.2012.6397411 pp. LS2b.4-1–LS2b.4-7.

[9] S. Corsi, F. De Villiers, R. Vajeth, Power system stability increase by secondary voltage regulation applied to the South Africa transmission grid, IREP Symposium Bulk Power System Dynamics and Control—VIII (IREP), August 2010, pp. 1–18, https://doi.org/10.1109/IREP.2010.5563264.

[10] S. Corsi, C. Sabelli, General blackout in Italy Sunday September 28, 2003, h. 03:28:00, IEEE PES General Meeting, 2004, pp. 1691–1702.

[11] M. Chiandone, G. Sulligoi, S. Massucco, F. Silvestro, Hierarchical voltage regulation of transmission systems with renewable power plants: an overview of the Italian case, Renewable Power Generation Conference (RPG), September 2014, pp. 1–5, https://doi.org/10.1049/cp.2014.0861.

[12] L. Yu, D. Czarkowski, F de Leon, Optimal distributed voltage regulation for secondary networks with DGs, IEEE Trans. Smart Grid 3 (2) (2012) 959–967, https://doi.org/10.1109/TSG.2012.2190308.

[13] M.E. Moursi, G. Joos, C. Abbey, A secondary voltage control strategy for transmission level interconnection of wind generation, IEEE Trans. Power Electron. 23 (3) (2008) 1178–1190, https://doi.org/10.1109/TPEL.2008.921195.

[14] W. Xiao, K. Torchyan, M.S. El Moursi, J.L. Kirtley, Online supervisory voltage control for grid interface of utility-level PV plants, IEEE Trans. Sustain. Energy 5 (3) (2014) 843–853, https://doi.org/10.1109/TSTE.2014.2306572.

[15] S. Peyghami, H. Mokhtari, P. Davari, P.C. Loh, F. Blaabjerg, On secondary control approaches for voltage regulation in DC microgrids, IEEE Trans. Ind. Appl. 53 (5) (2017) 4855–4862, https://doi.org/10.1109/TIA.2017.2704908.

[16] Y. Xu, H. Sun, W. Gu, Y. Xu, Z. Li, Optimal distributed control for secondary frequency and voltage regulation in an islanded microgrid, IEEE Trans. Ind. Inf. 15 (1) (2019) 225–235, https://doi.org/10.1109/TII.2018.2795584.

[17] J. von Appen, M. Braun, T. Stetz, K. Diwold, D. Geibel, Time in the sun: the challenge of high PV penetration in the German electric grid, IEEE Power Energy Mag. 11 (2) (2013) 55–64, https://doi.org/10.1109/MPE.2012.2234407.

[18] J. Arrinda, J.A. Barrena, M.A. Rodriguez, A. Guerrero, Analysis of massive integration of renewable power plants under new regulatory frameworks, International Conference on Renewable Energy Research and Application (ICRERA), October 2014, pp. 462–467, https://doi.org/10.1109/ICRERA.2014.7016428.

[19] R.K. Varma, M. Salama, Large-scale photovoltaic solar power integration in transmission and distribution networks, IEEE PES General Meeting, July 1932-55172011, pp. 1–4, https://doi.org/10.1109/PES.2011.6039860.

[20] Norma Italiana CEI, Norma Italiana CEI 0-16, 2012. Tech. Rep.

[21] C.E.I. Norma Italiana, Norma Italiana CEI 0-21, 2012. Tech. Rep.

[22] AISBL, ENTSOE, Entso-e network code for requirements for grid connection applicable to all generators, ENTSO-E AISBL, Brussels, Belgium, 2012.

[23] F. Delfino, R. Procopio, M. Rossi, G. Ronda, Integration of large-size photovoltaic systems into the distribution grids: a P-Q chart approach to assess reactive support capability, IET Renew. Power Gener. 4 (4) (2010) 329–340, https://doi.org/10.1049/iet-rpg.2009.0134.

[24] H.K. Tyll, F. Schettle, Historical overview on dynamic reactive power compensation solutions from the begin of AC power transmission towards present applications, IEEE PES Power Systems Conference and Exposition, March 2009, pp. 1–7, https://doi.org/10.1109/PSCE.2009.4840208.

[25] A.C. Franklin, D.P. Franklin, The J&P Transformer Book: A Practical Technology of the Power Transformer, Elsevier, Oxford, 2016.

[26] D. Dohnal, On-load tap-changers for power transformers, Maschinenfabrik Reinhausen GmbH, Regensburg, Germany, 2013 Tech. Rep.

[27] F. Milano, Power System Modelling and Scripting, Springer Science & Business Media, Germany, 2010.

[28] M.S. Calovic, Modeling and analysis of under-load tap-changing transformer control systems, IEEE Trans. Power Appar. Syst. (7) (1984) 1909–1915.

[29] Q. Wu, D.H. Popovic, D.J. Hill, M. Larsson, Tap changing dynamic models for power system voltage behaviour analysis, Power Systems Computation Conference (PSCC)1999.

[30] I.A. Hiskens, P.J. Sokolowski, Systematic modeling and symbolically assisted simulation of power systems, IEEE Trans. Power Syst. 16 (2) (2001) 229–234.

[31] IEEE PSDP Committee Power System Stability Subcommittee Test Systems for Voltage Stability and Security Assessment Task Force, Test systems for voltage stability analysis and security assessment, Technical Report PES-TR19, 2015 Tech. Rep.

[32] C. Gardiner, Stochastic Methods: A Handbook for the Natural and Social Sciences, fourth ed., Springer, Germany, 2009.

[33] F. Milano, R. Zárate-Miñano, A systematic method to model power systems as stochastic differential algebraic equations, IEEE Trans. Power Syst. 28 (4) (2013) 4537–4544.

[34] R. Zárate-Miñano, F.M. Mele, F. Milano, SDE-based wind speed models with Weibull distribution and exponential autocorrelation, IEEE PES General Meeting IEEE, 2016, pp. 1–5.

[35] C. Murphy, A. Keane, Local and remote estimations using fitted polynomials in distribution systems, IEEE Trans. Power Syst. 32 (4) (2017) 3185–3194.

[36] M.A.A. Murad, F.M. Mele, F. Milano, On the impact of stochastic loads and wind generation on under load tap changers, 2018 IEEE Power Energy Society General Meeting (PESGM), August 2018, pp. 1–5, https://doi.org/10.1109/PESGM.2018.8585857.

[37] J.L. Sancha, J.L. Fernandez, A. Cortes, J.T. Abarca, Secondary voltage control: analysis, solutions and simulation results for the Spanish transmission system, IEEE Trans. Power Syst. 11 (2) (1996) 630–638, https://doi.org/10.1109/59.496132.

[38] G. Sulligoi, M. Chiandone, V. Arcidiacono, NewSART automatic voltage and reactive power regulator for secondary voltage regulation: design and application, 2011 IEEE Power and Energy Society General Meeting, July 2011, pp. 1–7, https://doi.org/10.1109/PES.2011.6039618.

[39] R. Campaner, M. Chiandone, V. Arcidiacono, G. Sulligoi, F. Milano, Automatic voltage control of a cluster of hydro power plants to operate as a virtual power plant, IEEE International Conference on Environment and Electrical Engineering (EEEIC), June 2015, pp. 2153–2158.

[40] M. Chiandone, R. Campaner, V. Arcidiacono, G. Sulligoi, F. Milano, Automatic voltage and reactive power regulator for wind farms participating to TSO voltage regulation, IEEE Eindhoven PowerTech, June 2015, pp. 1–5.

[41] M. Chiandone, R. Campaner, A.M. Pavan, V. Arcidiacono, F. Milano, G. Sulligoi, Coordinated voltage control of multi-converter power plants operating in transmission systems. The case of photovoltaics, International Conference on Clean Electrical Power (ICCEP), June 2015, pp. 506–510.

[42] P.M.S. Carvalho, P.F. Correia, L.A.F.M. Ferreira, Distributed reactive power generation control for voltage rise mitigation in distribution networks, IEEE Trans. Power Syst. 23 (2) (2008) 766–772, https://doi.org/10.1109/TPWRS.2008.919203.

[43] J.C. Vasquez, R.A. Mastromauro, J.M. Guerrero, M. Liserre, Voltage support provided by a droop-controlled multifunctional inverter, IEEE Trans. Ind. Electron. 56 (11) (2009) 4510–4519, https://doi.org/10.1109/TIE.2009.2015357.

[44] A. Cagnano, E. De Tuglie, M. Liserre, R.A. Mastromauro, Online optimal reactive power control strategy of PV inverters, IEEE Trans. Ind. Electron. 58 (10) (2011) 4549–4558, https://doi.org/10.1109/TIE.2011.2116757.

[45] R. Campaner, M. Chiandone, G. Sulligoi, F. Milano, Automatic voltage and reactive power control in distribution systems: dynamic coupling analysis, IEEE International Conference on Renewable Energy Research and Applications (ICRERA) 2016, pp. 934–939, https://doi.org/10.1109/ICRERA.2016.7884472.

Chapter 11

Dynamic voltage stability

Sriram K. Gurumurthy and Antonello Monti
Institute for Automation of Complex Power Systems, RWTH Aachen University, Aachen, Germany

This chapter aims to address the challenges and solutions that are developed to address the dynamic voltage stability issues arising in distribution networks as the number of power-electronic interfaced renewable energy sources (RES) increases. This chapter is based on the research outcomes of the EU-funded project under the H2020 scheme RESERVE.

11.1 Voltage stability issues in futuristic distribution grids

In a conventional power system, frequency is the variable of interest for power grid operators specifically in the high voltage (HV) transmission network since the frequency deviation denotes a power imbalance between generation and demand. Furthermore, the rate of change of frequency (ROCOF) has a direct relationship with the inertia in the power grid as dictated by the swing equation. Australian grids are already experiencing low inertia conditions at the time when photovoltaic (PV) production and HV direct current (HVDC) imports are high [1, 2]. The challenges encountered when moving towards a fully RES power-electronic driven grid are numerous such as loss of inertia and the resulting new and faster dynamics. Such challenges are addressed in the previous chapters and novel solutions such as linear swing dynamics (LSD) are proposed [3].

In a conventional power system, at low voltage (LV) and medium voltage (MV) distribution grids, voltage is of major concern, and the system voltage is regulated through reactive power control. The reactive power flow is done directly or indirectly to maintain the necessary nominal voltage. Typically, in distribution grids, voltage and reactive power control are done mainly through on-load-tap-changing (OLTC) transformer, which is hosted in the secondary substation automation unit (SSAU). Buses, where high loads are connected, may use the switching of capacitor banks to enable reactive power injection for compensating for the voltage drop in the line. Flexible alternating current (AC) transmission (FACTS) devices are mainly used in MV and HV grids for the control of voltage and reactive power. FACTS devices such as

Converter-Based Dynamics and Control of Modern Power Systems
https://doi.org/10.1016/B978-0-12-818491-2.00011-0

static synchronous compensator (STATCOMs) and static VAR compensators (SVCs) and OLTC transformer are some examples of FACTS devices that are widely used [4]. Some of these methods directly control the steady state voltage magnitude and the reactive power flow is a resultant of voltage levels adopted in the network whereas some of these methods indirectly regulate the voltage by controlling reactive power. Furthermore, the implementation of these techniques in LV or MV or HV depends on the economic viability. A summary of conventional steady-state voltage control methods is provided in Table 11.1.

Conventional distribution grids mainly consisted of passive elements and loads, active loads such as constant power loads existed sparsely at the distribution level. Since there was not much power electronic interfaced distributed energy resources (DERs), fast dynamics in feeder voltage was not a major concern from a stability point of view [4]. This chapter mainly covers the control aspects of LVAC distribution grids, especially the 400 V consumer grid since this is where a large number of PV inverters are being integrated. This chapter will tackle the following questions:

1. What are the changes brought about by high penetration of power electronics in LVAC grids from a dynamic standpoint?
2. Can existing voltage control methods handle this transition?
3. What are some of the system-level monitoring and control strategies that are being proposed for the futuristic power system?

The present grid current supports the integration of RES and this trend will increase, grid code modifications are key enablers for such a transition.

- Loads such as implantable medical devices (IMDs) and light-emitting diode (LED) lighting, DC home technologies are not far away
- Recent grid codes are supporting extensive PV integration

TABLE 11.1 Summary of conventional steady-state voltage control methods.

Voltage control method	Direct/ indirect	LV cost-effective?	MV/HV cost-effective?
OLTC	Direct	Yes	Yes
Static VAR compensator	Indirect	No	Yes
Capacitor banks	Indirect	Yes	No
STATCOM	Indirect	No	Yes

Fig. 11.1 shows an exemplary voltage distribution across the feeder length for conventional loading and under high RES. OLTCs can only adjust the voltage on the LV side voltage level at the bus adjacent to the OLTC and does not have complete controllability over every bus. Typically, capacitor banks are deployed at end nodes to manage the voltage levels locally without using tap changers. A straightforward problem with this type of control is the lack of controllability on every bus. This means when many PV inverters are integrated into the system, voltage violations such as overshoot, undershoot and voltage unbalance across different phases can occur. Conventional grid codes allow inverters to operate with a strictly lagging power factor, meaning that reactive power needs to be absorbed by the inverters. For example, the Irish distribution grid code expects the inverter to operate with 0.95 power factor lagging [4, 5]. Many countries have more relaxed grid codes such as VDE-AR-N-4105 in Germany, where the inverter can also inject reactive power into the grid [5, 6]. This enables the inverters to participate in voltage regulation or support.

This chapter is organized as follows: First, a general outlook on voltage stability is presented and the new dynamics that need to be considered for voltage stability in high RES context are explained. Then, a new concept in modern power systems called harmonic stability is presented. Harmonic stability concepts are presented through the impedance approach for both single and three-phase inverters. New system dynamics would imply and call for new stability analysis methods that can characterize the state of the power system both locally and globally. Various methods of stability analysis are reviewed from the literature. It is understood that the stability of the power electronic interface depends on the grid impedance and therefore noninvasive measurement of such grid impedances becomes critical for monitoring stability and the control of converters to mitigate instabilities from interaction among power converters is increasingly important. Thus, a section is dedicated to explaining the wideband grid impedance measurement concept and standalone impedance measurement and stability monitoring device that can be used by distribution system

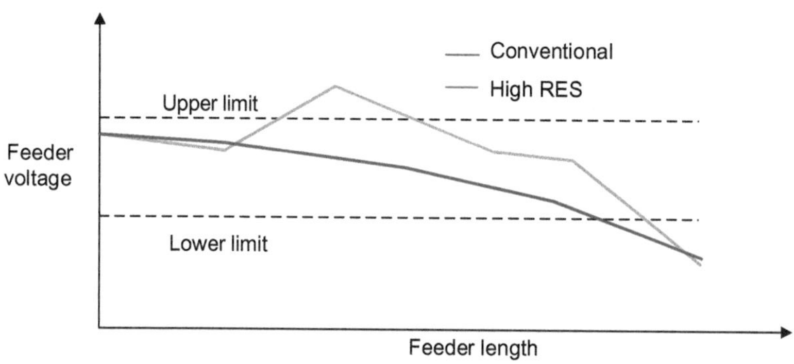

FIG. 11.1 Exemplary graph of voltage vs feeder length.

operators (DSOs). Harmonic instability mitigation techniques are then explained with their mathematical modeling. In project RESERVE, a decentralized stability monitoring technique was proposed for high RES context, known as dynamic voltage stability monitoring (DVSM), where the local information such as impedance is extracted from the power converters to analyze the stability of the network and then corrective actions can be synthesized. This method explores the degrees of freedom offered by power electronic converters. The converters provide ancillary services to the grid by providing the local impedance information to the central substation. Solid-state transformers (SSTs) are researched quite extensively as they offer increased flexibility and degrees of freedom and have the potential to replace the OLTCs in the future. The techniques proposed in RESERVE are extendable to the converters in the SST to guarantee stable operation and these extensions are briefly covered in the penultimate section of this chapter.

It must be noted that the techniques developed in RESERVE are also applicable to DC/DC converters and DC grids since the dynamic voltage stability concept has its origins in DC systems.

11.2 Voltage stability—An impedance approach

This section introduces the impedance-based approach for studying dynamic voltage stability. The methods for DC systems are discussed following by the extension for the AC system.

11.2.1 Middlebrook stability criterion

A new notion was introduced by Prod. R.D. Middlebrook in the early 1970s, which was a design-oriented approach, in which the input filters of DC converter systems can be designed such that the interconnection of the filter and the converter is stable [7, 8]. Such methods are "*stable by design*" and are highly desirable since the system is directly designed from a stability standpoint. Consider a generic source-load subsystem interconnected in cascade as shown in Fig. 11.2. Let P denote the point of interconnection from which the source side converter can be modeled as an impedance transfer function $Z_s(s)$ and similarly the load side converter/system can be modeled as an impedance transfer function $Z_l(s)$. This is represented by the red dotted line (black dotted line in print version) indicating subsystem partition. Let V_l represent the voltage at node P and this voltage can be expressed using Kirchhoff's Voltage Law (KVL) as shown in Eq. (11.1). The equivalent circuit representation of the cascade system is shown in Fig. 11.3.

$$V_l = \frac{V_s}{Z_s + Z_l} Z_l \tag{11.1}$$

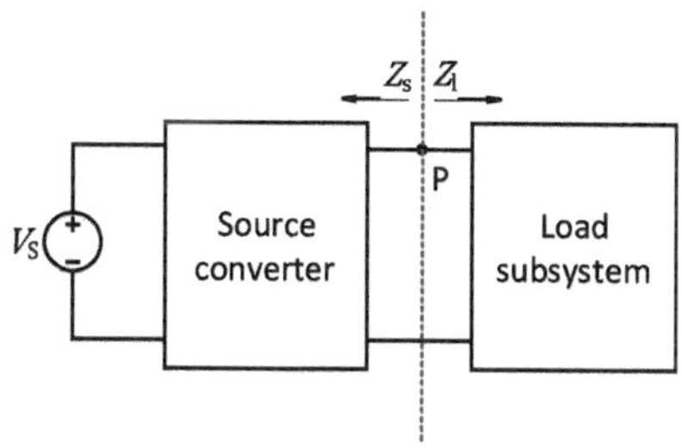

FIG. 11.2 Source-load cascade model.

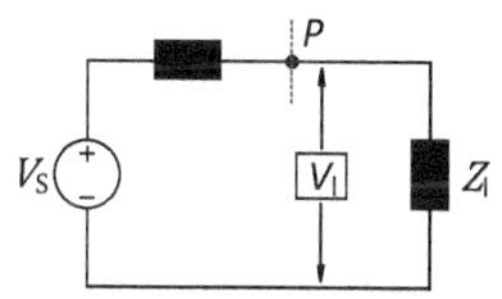

FIG. 11.3 Equivalent circuit of the cascade model.

The above equation can be written in an alternate form as shown in Eq. (11.2) and notice that the equation takes up the standard form of a single-input single-output (SISO) feedback system, with unity gain in the forward pass and the impedance ratio Z_s/Z_l in the feedback pass. The loop transfer function or the minor loop gain (MLG) is Z_s/Z_l.

$$V_l = \frac{1}{1+\dfrac{Z_s}{Z_l}} V_s \tag{11.2}$$

According to the Middlebrook criterion, the ratio of the magnitude of source-side impedance and load side impedance needs to be less than 1. In other words, the Nyquist plot of the MLG needs to lie inside the unit circle of the complex plane. The Middlebrook criterion is summarized by Eq. (11.3).

$$\left|\frac{Z_s(j\omega)}{Z_l(j\omega)}\right| \ll 1 \tag{11.3}$$

11.2.2 Nyquist stability criterion

The application of the Middlebrook criterion leads to a conservative design. For a closed-loop system, as shown in Fig. 11.4, the stability can be adjudged by applying the Nyquist criterion to the MLG. If voltage sources/current sources present in the source and load subsystem are stable and provided that the source and load subsystems are standalone stables, then applying the Nyquist criterion to MLG is valid. The unstable poles of the MLG are given by the right-half

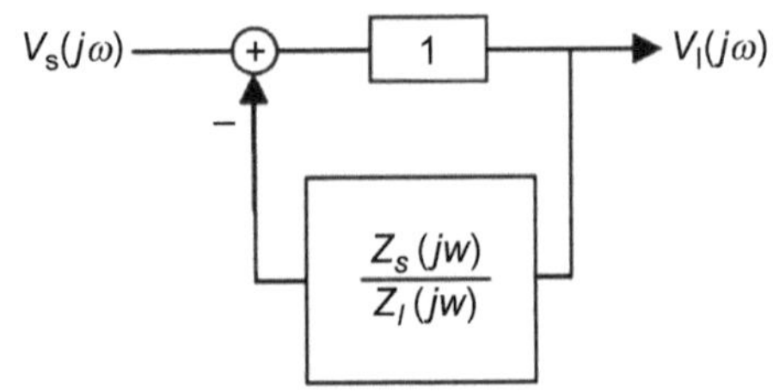

FIG. 11.4 Closed loop representation of cascade interconnection.

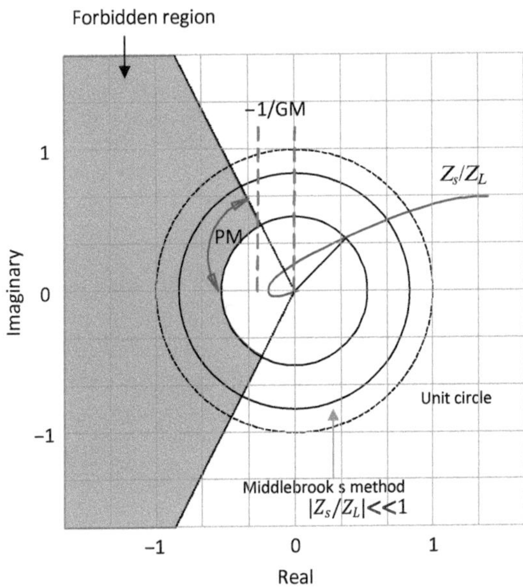

FIG. 11.5 Impedance based stability criterion for the DC system.

plane (RHP) poles of the source subsystem and RHP zeros of the load subsystem. Let these unstable poles of the MLG be denoted by P. The number of anticlockwise encirclements of the MLG around the critical point $-1+j0$ be N. Then, by the Nyquist stability criterion, the number of unstable closed-loop poles Z is given by Eq. (11.4) [9]. For closed stability, Z should be equal to zero and unstable for all integer values greater than zero.

$$Z = N - P \tag{11.4}$$

The application of the Middlebrook criterion leads to a conservative design. The interpretation of various stability criteria is explained in Fig. 11.5. An exemplary locus of the MLG is plotted and the system is stable according to Nyquist which is the correct small-stability interpretation for this exemplary system and unstable according to the Middlebrook criterion. For this system to satisfy the Middlebrook criterion, either the source side impedance needs to be decreased or the load side impedance to be increased. The load side impedance can be increased by reducing the bandwidth and using bulky filter design which is why methods that employ the classical Middlebrook criterion leads to artificial conservativeness. The Middlebrook approach is a pessimistic approach from a stability standpoint where the bandwidth achievable is very low compared to using other methods. The phase margin (PM) in the Nyquist plot is determined by the intersection of the Nyquist plot with the unit circle closer the Nyquist plot intersects with the unit circle with respect to the critical point $-1+j0$, lower is the PM. Low PM indicates tendency to enter into an oscillatory mode and therefore high PM is desirable. Therefore, by imposing a minimum required PM and gain margin (GM), a forbidden region is formulated. The source-side impedance and load side impedance need to be designed such that the Nyquist plot does not enter the forbidden region. This criterion is known as the GMPM criterion which is relaxed compared to the Middlebrook criterion and not very optimistic like the Nyquist criterion itself where low PMs may be implied [8].

11.2.3 Passivity-based stability criterion

The passivity-based stability criterion (PBSC) is valid only for radial networks. In a radial network, all sources and loads are connected in parallel. Passivity implies that the Nyquist plot lies on RHP and does not enter the left-half plane (LHP). For example, the exemplary plot of MLG in Fig. 11.5 is not passive since the Nyquist plot enters the LHP. In the Bode diagrams, a passive system has its phase strictly between −90 and +90 degrees. A system that satisfies passivity has strong disturbance rejection properties and remains stable. The PBSC criterion is based on the principle that when passive subsystems are interconnected in a parallel manner, passivity property is preserved [8, 10]. Thus, when all sources and loads are designed such that their output and input impedances are passive respectively, the equivalent network impedance which is the parallel of all sources and loads is passive as represented in Fig. 11.6. Controllers that guarantee closed-loop passivity for converters are suitable and adopting such an approach for all sources and loads enables the entire radial grid to satisfy the PBSC criterion [10, 11].

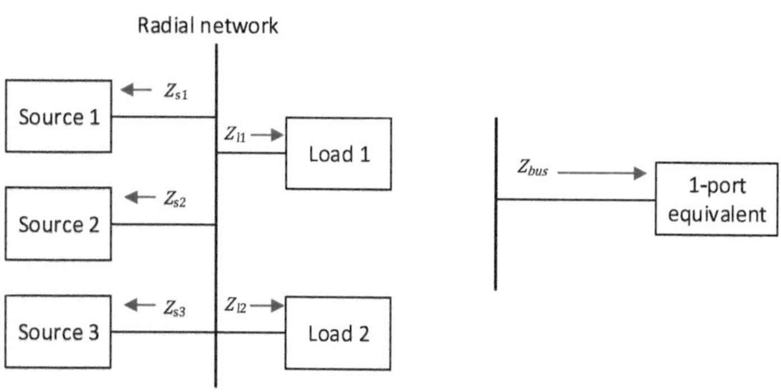

FIG. 11.6 PBSC concept.

11.2.4 Generalized Nyquist criterion

To study the interaction between two subsystems in three phase AC systems, direct-quadrature (*dq*) domain or sequence domain (with positive and negative sequence) modeling are often adopted. Some source/load subsystems may have strong coupling between two axes and therefore it is not possible to neglect the coupling and treat the system as two independent SISO systems. The system needs to be considered as a multiple-input multiple-output (MIMO) system and MIMO techniques need to be applied to study the stability of such systems [12, 13]. The Nyquist stability criterion covered in Section 11.2.2 is covering only SISO systems and this section is the extension for MIMO systems which is called the generalized Nyquist criterion (GNC).

Consider a three-phase grid-connected inverter modeled in the DQ domain. The exact closed-loop modeling shall be explained in the next section, however, consider the output impedance matrix of the inverter is given by the mapping from the *dq* output currents to the *dq* output voltage at the point of common coupling (PCC).

$$\begin{bmatrix} v_d \\ v_q \end{bmatrix} = \underbrace{\begin{bmatrix} z_{dq} & z_{dq} \\ z_{qd} & z_{qq} \end{bmatrix}}_{Z_{inv}} \begin{bmatrix} i_d \\ i_q \end{bmatrix} \tag{11.5}$$

From KCL we have,

$$\begin{aligned} (I_{inv} + I_2) &= Y_{inv} V_{pcc} \\ \left(V_{grid} - I_2 Z_{grid}\right) &= V_{pcc} \\ I_2 &= Z_{grid}^{-1}\left(V_{\text{grid}} - \text{V}_{\text{pcc}}\right) \end{aligned} \tag{11.6}$$

From the above, the PCC voltage can be expressed as follows:

$$V_{pcc} = \left(I + Z_{grid} Y_{inv}\right)^{-1} \left(Z_{grid} I_{inv} + V_{grid}\right) \tag{11.7}$$

Here, I is the identity matrix of size 2×2. Considering that the inverter is designed to standalone stable with an ideal grid and the grid is designed to be stable, then the stability can be adjudged by the return ratio matrix or the MLG matrix $Z_{grid}Y_{inv}$.

$$L_{dq}(s) = Z_{grid}(s) Y_{inv}(s) \tag{11.8}$$

Since $L_{dq}(s)$ is a transfer function matrix which is a function of frequency, its two eigenvalues $\lambda_1(s)$, $\lambda_2(s)$ are also functions of frequency. According to the GNC as defined in Ref. [14], "Let the multivariable system have no open loop unobservable or uncontrollable modes whose corresponding characteristic frequencies lie in the RHP, then the system will be closed-loop stable if and only if the net sum of anticlockwise encirclements of the critical point $(-1 + j0)$ by the set of characteristic loci (the eigenvalues $\lambda_1(s)$, $\lambda_2(s)$ in the complex plane) of $L_{dq}(s)$ is equal to the total number of RHP zeros of $Z_{grid}(s)$ and $Y_{inv}(s)$". It is very straightforward to know that the above rule is valid for the reciprocal of the eigen values, i.e., when we consider the inverse of $L_{dq}(s)$ matrix. Thus, the GNC can also be applied to the inverse MLG matrix and such a rule is known as the generalized inverse Nyquist stability criterion (GINC) [13, 14].

Steps for applying of GNC or GINC:

- Calculate the sum of open-loop RHP poles of of $Z_{grid}(s)$ and $Y_{inv}(s)$, let this be P.
- Draw the characteristic loci of of $L_{dq}(s)$, i.e. the locus of $\lambda_1(s)$, $\lambda_2(s)$ in the complex plane as the frequency $s = j\omega$ is varied from $\omega = \omega_{min}$ to $\omega = \omega_{max}$. Let the total number of anticlockwise encirclements from the locus of both $\lambda_1(s)$, $\lambda_2(s)$ around the critical point be N.
- Then, total number of closed-loop RHP poles Z, is given by the following equation:

$$Z = N - P \tag{11.9}$$

Example 11.1: Application of generalized Nyquist criterion

Consider a three-phase inverter with a closed admittance transfer function.

$$Y_{inv} = \begin{bmatrix} \dfrac{3333s}{s^2 + 558s + 1.79e05} & \dfrac{-1.046e06}{s^2 + 541.4s + 1.831e05} \\ \dfrac{1.046e06}{s^2 + 541.4s + 1.831e05} & \dfrac{3333s}{s^2 + 541.4s + 1.831e05} \end{bmatrix}$$

The grid impedance is an *RL* branch and the generic form of the *dq* domain transfer function for *RL* branch is given as:

$$Z_{grid} = \begin{bmatrix} R_g + sL_g & \omega L_g \\ -\omega L_g & R_g + sL_g \end{bmatrix}$$

Two cases are considered for the RL branch. Case 1: $R_g = 0.1\ \Omega$ and $L_g = 0.318$ mH; Case 2: $R_g = 0.45\ \Omega$ and $L_g = 9.5$ mH. In both these cases, the inverter admittance and the grid impedance are standalone stable systems which imply that $P = 0$. The characteristic loci are plotted considering the angular frequency to change from -3000 to 3000 rad/s.

Case 1: In this case, as observed from Fig. 11.7 there is no anticlockwise encirclement, $N = 0$. Thus, $Z = 0$ and system is stable.

Case 2: In this case, as observed from Fig. 11.8 there are 2 anticlockwise encirclements, $N = 2$. Thus, $Z = 2$ and system is unstable.

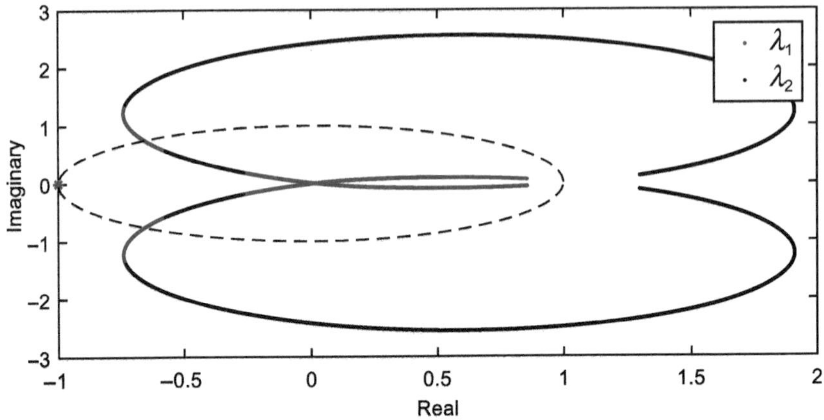

FIG. 11.7 Case 1 GNC example: stable case.

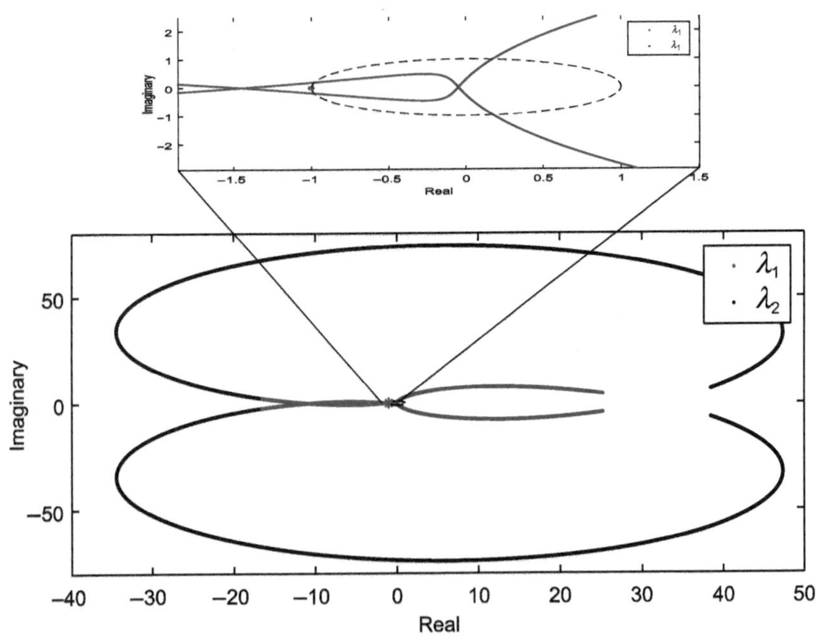

FIG. 11.8 Case 2 GNC example: unstable case.

11.3 Harmonic stability theory—An impedance phenomenon

Studies on power system harmonics conventionally have referred to the higher integer multiples of fundamental frequency which can occur due to the presence of FACTS devices such as STATIC compensators or rectifiers. For example, a thyristor-based rectifier feeding a DC machine can have a significant impact on higher-order odd harmonics such as 5th, 7th, and 11th harmonics [15–17]. The damping present in the network decides the impact of such harmonics. Weaker damping enables significant amplification and sustenance of such higher-order harmonics and vice-versa a stronger damping can mitigate or reduce the harmonic current magnitudes. Solutions such as phase-shifting transformers are adopted to interface such large rectification units with the AC grid so that the higher-order odd harmonics are canceled and eliminated [16].

With the advent of power electronic converters, many types of dynamics exist, and it is essential to differentiate them based on frequency range and study their cause and effect. The mapping of different converter dynamics is presented in Fig. 11.9. Subsynchronous oscillations emerge from active and reactive power flow dynamics and they occur at frequencies much less than 50 Hz owing to the low bandwidth of power controller. Near synchronous oscillations occur at frequencies close to 50 Hz up to 100 Hz. They occur due to unbalances or small changes in the PCC voltage or change of reference current to which the inner controller and phase-locked loop (PLL) react. Supersynchronous or Harmonic Stability occurs between 100 Hz to half the switching frequency. This stability is subjected to the impedance of inverter and grid; typically caused due to negative damping at the resonance frequency of the inverter-grid tied system [18–21]. At multiples of switching frequency and its sidebands, harmonics exist due to pulse width modulation (PWM) and the nature of harmonics is dependent on the PWM type. This section focuses only on one of the most complex medium frequency interactions between subsystems, which is harmonic stability.

Harmonic stability can be characterized by the impedance overlap that occurs between in the inverter admittance and the grid impedance [4]. When the magnitudes of Y_{inv} and Z_{grid} intersect; there exists a resonance in the inverter-grid tied system. If the phase angle difference between Y_{inv} and Z_{grid} at this resonance frequency is less than 180 degrees, then the resonance is stable. The harmonic stability condition is written as Eq. (11.10).

$$\arg(Y_{inv}) - \arg(Y_{grid}) < 180\,\text{degrees} \tag{11.10}$$

If the phase angle difference between Y_{inv} and Z_{grid} at this resonance frequency is greater than or equal to 180 degrees, then the system is unstable.

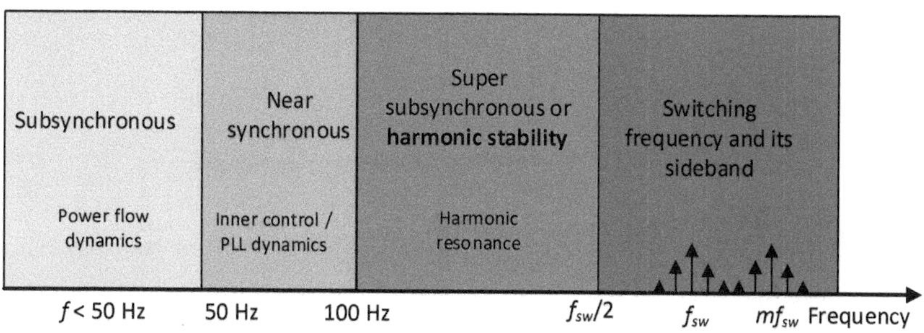

FIG. 11.9 Mapping of dynamics from power electronics.

An angle greater than 180 degrees indicates negative damping in the system and angle less than 180 degrees indicates net positive damping. Under negative damping, the harmonic current at the resonance frequency increases, and the rate of increase depends on the magnitude of negative damping in the system. Based on the nature of the control loop used in the inverter system or changes happening in the grid/load side system due to this harmonic current flow, there may be a change of impedance in the inverter or grid side where the negative damping is nullified to zero damping. In such a case, the harmonic current stabilizes and thus there is a stable yet distorted 50Hz current and voltage. Even this condition is referred to as harmonic instability due to sustained oscillations. Converters need to have virtually emulated damping such that the damping frequency can be varied to match the resonance frequency between the inverter-grid tied system. This concept is called as virtual output impedance (VOI) or active damping (AD) and it is covered in Section 11.5.

Let us consider an exemplary distribution grid as shown in Fig. 11.10. For simpler analysis, we neglect the cable impedances. The inverter is modeled as a current source with shunt admittance and the grid is modeled as a voltage source with series impedance.

Consider the stability at node P_1. The grid or load side impedance is Z_{grid} and source-side admittance is the sum of all inverter admittances since they are in parallel. Therefore, MLG at P_1 is

$$L_{P1} = (Y_{inv1} + Y_{inv2} + Y_{inv3})Z_{grid}$$

If Z_{grid} is zero at all frequencies, then L_{P1} is zero and the system is always stable. The harmonics injected by the inverters sink through the stiff grid. However, if Z_{grid} is high at a certain frequency, then that frequency is prone to harmonic instability. In such a case, we can only comment on the aggregated stability of all inverters with the grid impedance. The internal interaction cannot be judged.

Consider the stability at node P_2. The source-side admittance is Y_{inv2} and grid/load side impedance is the inverse of the sum of inverter admittances Y_{inv1}, Y_{inv3} and gird admittance $Y_{grid} = 1/Z_{grid}$ since they are in parallel. Therefore, MLG at P_2 is

$$L_{P2} = Y_{inv2}\frac{1}{(Y_{inv1} + Y_{inv3} + Y_{grid})}$$

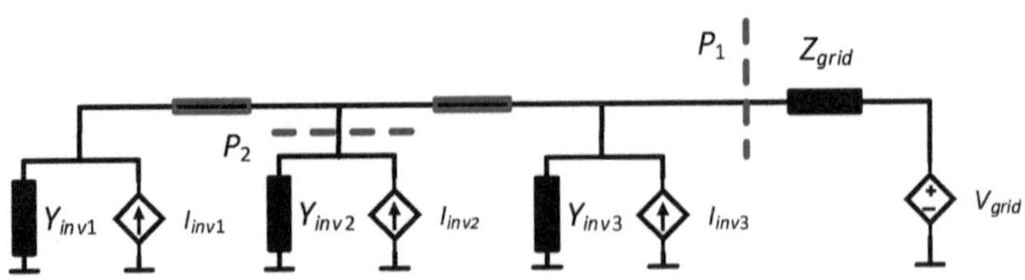

FIG. 11.10 Harmonic stability assessment in the distribution grid.

Similarly, if Z_{grid} is zero or Y_{grid} is infinity at all frequencies, then L_{P2} is zero and the system is always stable. However, the grid is not ideally stiff, and the impedance of the grid is a function of frequency and there could also be resonance in the grid impedance. In such a scenario, there could be multiple resonance frequencies of the MLG, and the stability of the system depends on the stability at each resonance frequency. Positive damping needs to be present at each of those resonance frequencies. Positive damping is characterized by a low impedance shunt path to the ground or neutral at the resonance frequency or a high impedance series element at the resonance frequency. The example presented in Fig. 11.10 with cable impedance being omitted is only for the ease of understanding the theory of a multisource radial system. When cable impedances are included, the analysis becomes complex, and stability results can be concluded only by deriving the transfer functions at the breakaway point. A detailed example covering the impact of cable impedance is presented in Section 11.6 where the conclusions drawn from the impact of cable impedances are discussed.

As explained above, since harmonic stability is an impedance phenomenon, it is essential to correctly model the closed-loop impedance of power electronic converters.

11.3.1 Impedance modeling of single-phase inverter

Consider a single-phase inverter with the LCL filter as shown in Fig. 11.11A and the equivalent circuit of the AC side is shown in Fig. 11.11B. Since various passive damping options are available, the modeling of each component is kept as generic transfer functions. The open-loop plant model of the inverter is shown in Fig. 11.11C and the blocks represent the transfer function of an individual passive element of the LCL filter. Y_L and Y_G represents the admittances of the converter side and grid side inductance and Z_{sh} represents the impedance of the filter capacitor. When complex damping circuits are designed, typically in shunt, Z_{sh} gets modified accordingly. For an LCL filter, considering internal resistances for the inductances, we have the following transfer functions:

$$Y_L(s) = \frac{1}{r_c + sL_c}$$
$$Z_{sh}(s) = \frac{1}{sC_o} \tag{11.11}$$
$$Y_G(s) = \frac{1}{r_g + sL_g}$$

From the above block diagram, the open-loop control transfer function and open-loop admittance of the inverter can be derived as shown below.

$$\frac{i_g(s)}{v_{inv}(s)} = G_{id}(s) = \frac{Y_L(s)Z_{sh}(s)Y_G(s)}{1 + Z_{sh}(s)(Y_L(s) + Y_G(s))} \tag{11.12}$$

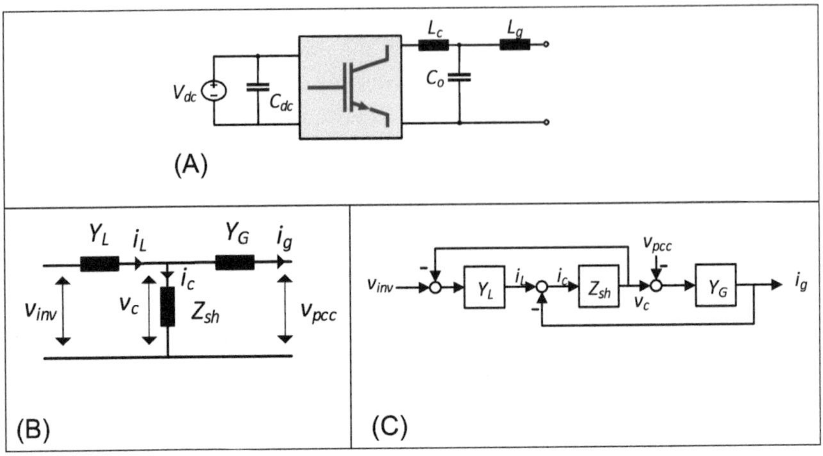

FIG. 11.11 Single phase inverter—Open loop model. (A) Single phase inverter with LCL filter, (B) equivalent circuit, and (C) open loop plant model.

$$\frac{-i_g(s)}{v_{pcc}(s)} = Y_{ol}(s) = \frac{Y_G(s)(1+Z_{sh}(s)Y_L(s))}{1+Z_{sh}(s)(Y_L(s)+Y_G(s))} \tag{11.13}$$

The single-phase inverter is considered with the current controller $K_c(s)$ to track the reference current and a grid voltage feedforward is also considered. Since the controllers are implemented in the digitally, delays due to sampling and hold are also considered $G_{del}(s)$. Closed-loop block diagram of the inverter is shown in Fig. 11.12.

The closed-loop reference tracking transfer function of the inverter if given by Eq. (11.14).

$$\frac{i_g(s)}{i_{g,ref}(s)} = T_{cl}(s) = \frac{G_{id}(s)G_{del}(s)K_c(s)}{1+G_{id}(s)G_{del}(s)K_c(s)} \tag{11.14}$$

The closed-loop output admittance of the inverter that considers the influence of control loops can be shown to be Eq. (11.15).

$$\frac{-i_g(s)}{v_{pcc}(s)} = Y_{cl}(s) = \frac{Y_{ol}(s) - G_{id}(s)G_{del}(s)}{1+G_{id}(s)G_{del}(s)K_c(s)} \tag{11.15}$$

11.3.2 DQ domain impedance modeling of three phase inverter

The three-phase inverter is often modeled in *dq* domain so that three-phase AC system is transformed into a DC system equivalent and hence first order controllers such as PI controller can be used for control. Due to the coupling

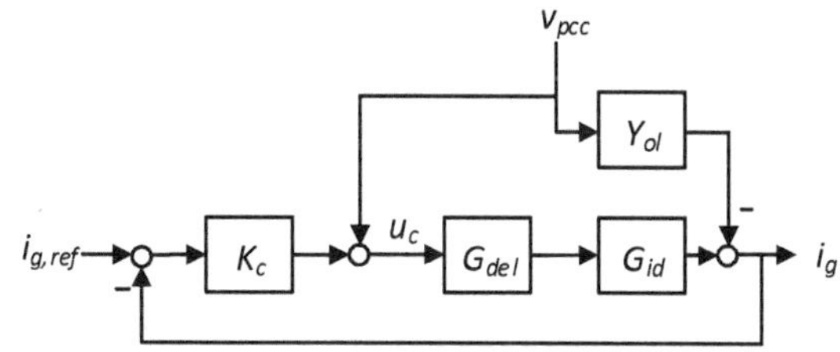

FIG. 11.12 Single phase inverter—Closed loop model.

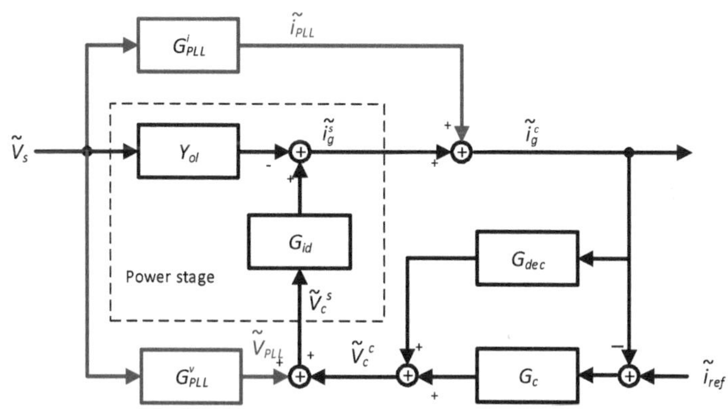

FIG. 11.13 Control block diagram of three-phase inverter [22].

between d and q-axis, every transfer function is a MIMO system. The control block diagram of the three-phase inverter in the DQ domain is shown in Fig. 11.13. Firstly, we neglect the PLL dynamics as marked in red. The closed-loop DQ admittance of the inverter takes the form Eq. (11.16), where I is the identity matrix of size 2.

$$Y_{cl} = [I + G_{id}(G_{ci} - G_{dec})]^{-1}[Y_{ol} - G_{id}]$$
$$-i_g(s) = v_{pcc}(s)Y_{cl}(s) \tag{11.16}$$

PLL is used to synchronize the inverter with the grid. Either the d or q-axis is chosen as a reference for the PLL. A PI controller is used to vary the frequency input to the sawtooth oscillator that generates the instantaneous phase angle of the inverter, such as the chosen reference axis voltage is driven to zero. The power flow convention is dependent on the choice of the reference axis. If q-axis voltage is chosen and reference, then the PLL tries to generate the instantaneous angle to drive v_q at PCC to zero. In this sign convention, positive values of d and q currents indicate the injection of active and reactive power from the inverter into the grid respectively. The PLL acts as a loyal grid follower, thus, any change in grid frequency or skip in the phase in the measured grid voltage, causes the PLL to react. The phase angle dynamics of the PLL propagate into the grid voltage feedforward path, denoted by G^{v}_{pll} and since the current measurements use a dq transformation, the PLL dynamics also propagate into current measurements, which is denoted as G^{i}_{pll} as shown in Fig. 11.13. Detailed modeling on these transfer functions can be found in [12, 13, 22, 23]. The closed-loop admittance of the inverter considering PLL dynamics is given by Eq. (11.17) [23]. Due to the PLL, the impedance is shaped as a negative resistance in frequencies closer to fundamental and therefore is prone to instabilities under sudden frequency fluctuations. However, a stabilizing PLL design can be achieved by designing with time-domain constraints such as settling time and frequency domain constraints such as damping factor since the PLL is a second-order system [24].

$$Y_{cl,pll} = [I + G_{id}(G_{ci} - G_{dec})]^{-1}[Y_{ol} - G_{id}\{G_{v,pll} - (G_{ci} - G_{dec})G_{i,pll}\}] \quad (11.17)$$

Example 11.2: Harmonic stability in grid connected inverter

The converter parameters used in this example are provided in Table 11.2. An RL grid impedance is considered and the values for the strong grid are $R_{grid}=0.001\ \Omega$ and $L_{grid}=3.18\ \mu H$; the values for weak grid are $R_{grid}=0.45\ \Omega$ and $L_{grid}=0.79$ mH.

TABLE 11.2 Converter parameters of three phase inverter.

Converter parameters	Values
DC-link voltage	700 V
Grid voltage and frequency	400 V, 50 Hz
Switching frequency	20 kHz
Converter side choke	1.0 mH, 0.03 Ω
Grid side choke	0.4 mH, 0.03 Ω
Filter capacitance *Co*	9.98 μF
Current control parameters (K_p, K_i)	0.039, 3.554

For simplicity, we do not consider the cross-coupling terms and therefore the inverter closed-loop model is equivalent to that of a single-phase inverter as it corresponds to the impedance Z_{dd}. The closed-loop admittance of the inverter is calculated using the equations defined above. The grid impedance transfer function is $Z_{grid}=R_{grid}+sL_{grid}$. Let us consider a bode plot with the inverter closed-loop admittance $Y_{inv,cl}$ and the grid admittance $Y_{grid}=\frac{1}{Z_{grid}}$ as shown in Fig. 11.14. Notice all the intersections from the magnitude plot of $Y_{inv,cl}$ with

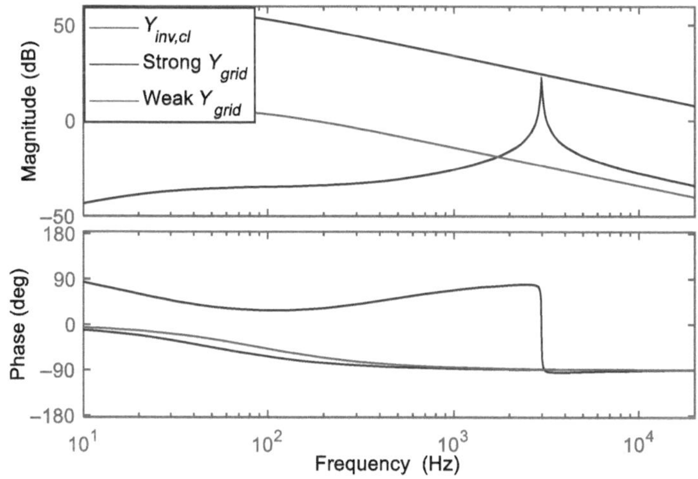

FIG. 11.14 Harmonic stability analysis—Bode plot.

Y_{grid} and calculate the phase difference between $Y_{inv,cl}$ and Y_{grid} at these intersecting frequencies in the phase plot. The magnitude plots do not intersect in the strong grid case but in the weak grid case, the magnitude plots intersect, and the angle is close to 180 degrees (164 degrees to be exact), and therefore the system is prone towards harmonic instability.

This is verified from the Nyquist plot of the impedance ratio $L = Y_{inv,cl}Z_{grid}$ as shown in Fig. 11.15. In this chosen example for a strong grid, the inverter admittance magnitude is smaller than the grid admittance at all frequencies, therefore the grid-tied system also satisfies the Middlebrook stability criterion. Hence, the Nyquist plot lies within the unit circle. For the weak grid condition, the Nyquist plot intersects the unit circle close to the critical point. The PM is only 16 degrees for the system, the system is not unstable however the system can enter oscillatory modes.

11.4 Wideband grid impedance measurement techniques

The previous sections have underlaid the foundation that Harmonic stability can be characterized and modeled as an impedance phenomenon. Therefore, measuring such an impedance over a wide range of frequencies becomes critical to perform diagnostics on the PEDGs. The first methods of impedance measurement are offline techniques i.e., when the power is not flowing in the network. Network analyzers or frequency analyzers were used to apply a sinusoidal voltage perturbation across the terminals of the network where the impedance is to be measured and the current drawn by the network is recorded. Comparing the magnitude and phase of the current with respect to the applied voltage, the impedance of the network is extracted. The modern-day network analyzers

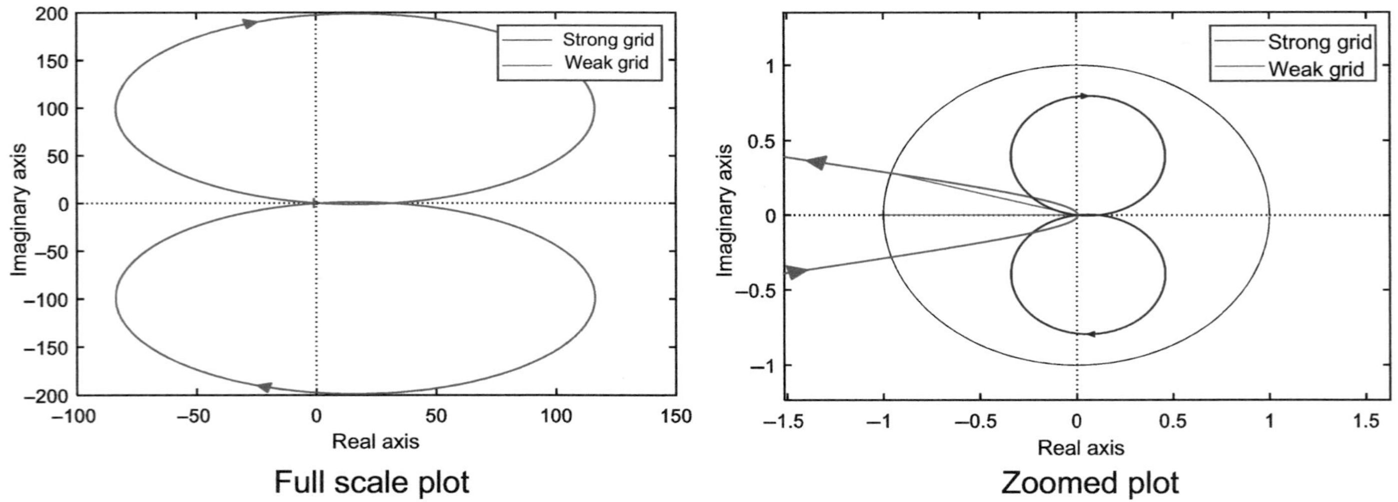

FIG. 11.15 Harmonic stability analysis—Nyquist criterion.

are equipped to perform these measurements online. A linear amplifier acts as a voltage source which is coupled to the primary of a high-frequency transformer and the secondary coil is connected to the network terminals across which the impedance is measured. Such an arrangement enables galvanic isolation between the high power network and the low power signal generator/linear amplifier. Such small-signal methods based on sinusoidal perturbation take several minutes to measure wideband impedance with high resolution in frequency [25]. Large signal methods are those that cause transient disturbance to the grid during the measurement process. Methods such as Dynamic Load Switching use square wave switching patterns to turn loads ON and OFF causing a transient disturbance and by recording the voltage and current, the impedance is extracted [26]. Such methods are also noninvasive and cannot be integrated into inverters.

In this section, a small signal method that is noninvasive and yet has a short measurement duration is presented.

11.4.1 Wideband system identification technique

Instead of measuring one frequency at a time, a wideband of frequencies can be excited simultaneously so that the impedance is measured at all those frequencies at the same time. Instead of sinusoidal, white noise perturbation can be used. White noise approximation can be generated digitally using feedback shift registers and they are referred to as pseudo random binary sequence (PRBS) [27–29]. A typical PRBS generator uses an N-bit shift register, where bit N and bit N-1 are applied to an XOR gate and the output of XOR gate is fed back to 1st bit and the output of PRBS is generator corresponds to bit N. One cycle of PRBS consists of $2^N - 1$ samples. For a single-phase inverter, the PRBS signal is sufficient to extract the grid impedance. By injecting the PRBS signal in the duty cycle, the inverter being a current source injects current perturbations into the grid. By recording the voltage at PCC before injection and during injection, it is possible to know the change in PCC voltage caused by the injected current and the impedance is calculated using Eq. (11.18). Here, an assumption is that the grid impedance does not change during measurement time.

$$Z_{grid}(j\omega) = \frac{V_{pcc,PRBS}(j\omega) - V_{pcc,NO,PRBS}(j\omega)}{I_{grid,PRBS}(j\omega) - I_{grid,NO,PRBS}(j\omega)} \tag{11.18}$$

As explained in the previous section, the three-phase Inverter in DQ domain is a MIMO system where the transfer function elements are the Fourier transform of the cross-correlation between the input and output as given in Eq. (11.19). PRBS cannot be injected in both d and q axis simultaneously due to the inherent coupling between d and q axis and therefore the measurement could only be done sequentially. Between the d-axis and q-axis injection, an idle time t_{idle} is required for the d-axis dynamics to settle. The measurement period increases by the idle time which generally 1–10 cycles depending on the PRBS amplitude.

$$y_n(m) = \sum_{i=1}^{q} \sum_{k=1}^{\infty} g_{ni}(k) u_i(m-k)$$

$$R_{u_p y_n}(m) = \alpha g_{np}(m) \quad (11.19)$$

$$G(j\omega) = \frac{1}{\alpha} \mathcal{F}\left[R_{u_p y_n}(m)\right]$$

To measure all impedance elements in the transfer function matrix simultaneously, then orthogonal excitation signals should be used. Orthogonal signals do not have the same frequency content and thus they can be injected into a MIMO system due to their uncorrelated nature and all elements can be measured. An orthogonal sequence referred to as inverse repeat sequence (IRS) can be generated by considering 2 rounds of PRBS and inverting every alternate output of the PRBS signal [30]. An example of PRBS and IRS orthogonal generation is explained below. A comparison of measurement time and frequency resolution is provided in Table 11.3. These tools can be integrated into existing grid forming and grid feeding converters to include the diagnostic capabilities. The impedance data obtained can be converted to a transfer function through complex curve fitting techniques. The numerator and denominator order of the unknown grid impedance transfer function needs to be assumed [27, 28].

Example 11.3: PRBS and IRS signals

Consider a $N=11$ bit shift registers for generating PRBS signal. This corresponds to $2^N-1=2047$ number of samples for 1 round of PRBS. However, we also need to generate the IRS sequence such that both PRBS and IRS have the same energy level. Thus, we consider 2 rounds of PRBS signal which consists of 4094 samples. Every alternate output of the PRBS signal is inverted to generate the IRS sequence. Fig. 11.16 shows the spectrum of PRBS and IRS signals, notice that the frequency content of both signals is not overlapping, thus they can be used to excite both d and q-axis without any interference provided that the MIMO system is linear.

TABLE 11.3 Comparison of PRBS and orthogonal PRBS/IRS injection.

Measurement aspects/excitation type	PRBS	Orthogonal PRBS and IRS
d-axis measurement	2^N-1	$2(2^N-1)$
q-axis measurement	2^N-1	
Idle time	t_{idle}	0
Frequency resolution	$\frac{1}{2^N-1}$	$\frac{1}{2(2^N-1)}$

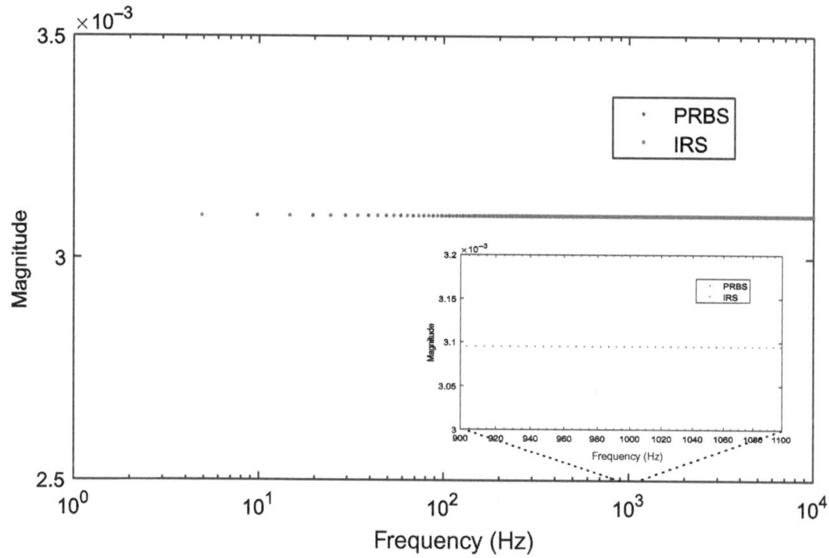

FIG. 11.16 Orthogonal wideband perturbation signals.

11.4.2 Wideband-frequency grid impedance device

Various impedance measurement techniques were discussed previously. Noninvasive wide techniques are best suited for measurement of grid impedance in a short time. The wideband system identification (WSI) technique is proposed as a tool that could be integrated into RES inverters. However, considering a distribution grid, harmonic stability assessment becomes critical not only at the inverter's terminal but also in a few selected buses along the radial feeder. Thus, a wideband-frequency grid impedance (WFZ) measurement device is proposed as a standalone impedance measurement and stability monitoring device [27, 31]. The proposed device does not require a DC power supply and thus offers high plug-play capability. This device can be used by DSOs and can be installed in multiple buses to continuously monitor the impedance of the grid as observed from the buses and monitor the stability margins. Such a device can also be extended to MV range by using a step-up transformer or by adopting multilevel converter design. Transmission system operators (TSOs) can use such a device to measure and monitor the impedance of large PV parks or wind farms. The black-box approach from the WFZ device means that RES power plant operators need not share confidential data required by the TSO operator for modeling.

The circuit diagram of the WFZ device is shown in Fig. 11.17. It consists of ta DC link capacitor, three phase two-level converter, and an LCL output filter. The software part of the device consists of a digital controller and WSI tool in an FPGA. The converter uses DC link voltage control to maintain nominal voltage. Perturbations are injected in the duty cycle to inject perturbation currents.

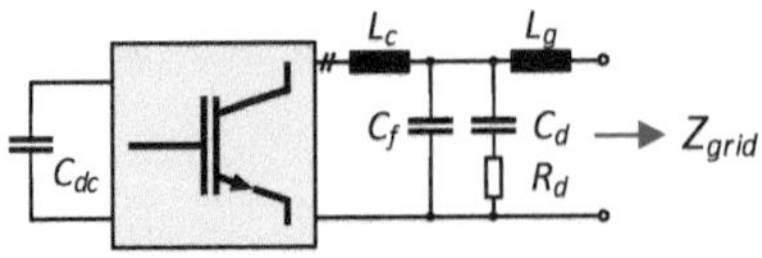

FIG. 11.17 WFZ device.

FIG. 11.18 Prototype of the WFZ device.

As explained previously, the impedance is extracted by the change in voltage at PCC before and during the injection. The prototype of the measurement device is shown in Fig. 11.18. A detailed measurement mechanism and uncertainty analysis can be found in [27].

Example 11.4: Wideband impedance measurement

An experiment based on the WFZ device is presented here to demonstrate wideband grid impedance extraction. Fig. 11.19 shows the experimental setup which consists of the WFZ device connected to a grid emulator through a known passive impedance. The parameters of the WFZ device are given in Ref. [27] and the parameters of the grid impedance are shown in Fig. 11.19. The WFZ device is operated at 20 kHz and uses a 11-bit shift register to generate the PRBS. The three-phase grid emulator acts as an ideal voltage source and it is programmed to produce a grid voltage of 40 V RMS. Since it is the initial testing phase, the WFZ device is supported by a DC power supply of 160 V.

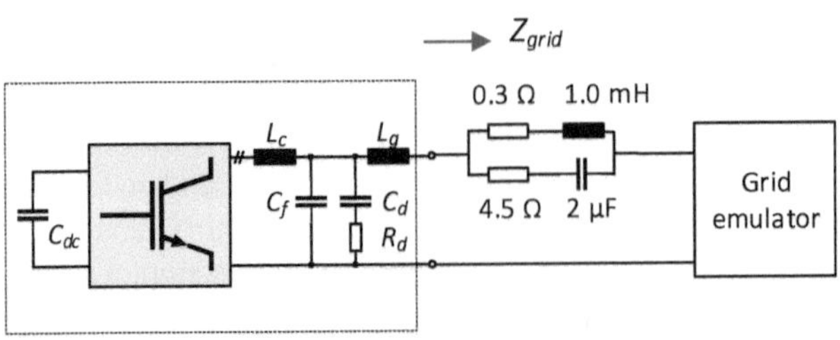

FIG. 11.19 Experimental setup.

The measurement consists of two parts, the first part is the scanning period when no noise is injected, and the second part is the perturbation period where the PRBS is injected onto the duty cycle. The PRBS is perturbing the d-axis duty cycle with an amplitude of 0.15. In this experiment, IRS is not implemented, so only a sequential measurement of the impedance matrix elements can be made. Fig. 11.20 shows the original sequence grid current and PCC voltage waveform and the PLL instantaneous angle which is obtained from the controller is plotted in Fig. 11.21. Fig. 11.21 also shows the DC link voltage fluctuation during PRBS injection. The PLL angle is used to transform the *abc* voltages and currents into *dq* domain and they are plotted alongside the *abc* quantities in Fig. 11.20. Using fast Fourier transform (FFT), the time domain current, voltage quantities are transformed into frequency domain. Fig. 11.20 shows the difference in the spectrum before and during PRBS is injected.

Using Eq. (11.18), the impedance spectrum can be extracted and in this case, since the external grid impedance is a known passive branch, we can analytically model or derive the grid impedance transfer function which then enables comparison and aids in uncertainty analysis. Fig. 11.22 shows the comparison of extracted nonparametric impedance data of the WFZ device with the analytical transfer function. The device is able to accurately measure the impedance up to 8 kHz and it is also able to capture the resonance frequency present in the 2nd order grid impedance. The accuracy is low at those frequencies where the impedance is small. A larger current is required to accurately measure small impedances since a larger current is required to produce an accurately measurable voltage change at the PCC than at a high impedance case where a small current is enough to produce a measurable voltage change at PCC. Thus, the minimum impedance that is of interest to grid operators needs to establish before designing the WFZ device. Based on the minimum impedance to be measured, the magnitude of current, resolution of voltage measurement unit needs to be fixed, following which the DC link capacitance required will be estimated depending on the allowable voltage ripple in the DC side.

11.5 Virtual output impedance control techniques

VOI control has existed in literature since the introduction of converter dominated AC microgrids. VOI control loops are generally used in the outer loops for grid supporting converters so that active and reactive power can be shared among converters. So, they are conventionally designed to be in the droop control loop where power control is carried out. The ratio of power-sharing is dependent on the magnitude of virtual impedance values used by every converter. The slope of the drooping line is dependent on the virtual impedance magnitude. Analogously, a virtual resistance control logic is used in the secondary droop control loop in the MVDC shipboard power system through which

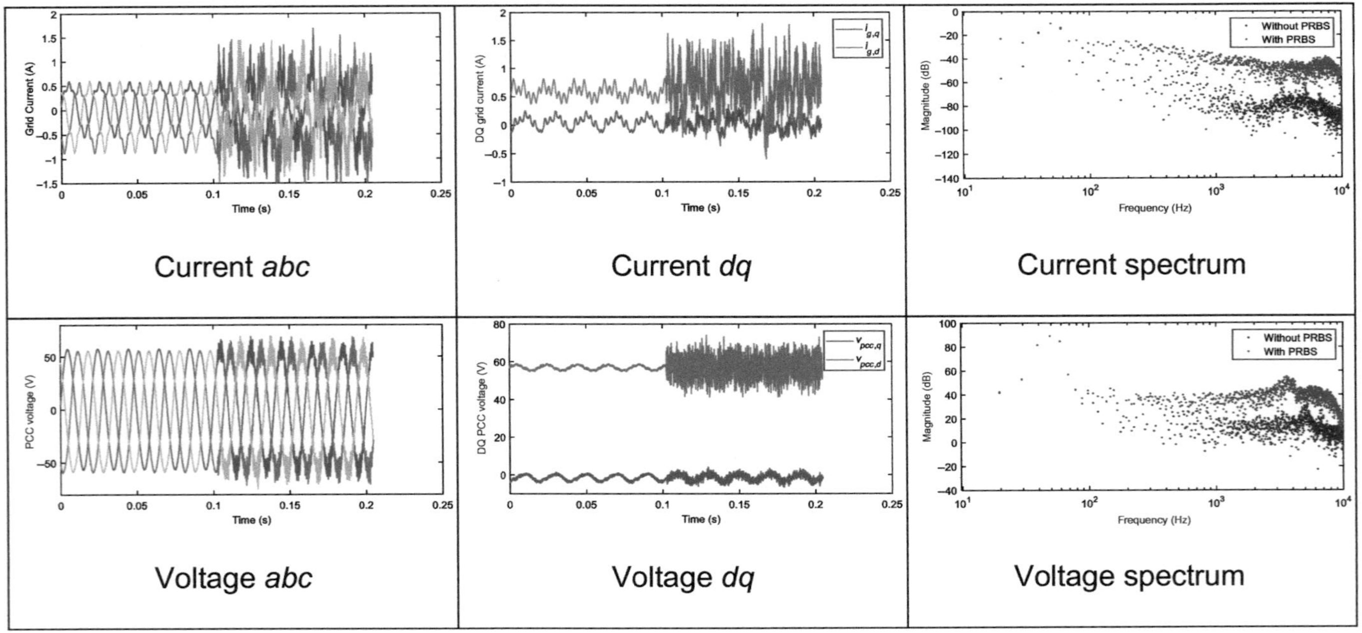

FIG. 11.20 Grid current, voltage waveform and their spectrum.

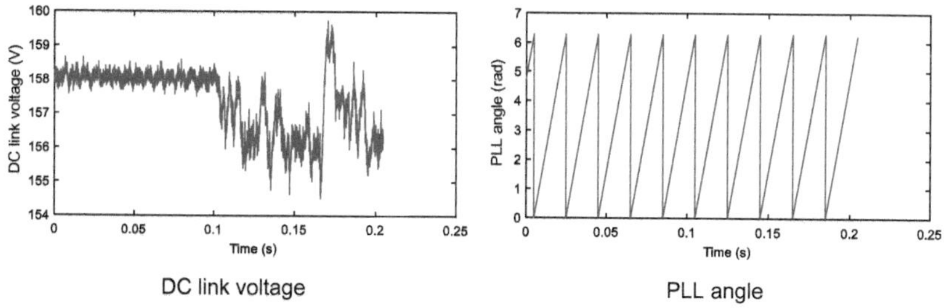

FIG. 11.21 DC link voltage and PLL angle.

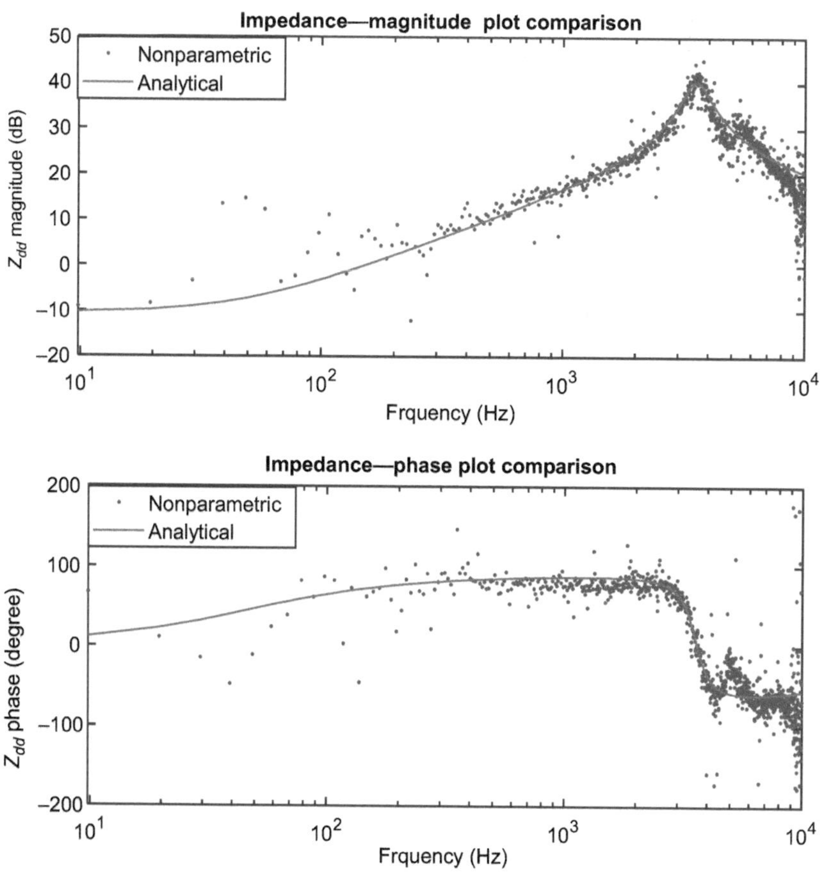

FIG. 11.22 Comparison of extracted impedance and analytical grid impedance transfer function.

power-sharing is attained. Since the harmonic stability issue started to gain importance, the inverters' output impedance needs to be reshaped such that it is stable for the existing grid impedance. This is achieved by modifying the inner current/voltage control loops, unlike the power control loops. Harmonic resonances are damped and mitigated through such control loops and generally, they are designed to emulate passive damping elements that could be connected across the filter capacitor. Filter capacitor voltage or current is required to emulate the damping scheme and these controllers are referred to as active damping (AD) controllers since they replace passive damping functionality and they are also referred to as VOI controllers since they modify the inverters output impedance [32, 33]. LCL filter design and passive damping method is shortly covered to explain the role of shunt impedance before deriving transfer functions for AD methods.

11.5.1 Passive damping

The open-loop plant model of the inverter is shown in Fig. 11.23 and the blocks represent the transfer function of the individual passive element of the LCL filter. Y_L and Y_G represents the admittances of the converter side and grid side inductance and Z_{sh} represents the impedance of the filter capacitor. When complex damping circuits are designed, typically in shunt, Z_{sh} gets modified accordingly. For an LCL filter, considering internal resistances for the inductances, we have the following transfer functions:

$$
\begin{aligned}
Y_L(s) &= \frac{1}{r_c + sL_c} \\
Z_{sh}(s) &= \frac{1}{sC_o} \\
Y_G(s) &= \frac{1}{r_g + sL_g}
\end{aligned}
\tag{11.20}
$$

The converter side inductance (L_c) is designed based on the maximum acceptable current ripple when the converter operates at peak power. The formula for determining L_c is dependent on the type of PWM, DC link voltage, maximum current ripple, and switching frequency. For space-vector PWM, the current ripple formula is provided in [34]. By assuming the apparent power

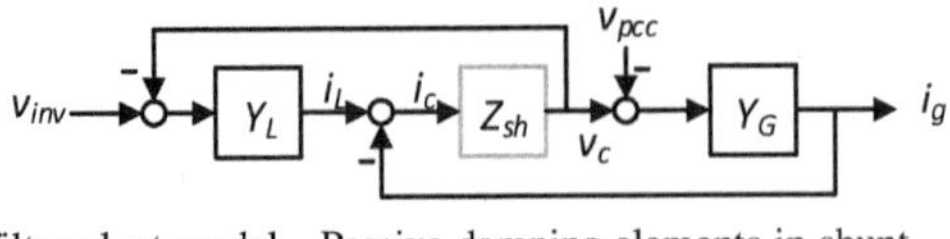

FIG. 11.23 LCL filter plant model—Passive damping elements in shunt.

rating of the inverter as the base power, the base impedance can be determined. The capacitor C_o is designed to be 5% of the base impedance. Designing the grid side inductance is mainly based on harmonic pollution recommendation from grid codes. However, an optimal selection can be made when the total percentage of converter plus grid side inductance is less than the capacitive branch percentage impedance which is 5% [34]. An exemplary design is shown in Fig. 11.24 where the grid side inductance can be chosen by first plotting the variation of resonance frequency and total percentage impedance of L_c and L_g as L_g is varied.

Without considering any feedback loop introduced by the controller, the open-loop admittance transfer function can be derived as Eq. (11.21).

$$\frac{i_g(s)}{v_{pcc}(s)} = Y_{ol}(s) = \frac{Y_G(s)(1+Z_{sh}(s)Y_L(s))}{1+Z_{sh}(s)(Y_L(s)+Y_G(s))} \quad (11.21)$$

Consider $u_c(s)$ as the control signal as shown in Fig. 11.27. There is always a certain delay introduced by zero-order hold method of sampling and update. Typically, in almost all power-electronics application, the sampling is done in the middle of the period so that the current/voltage is sampled in the middle of the ripple. The average of ripple is recovered through such sampling. If the control calculation can be completed within half a period, then the PWM is updated at the start of the immediate period. Let $G_{del}(s)$ be the delay transfer function. For the above case with half a period delay, the delay transfer function is

$$G_{del}(s) = e^{-0.5sT_{sw}} \quad (11.22)$$

If the controller computation takes longer than half the period, then the control can be updated 1.5 periods later. Then the delay transfer function is

$$G_{del}(s) = e^{-1.5sT_{sw}} \quad (11.23)$$

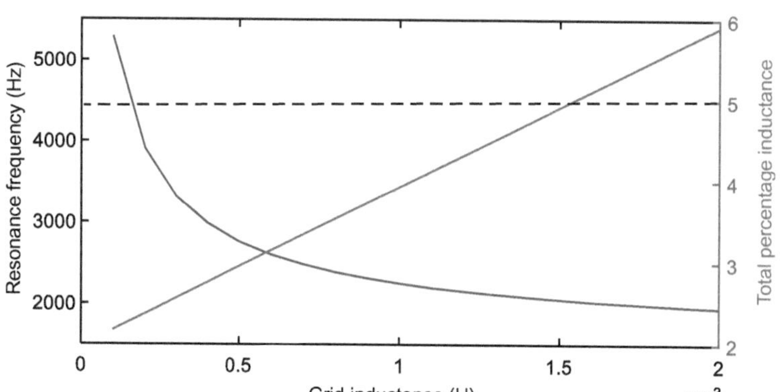

FIG. 11.24 Grid side inductance design.

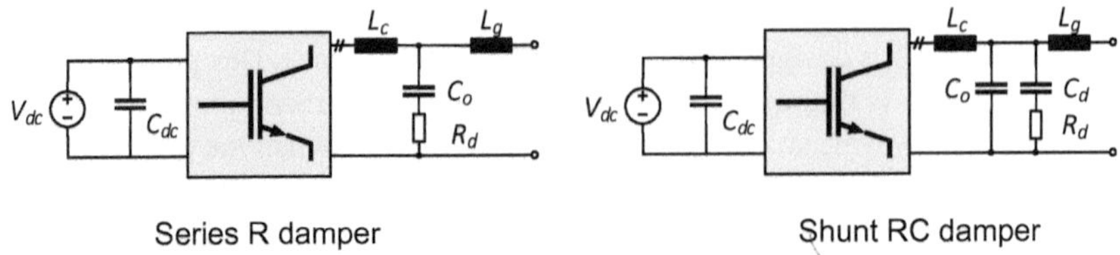

FIG. 11.25 Basic passive damping filters.

The open-loop control transfer function considering delay can be shown as Eq. (11.24).

$$\frac{i_g(s)}{u_c(s)} = G_{id}(s) = \frac{Y_G Z_{sh} Y_L G_{del}}{1 + Z_{sh}(Y_L + Y_G)} \tag{11.24}$$

The shunt elements are designed with damping circuits so that higher-order harmonics flow through the shunt path. Designing damping circuit in series elements leads to power losses and voltage drop. A wide range of passive damping design can be found in Ref. [34]. For ease of understanding, we select the most basic damping types: Series R damping and Shunt RC damping, and their topology is shown in Fig. 11.25. The impedance transfer function of the shunt element for these damping topologies is given in Eq. (11.25).

$$\begin{aligned} \text{Series R: } Z_{sh}(s) &= \frac{sC_oR_d + 1}{sC_o} \\ \text{Shunt RC: } Z_{sh}(s) &= \frac{sC_dR_d + 1}{s(sC_dC_oR_d + C_o + C_d)} \end{aligned} \tag{11.25}$$

Example 11.5: Passive damping

The parameters of the three-phase inverter used for this example are given in Table 11.2. By following the approach in Ref. [34], a shunt RC damper is designed with a split ratio of $n=1$. For the choice, $n=1$, the optimal quality factor is $Q=9$. Thus, the characteristic impedance is $Z_o=5.36\ \Omega$ and the damping impedance $R_d=48.27\ \Omega$ and $C_d=C_o=4.97\ \mu F$. From these new values, Z_{sh} can be formulated and hence the set of open loop transfer functions can be formulated. Fig. 11.26 shows the resultant bode plots of the open loop transfer function. Notice that with Shunt RC damper, the resonance peak is reduced, and the resonance is shifted to a higher frequency. Both these damping topologies perform in a nearly identical manner and the only difference is increased losses in Series R damper [34]. The split ratio and quality factor can be experimented to determine better peak minimization, or a higher-order passive damping circuit can be designed.

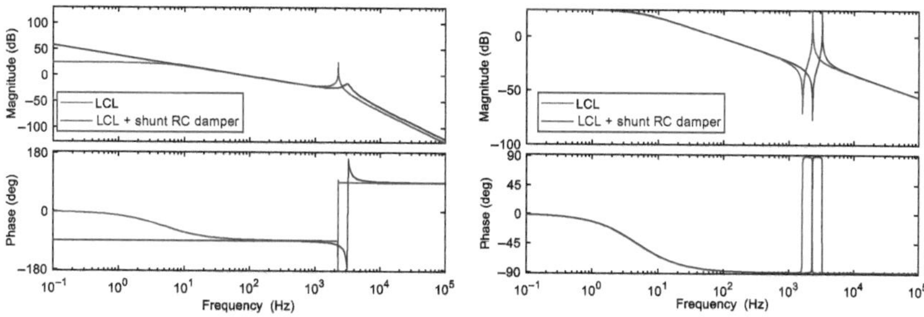

Open loop control transfer function G_{id} s Open loop admittance $Y_{ol}(s)$

FIG. 11.26 Passive damping result.

11.5.2 Active damping/VOI control

To emulate such damping behavior as discussed in the previous section, a feedback loop is introduced that modifies the control signal $u_c(s)$. Since the shunt element needs to be emulated, the measurement of the current through the shunt branch $i_c(s)$ or the node voltage across the shunt branch $v_c(s)$ is sufficient for implementing the virtual damping mechanism. Such an approach is known as AD or VOI. $K_{voi}(s)$ is the VOI controller transfer function.

Fig. 11.27 shows the open-loop control block diagram considering the VOI controller. It is referred to as open-loop since the reference tracking current controller $K_c(s)$ is not included in this analysis. Fig. 11.27 shows both capacitor current feedback and capacitor voltage feedback-based VOI control. The open-loop admittance transfer function with capacitor current feedback is shown in Eq. (11.26).

$$\frac{-i_g(s)}{v_{pcc}(s)} = Y_{ol}(s) = \frac{Y_G[1 + Y_L(Z_{sh} - K_{voi}G_{del})]}{1 + Y_L(Z_{sh} - K_{voi}G_{del}) + Z_{sh}Y_G} \tag{11.26}$$

Open-loop admittance transfer function with capacitor voltage feedback can be derived as Eq. (11.27).

$$\frac{-i_g(s)}{v_{pcc}(s)} = Y_{ol}(s) = \frac{Y_G[1 + Y_L Z_{sh}(1 - K_{voi}G_{del})]}{1 + Y_L Z_{sh}(1 - K_{voi}G_{del}) + Z_{sh}Y_G} \tag{11.27}$$

In a similar manner, the open-loop control transfer functions with capacitor current feedback are derived as Eq. (11.28),

$$\frac{i_g(s)}{u_c(s)} = G_{id}(s) = \frac{Y_G Z_{sh} Y_L G_{del}}{1 - K_{voi}G_{del}Y_L + Z_{sh}(Y_L + Y_G)} \tag{11.28}$$

and the open-loop control transfer function with capacitor voltage feedback is derived as Eq. (11.29)

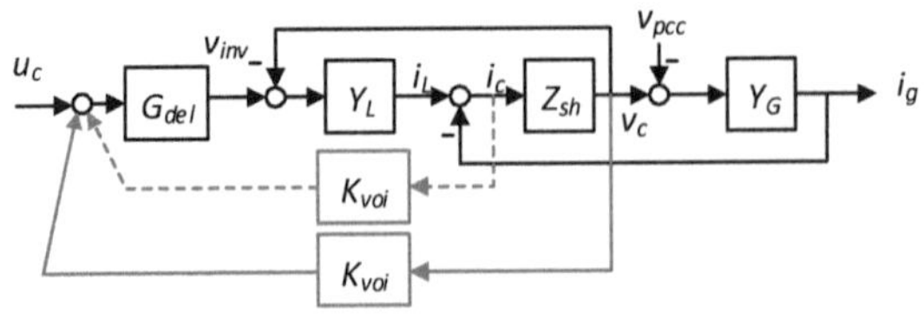

FIG. 11.27 Active damping/VOI—Open loop shaping.

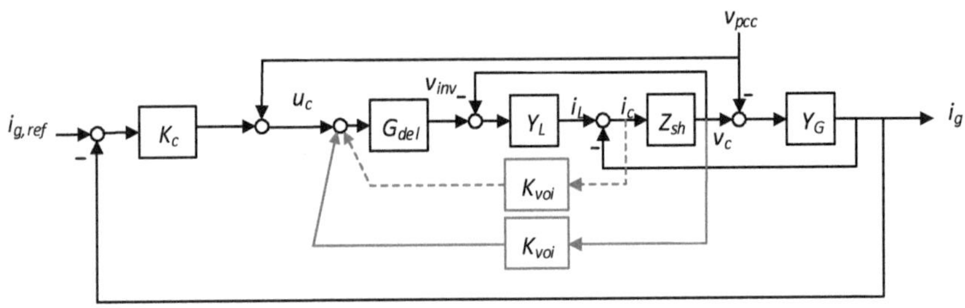

FIG. 11.28 Active damping/VOI—Closed loop shaping.

$$\frac{i_g(s)}{u_c(s)} = G_{id}(s) = \frac{Y_G Z_{sh} Y_L G_{del}}{1 - K_{voi} G_{del} Y_L Z_{sh} + Z_{sh}(Y_L + Y_G)} \tag{11.29}$$

Upon closing the loop with the current controller $K_c(s)$, the closed-loop admittance transfer function can be derived by reducing the block diagram in Fig. 11.28. The closed-loop admittance transfer function considering capacitor current feedback is derived as Eq. (11.30),

$$\frac{-i_g(s)}{v_{pcc}(s)} = Y_{cl}(s) = \frac{Y_G(1 + Y_L Z_{sh}) - G_{del} Y_L Y_G (K_{voi} + Z_{sh})}{1 + Y_L G_{del}(K_c Y_G Z_{sh} - K_{voi}) + Z_{sh}(Y_L + Y_G)} \tag{11.30}$$

and the same when considering capacitor voltage feedback is derived as Eq. (11.31).

$$\frac{-i_g(s)}{v_{pcc}(s)} = Y_{cl}(s) = \frac{Y_G(1 + Y_L Z_{sh}) - G_{del} Y_L Y_G Z_{sh}(K_{voi} + 1)}{1 + Y_L G_{del} Z_{sh}(K_c Y_G - K_{voi}) + Z_{sh}(Y_L + Y_G)} \tag{11.31}$$

A suitable transfer function candidate for $K_{voi}(s)$ to emulate the Series R or Shunt RC type of damping is a derivative controller with a first-order low pass filter of the form:

$$K_{voi}(s) = K_d \frac{s}{s + \alpha'} \tag{11.32}$$

where the parameter $\alpha = \frac{1}{R_d C_d}$ of the shunt branch. K_d gain can be chosen to have enough damping of the resonance peak. With increased K_d, the damping of the system is improved however the controller is sensitive to noise due to the derivative action. Therefore, the optimal choice of K_d needs to be made from both time and frequency domain analysis.

Example 11.6: Active damping/VOI control

In this example, we consider the capacitor voltage measurement as feedback for implementing the damping. The parameters of the three-phase inverters are the same as that of Example 11.5 (refer Table 11.2). The low pass filter parameter of the VOI controller is chosen as $\alpha = \frac{1}{R_d C_d} = 4.16E3$. The derivative gain can be increased to have higher damping. Fig. 11.29 shows the open-loop shaping of control and admittance transfer function as the derivative gain is increased, notice that the resonance peaks are damped. These transfer functions are evaluated using Eqs. (11.27), (11.29).

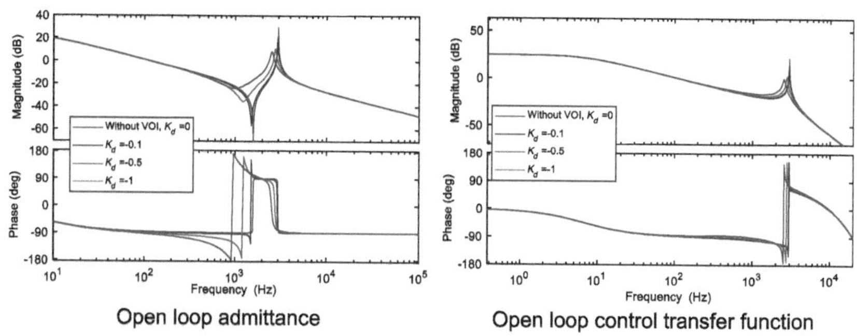

FIG. 11.29 Active damping—Open loop transfer functions.

Considering the presence of the controller, the closed-loop admittances for the above AD scenarios are presented in Fig. 11.30. In the same plot, the grid impedance is also plotted. For the case without VOI, the grid impedance magnitude plot intersects twice with a magnitude plot of inverter admittance during which the phase of the inverter admittance is close to −90 degrees and the phase of grid impedance is close to +80 degrees. The difference between the phases is close to 180 degrees and thus the stability margins are low. On the other hand, with VOI damping, the resonance peak is reduced significantly and therefore there is no intersection of the inverter admittance and grid impedance. Hence the inverter-grid tied system will not interact at that frequency with the presence of VOI.

The Nyquist criterion applied to the ratio of inverter admittance and grid impedance is shown in Fig. 11.31. We consider both strong and weak grid conditions and it can be observed that the stability margins are significantly reduced without the VOI control and the grid-tied inverter system is prone to harmonic oscillations. With resonance peak clipping, there is no intersection of magnitude plots from the Bode's perspective and the equivalent Nyquist plots of MLG show an almost passive behavior. In this case, the inverter with VOI is also Middlebrook stable since the entire Nyquist plot lies within the unit circle, however, this may not be true always.

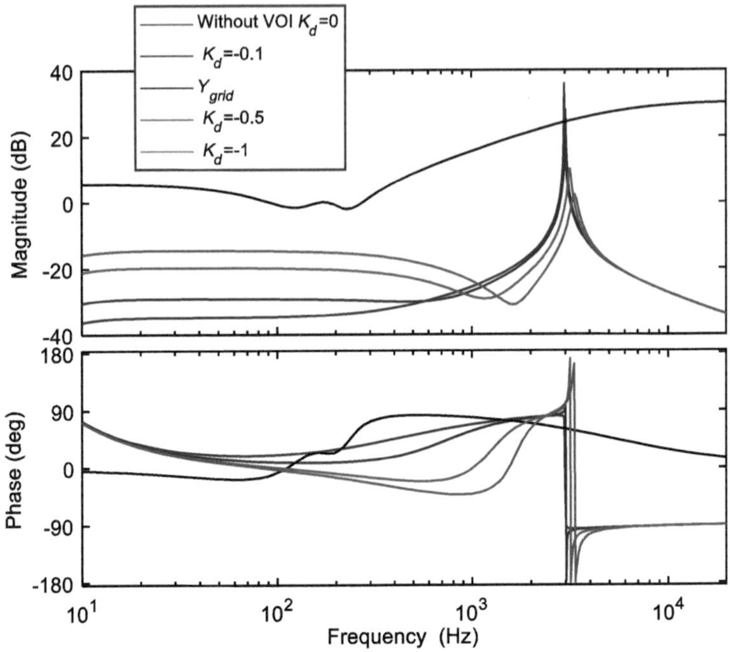

FIG. 11.30 Closed-loop admittance with VOI control.

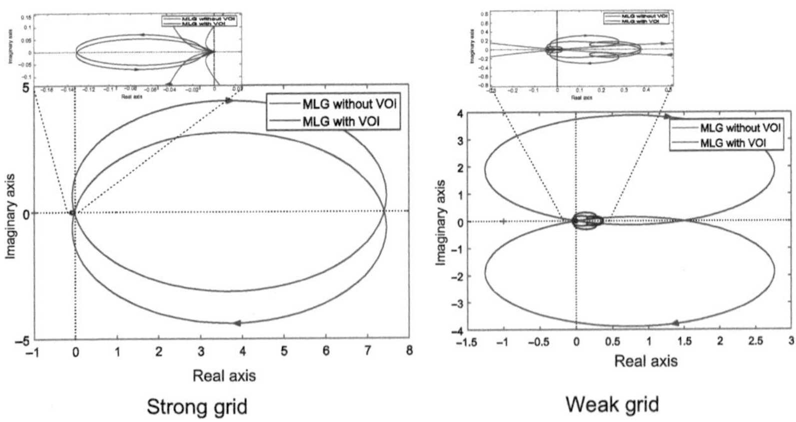

FIG. 11.31 Stability analysis with VOI controller under strong and weak grid conditions.

11.5.3 Generalized framework for VOI synthesis

In the previous section, the damping control was designed without considering the grid impedance model. Moreover, the stability analysis in the previous example was done by assuming a 2nd order grid impedance model and damping controller eliminates the resonance of the inverter. However, there could be a case where the grid impedance model itself has one or more resonances at a much lower frequency than the resonance frequency. Then the damping controller needs to be designed to damp at those lower frequencies as well. With the advent of the WSI technique, if the order of the grid impedance is known, a suitable grid impedance transfer function can be extracted from the impedance measurement process. Thus, the grid impedance transfer function knowledge can be used to design a stabilizing VOI controller. A generalized framework for the synthesis of VOI was first developed in Ref. [32]. An output feedback structure is chosen where the output grid current measurement is used for reference tracking and also for implementing the damping.

The closed-loop impedance of the inverter considering the presence of VOI can be reduced from the block diagram presented in Fig. 11.32. The closed-loop admittance is given by Eq. (11.33), notice the term in bracket containing the VOI controller transfer function Z_{voi} and current controller transfer function K_{ig}. This shows that the frequency domain behavior of the current controller is effectively modified by the VOI controller. Thus, the current controller can be designed for reference tracking and VOI controller can be designed to mitigate harmonic resonance.

$$Y_{cc,voi} = \left[I + G_{ig}K_v\left(I + G_{ig}K_v\right)^{-1}\left(K_{ig} - Z_{voi}\right)\right]^{-1}\left(Y_{cc,ol} + G_{ig}\right) \qquad (11.33)$$

The measured grid impedance transfer function, together with the impedance of grid side inductance form the total impedance denoted at Z_T. The total inductance relates the output current and the capacitor voltage v_c. Since the VOI controller needs to damp the harmonic resonance, which is a medium frequency

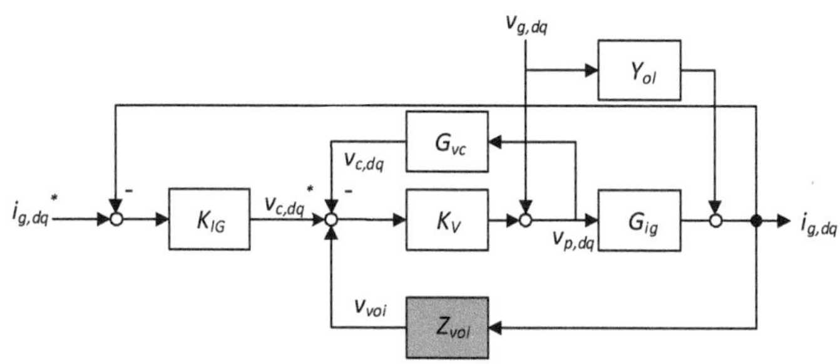

FIG. 11.32 Output feedback VOI control structure.

phenomenon, low pass behavior is required for the signal shape $u = v_{voi}$, hence a high pass filter is chosen as the weighting function W_u.

$$W_u = I_{2x2}\frac{s + {}^{\omega_u}/_{M_u}}{\varepsilon_u s + \omega_u} \tag{11.34}$$

For a similar reason, the weighting function W_y is also chosen as a high pass. Advanced options for the weighting function such as 2nd order bandstop filters can also be chosen to effectively eliminate resonances in a certain range of frequencies. Thus by augmenting the grid impedance Z_T and the necessary weighting function, we have formulated the generalized plant model (GPM) denoted as $P_{cc}(s)$ as shown in Fig. 11.33.

The GPM $P_{cc}(s)$ relates the exogenous inputs and outputs as expressed in Eq. (11.35).

$$\begin{bmatrix} z_y \\ z_u \\ v \end{bmatrix} = \underbrace{\begin{bmatrix} P_{wz_y} & P_{uz_y} \\ P_{wz_u} & P_{uz_u} \\ P_{wv} & P_{uv} \end{bmatrix}}_{P_{cc}(s)} \begin{bmatrix} w \\ u \end{bmatrix} \tag{11.35}$$

Thus, linear optimal control theory can be applied to minimize the H-Infinity norm of the $P_{cc}(s)$ and the controller Z_{voi} can be calculated. The derivation of transfer function elements of $P_{cc}(s)$ are provided in [32].

Example 11.7: VOI synthesis by applying generalized framework

An example is presented based on the work in Ref. [32], where the inverter parameters can be found. As discussed before, the weighting functions are high pass to achieve corresponding signal shape as low pass. In this example, we consider the impact of cut off frequency ω_u on the stability of inverter-grid tied system. As ω_u is increased, the bandwidth of the VOI controller is increased such that it can damp higher frequency effectively. This results in increased stability of the inverter-grid tied system as shown in Fig. 11.34. Notice that the characteristic loci shifts to the right and therefore the intersection with the unit circle occurs at a higher angle and therefore we have an increase in stability margins.

11.5.4 Invasive methods—Active impedance cancellation devices

The methods previously discussed were noninvasive methods that do not require additional passive hardware components or other power electronic devices connected at the output. The presence of grid impedance changes the PCC voltage dependent on the magnitude of current injected. Additionally, the harmonics that enter resonance depends on the control loops of inverter and the grid impedance. Thus, an invasive method can be invoked based on the concept of series active filter wherein a power electronic converter that acts

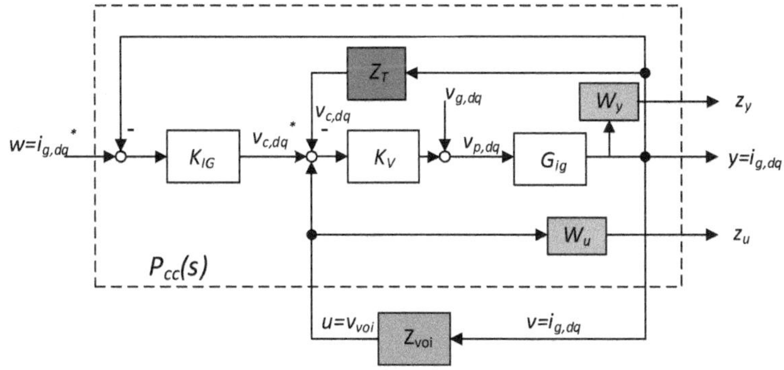

FIG. 11.33 Generalized plant model.

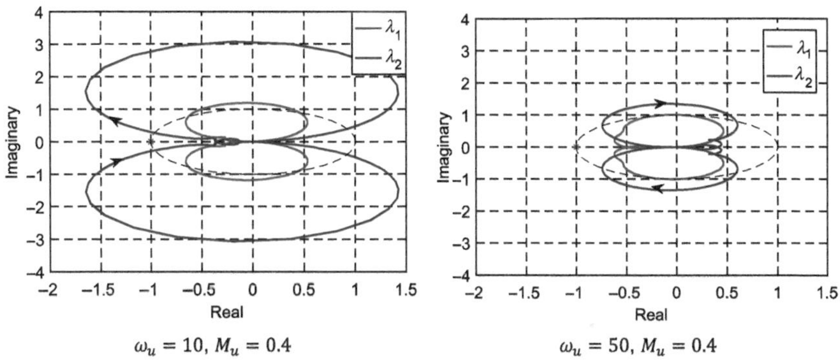

FIG. 11.34 Impact of weighting function on stability margins.

as a voltage source can be connected in series at the output of the inverter [35, 36]. An example of such a device is presented in Fig. 11.35 wherein a single-phase application is shown. At the output of the main single phase inverter, the impedance cancellation device is connected. The impedance cancellation device is itself a single-phase inverter with low power and voltage ratings. The device draws a small amount of active power to compensate for losses and DC link voltage control. The device acts as a harmonic voltage source as shown in Fig. 11.35 and the magnitude and phase of this voltage source is controlled such that it eliminated the presence of grid impedance and by which harmonic resonance is mitigated.

Effective comparison of the harmonic resonance mitigation methods is presented in Table 11.4.

11.6 Dynamic voltage stability monitoring

The dynamic voltage stability monitoring (DVSM) technique proposed within the EU Horizon 2020 funded project RESERVE is a decentralized approach for monitoring dynamic voltage stability for futuristic distribution grids. DVSM is

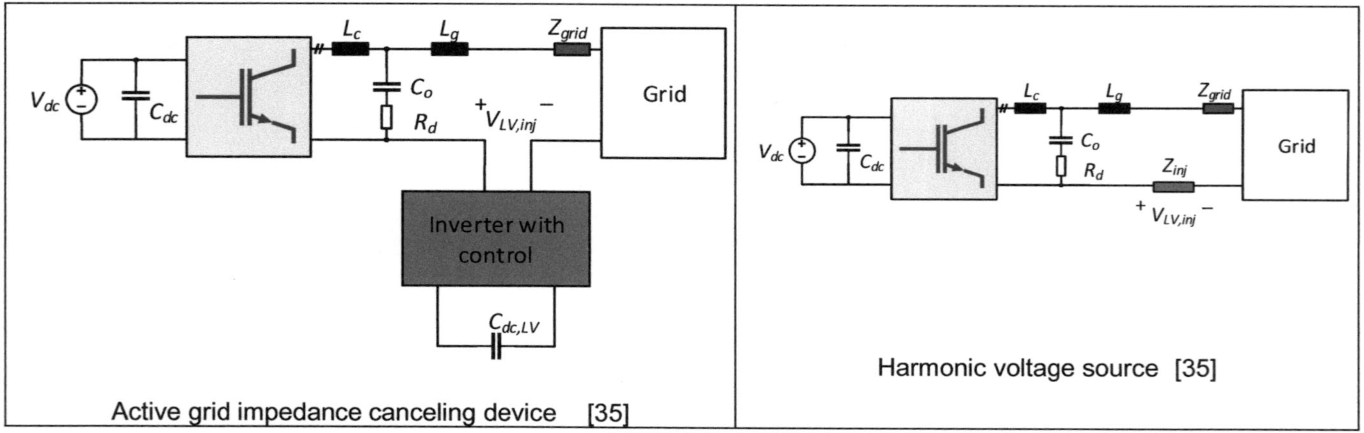

FIG. 11.35 Active grid impedance cancellation device—Proposed in [35].

TABLE 11.4 Comparison of inverter-grid harmonic resonance mitigation methods.

Desirable properties	Passive damping	Active damping/ VOI	Active impedance cancellation devices
Noninvasive	No	Yes	No, due to usage of a device external to the main inverter
Adaptability	No	Yes	Yes
Bandwidth/ sampling rate	Not applicable	VOI controller is limited by bandwidth and sampling rate of the main inverter	Independent of bandwidth or sampling rate of main inverter. The impedance cancellation device can have a much higher bandwidth and sampling rate
Power losses and complexity	Increased losses and complexity based on the passive damping structure	No losses specific to VOI functionality. Only the typical losses in the main inverter	Apart from losses in main inverter, switching and conduction losses in impedance cancellation device

a 3-step process that monitors and maintains the stability of a power-electronic dominated grid. The DVSM technique consists of an impedance measurement step followed by stability monitoring step and the last step is the synthesis of a virtual output impedance control [22, 37, 38]. Coordination between RES inverters and the SSAU is done through communication links with low latency. The SSAU monitors all the RES inverters sequentially. Since the stability monitoring between inverters and the grid is done one at a time, low latency communication such as 5G is ideal for the proposed technique. A detailed investigation on the communication requirements can be found in Ref. [39]. It has been estimated that on an average of 400 households under an SSAU can be coordinated or monitored within 5–10 min, i.e., the monitoring cycle for every RES inverter occurs between 5 and 10 min [37, 39].

The DVSM technique is shown in Fig. 11.36. RES inverters are required to have some changes in their software such as wideband system identification for measurement of grid impedance and VOI control functionality. RES inverters are also required to have communication modules due to the required data exchange between inverters and SSAU. Three steps here

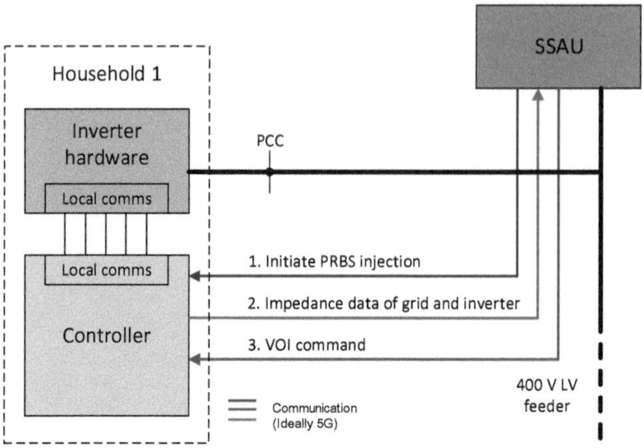

FIG. 11.36 DVSM technique.

Although inverters can perform the above mentioned three steps without the requirement of SSAU, the coordination done by the SSAU is important for the following reasons:

- Multiple inverters injecting PRBS noise leads to interference and the impedance measurement gets corrupted.
- The stability calculation as explained in Section 11.2.4 and VOI synthesis as explained in Section 11.5.3 can be computationally expensive for inverters. To minimize inverter hardware costs, these algorithms can be executed in SSAU.

Impedance measurement devices such as the proposed WFZ device can enable real-time stability monitoring at any given bus in the distribution grid. An example of the harmonic stability assessment for the distribution grid is provided below.

Example 11.8: Harmonic stability analysis of distribution grid

Consider a distribution grid model as shown in Fig. 11.37. The cable parameters and structure are adopted from an Irish LV network and modified to consider the presence of three-phase RES inverters on every bus. Z_{OLTC} represents the impedance of the OLTC transformer plus the low voltage referred to grid impedance of the medium voltage grid. Z_{ij} represents the cable impedance between node i and j, $Y_{L,\,i}$ represents the load admittance at bus i and $Y_{inv,\,i}$ represent the closed loop output admittance of the RES inverter. A stability analysis at Bus 2 is obvious since the stability depends on the strength of Z_{OLTC}, lower Z_{OLTC} more stable is Bus 2. To make the scenario interesting, consider that a device like WFZ is connected at Bus 6. If the currents to the left and right of node 6 are available to the WFZ device, then it can observe and measure the impedances Z_{6L} and Z_{6R} to the left and right respectively.

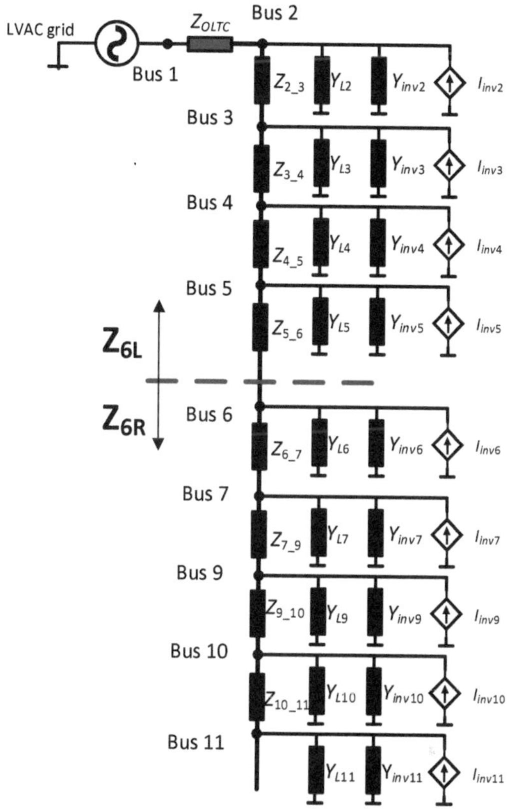

FIG. 11.37 Distribution grid with high RES proliferation [22].

Fig. 11.38 shows the harmonic stability analysis of the actual network at Bus 6. The cable lengths are the same as the original network. The stability margins can be obtained as explained previously in Section 11.2.2. The PM is around 68 degrees. Now the cable lengths are varied to study the interaction among converters for the fixed grid impedance.

Fig. 11.39 presents two stability results, where in the first case the network cables are scaled down by a factor of 10 and the second case where the cables are scaled up by a factor of 8. For cable length scaling of 0.1, the phase margins are much lower, around 46 degrees, which denotes an increased tendency towards oscillations. For cable length scaling of 8, the phase margins are much high, around 105 degrees and it can be observed that the entire GNC plot lies on the RHP. This indicates that the system is passive and stable. Distribution cables have a high R/X ratio due to which there is increased damping for harmonics.

The PM figures presented for the abovementioned scenario decreases when the grid impedance Z_{OLTC} increases.

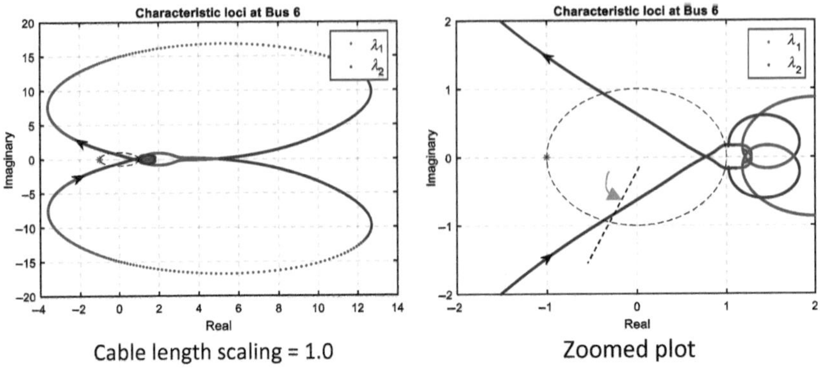

FIG. 11.38 Harmonic stability of actual LV distribution grid [22].

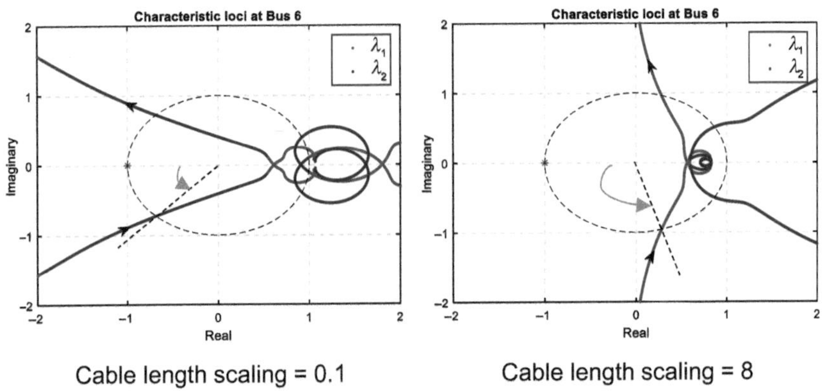

FIG. 11.39 Harmonic stability of scaled LV distribution grid [22].

The following general conclusions can be drawn:

- For a given cable length, increasing the grid impedance causes an increased tendency towards oscillations.
- For a given grid impedance, increasing the cable length, where the nature of cable has a high R/X ratio, then the network tends towards passivity. Thus, from the above two points, the stability can be concluded only after including both the grid and cable impedances in the modeling.
- For low cable lengths, the stability is critically dependent solely on the grid impedance. High grid impedance can cause inverters to interact with one another.

11.7 Role of solid-state transformer in futuristic distribution grids

As discussed previously, interaction among RES inverters would increase with a greater number of converters. Mitigation approaches from the point of view of the converter were discussed. However, an important factor playing a role in harmonic stability of distribution grids is the impedance or strength of the medium voltage grid. Conventional OLTCs do not address a solution for weak MV connection and harmonic resonance at a higher frequency due to a weak MV grid condition. This is due to the very low bandwidth of OLTCs as they only regulate steady voltage and not voltage dynamics. The typical reaction time of OLTCs last from 1 to 3 s. Hence, there is growing motivation to develop and replace OLTCs with power electronic-based transformer which is referred to as SST [40–42]. Since the role of SST is covered in this section.

The SST as shown in Fig. 11.40 consists of an AC/DC converter wherein the AC side of the converter connects the MV distribution grid and the DC side is an MV port. A DC/DC converter interfaces this MV DC port to an LV DC port. A suitable choice for such a converter is a dual active bridge (DAB) since it provides galvanic isolation through the medium frequency AC link transformer. In the final stage, we have a DC/AC converter that interfaces the LV DC side to the LV distribution grid. The AC/DC converter in the first stage operates in grid-following mode and regulates the DC link voltage in MV DC port, the DC/DC converter regulates the voltage in the LV DC port and DC/AC converter in the final stage operates in a grid forming mode and controls the AC side voltage and thus forms voltage reference for the LV grid [43].

As marked in Fig. 11.40, the stability at the interface nodes P_1, P_2, and P_3 are important for guaranteeing the stability of the SST. The first and last stage converter connects to AC grids, so for node P_1 and P_3, measuring the wideband grid impedance is required. Based on the measured impedance, the converters can modify their self-impedances using the VOI techniques previously discussed in Section 11.5.3. The Thevenin equivalent of the above concept is shown in Fig. 11.41. Stability at node P_2 depends on the DC/DC converter and

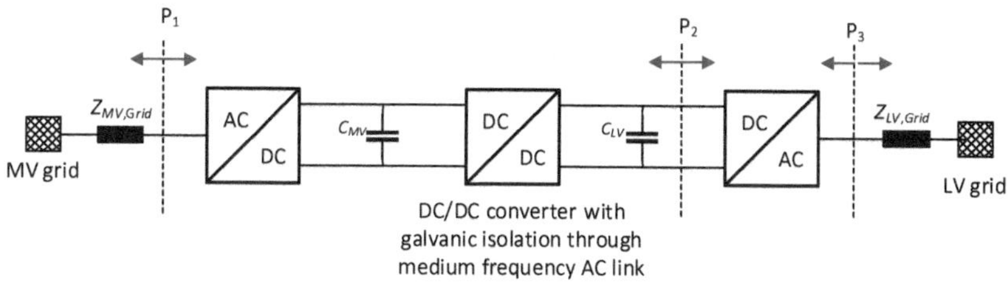

FIG. 11.40 Solid state transformer.

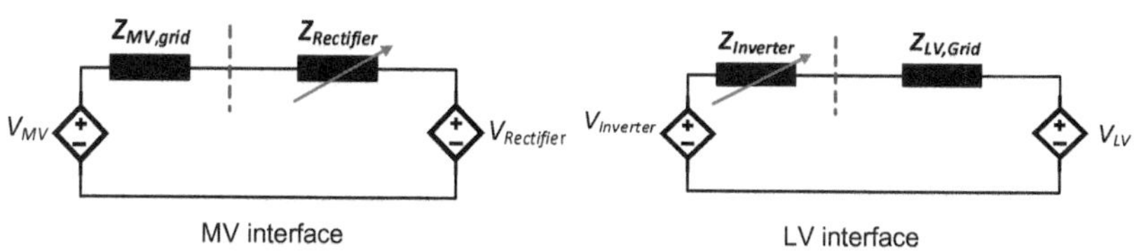

FIG. 11.41 VOI control for SST.

DC/AC converter interface. Since the control objective of the DC/AC converter is to stiffly regulate the AC voltage and handle fast power dynamics in the LV grid such as load increase or decrease, the DC/AC converter behaves like a constant power load (CPL) as seen by the DC/DC converter. Due to the CPL behavior, there is an inherent negative resistance behavior that causes a destabilizing effect. In order to tackle the presence of a CPL, port-controlled Hamiltonian (PCH) modeling and control approach is proposed for the DC/DC converter [11, 43]. An algebraic approach is known as interconnection and damping assignment passivity-based control (IDAPBC) is a simple yet elegant method to derive the PCH control law. Upon implementing the PCH control law, the closed-loop system is passive and stable. Sufficient virtual positive damping can be assigned to counter the negative resistance behavior of CPL.

Example 11.9: PCH control for DAB

This example is based on the work in Ref. [11], wherein the PCH control law based on the IDAPBC approach is derived for DAB DC/DC converter. The control law is given by Eq. (11.36).

$$\delta = \frac{\pi}{2} - \sqrt{\left(\frac{\pi}{2}\right)^2 - \left(\frac{\pi n_t \omega L' i_m v^*}{V_{in} v_m}\right) + \left(\frac{\pi n_t \omega L'}{V_{in}}\right) r_1 (v_m - v^*)} \tag{11.36}$$

State equation at the output of the DAB DC/DC converter is given by Eq. (11.37).

$$C_o \frac{dv}{dt} = \frac{4 V_{in} \delta}{n_t \omega L'} \left(1 - \frac{\delta}{\pi}\right) - i_L \tag{11.37}$$

By using Eq. (11.36) in (11.37), the closed-loop state-space equation can be established. The closed-loop equation needs to be linearized around the operating point and then the state space can be formulated such that the load current i_L is the input and the only state variable, which is the output voltage v is the output. From the state space representation, the impedance transfer function can also be evaluated. The Nyquist plot of the closed-loop impedance transfer function is shown in Fig. 11.42. The entirety of the Nyquist plot lies on the RHP; therefore, the DAB converter is passive and stable. In the third stage of the SST, the DC/AC converter and the LVAC grid can be modeled as an ideal CPL in parallel with resistive load. To analyses the stability between the DAB stage and the third stage at node P_2, the nature of the load is varied from being fully resistive to fully CPL. A total rating of 5 MW is assumed for the DAB converter since typically distribution OLTC transformers have 2–5 MVA capacity based on the size of the LVAC grid. Impedance based stability analysis is conducted at the output DC/DC converter at node P_2 and the results are plotted in Fig. 11.43. The interconnection is passive for fully resistive load and the passivity decreases as CPL contribution increases. However, for all the cases, the Nyquist plots are stable.

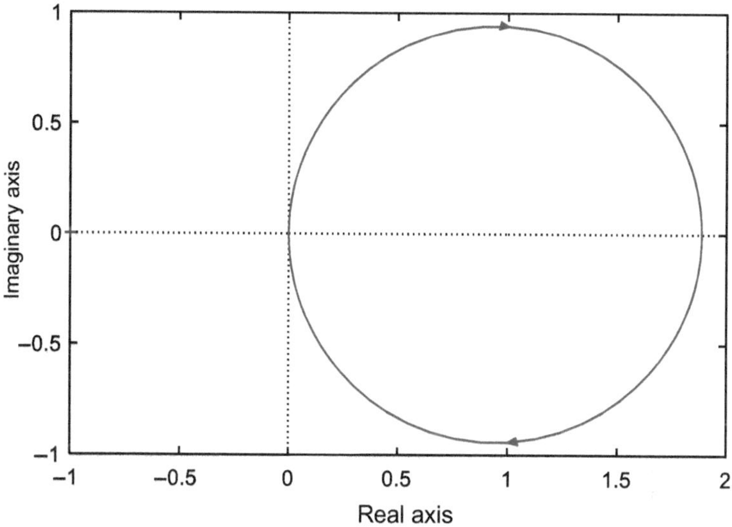

FIG. 11.42 Impedance of PCH controlled DAB.

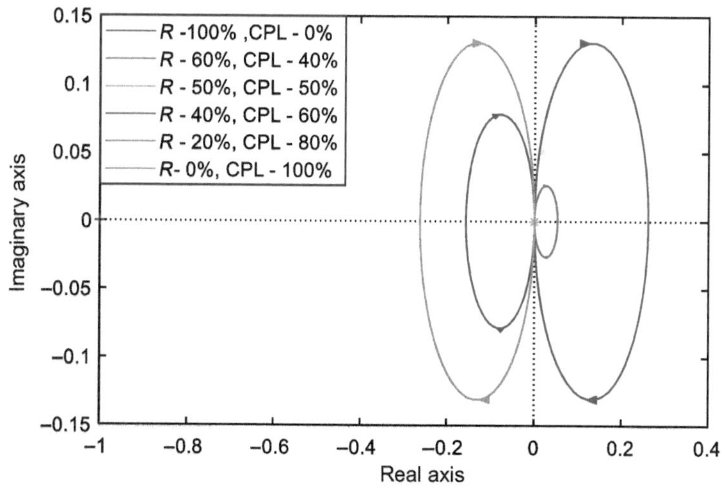

FIG. 11.43 MLG of DAB with variation in contribution from resistive load and CPL.

11.8 Summary

This chapter presents the new perspective of voltage stability and voltage control for power electronics dominated grids. Covering from the fundamental framework of the Middlebrook criterion and Nyquist criterion, extensions are provided to study the dynamics arising from the interaction of power electronic converters. A resonance phenomenon that occurs between the inverter and the grid is known as harmonic instability and the theory of harmonic instability has been described through an impedance-based formalism. Increased grid impedance or presence of resonance in the grid impedance can cause harmonic instabilities to the inverter-grid tied systems. Some of the most recent trends in the measurement of the unknown grid impedance are presented. WSI is a very important measurement and monitoring tool for power electronic converters. A standalone device based on this technique known as the WFZ device can be used as a mobile and real-time stability monitoring device by grid operators. Due to its low weight and no necessity for DC power supply, it offers high plug-play capability. As a solution to the harmonic resonance problem, VOI controllers are proposed to modify the inverters output impedance based on the measured grid impedance, such that the closed-loop system is stable. This chapter presents some of the conventional damping approaches and the recently proposed generalized framework for the synthesis of VOI controllers.

References

[1] N. Modi, R. Yan, Low inertia power systems: frequency response challenges and a possible solution, in: 2016 Australasian Universities Power Engineering Conference (AUPEC), 2016.

[2] F. Milano, F. Dörfler, G. Hug, D.J. Hill, G. Verbič, Foundations and challenges of low-inertia systems, in: 2018 Power Systems Computation Conference (PSCC), 2018.

[3] D. Raisz, A. Musa, F. Ponci, A. Monti, Linear and uniform swing dynamics, IEEE Trans. Sustain. Energy 10 (3) (2019) 1513–1522.

[4] S.K. Gurumurthy, A. Monti, C. Murphy, H. Gill, A. Keane, S. Walsh, Power electronic stability criteria for AC three phase systems, in: EU H2020 Project RESERVE, Deliverable 3.1, 2017.

[5] A. Nouri, A. Souroudi, A. Keane, S.K. Gurumurthy, A. Monti, D. Ryan, M. Ponce De Leon, N. Grant, D. Leniston, Z. Mehdi, R. Farac, R. Murphy, T. Clancy, A. Löwen, WP3 drafting of ancillary services and network codes definitions V2, in: EU H2020 Project RESERVE, Deliverable 3.9, 2019.

[6] Y. Bae, T.-K. Vu, R.-Y. Kim, Implemental control strategy for grid stabilization of grid-connected PV system based on German grid code in symmetrical low-to-medium voltage network, IEEE Trans. Energy Convers. 28 (3) (2013) 619–631.

[7] R.D. Middlebrook, V. Vorpérian, J. Lindal, The N extra element theorem, IEEE Trans. Circuits Syst. I Fund. Theor. Appl. 45 (9) (1998) 919–935.

[8] A. Riccobono, E. Santi, Comprehensive review of stability criteria for DC power distribution systems, IEEE Trans. Ind. Appl. 50 (5) (2014) 3525–3535.

[9] M. Stojic, D. Siljak, Generalization of Hurwitz, Nyquist, and Mikhailov stability criteria, IEEE Trans. Autom. Control 10 (3) (1965) 250–254.

[10] A. Riccobono, E. Santi, Stability analysis of an all-electric ship MVDC Power Distribution System using a novel passivity-based stability criterion, in: 2013 IEEE Electric Ship Technologies Symposium (ESTS), 2013.

[11] M. Cupelli, S.K. Gurumurthy, S.K. Bhanderi, Z. Yang, P. Joebges, A. Monti, R.W. De Doncker, Port controlled Hamiltonian modeling and IDA-PBC control of dual active bridge converters for DC microgrids, IEEE Trans. Ind. Electron. 66 (11) (2019) 9065–9075.

[12] B. Wen, D. Boroyevich, R. Burgos, P. Mattavelli, Z. Shen, Analysis of DQ small-signal impedance of grid-tied inverters, IEEE Trans. Power Electron. 31 (1) (2015) 675–687.

[13] B. Wen, D. Boroyevich, R. Burgos, P. Mattavelli, Z. Shen, Inverse Nyquist stability criterion for grid-tied inverters, IEEE Trans. Power Electron. 32 (2) (2016) 1548–1556.

[14] A.G.J. MacFarlane, I. Postlethwaite, The generalized Nyquist stability criterion and multivariable root loci, Int. J. Control. 25 (1) (1977) 81–127.

[15] J.D. Ainsworth, Harmonic instability between controlled static convertors and ac networks, in: Proceedings of the Institution of Electrical Engineers, 1967.

[16] P.A. Hazell, J.O. Flower, Theoretical analysis of harmonic instability in ac-dc convertors, in: Proceedings of the Institution of Electrical Engineers, 1970.

[17] G. Lemieux, Power system harmonic resonance—a documented case, IEEE Trans. Ind. Appl. 26 (3) (1990) 483–488.

[18] X. Wang, F. Blaabjerg, Harmonic stability in power electronic-based power systems: concept, modeling, and analysis, IEEE Trans. Smart Grid 10 (3) (2018) 2858–2870.

[19] J. Sun, G. Wang, X. Du, H. Wang, A theory for harmonics created by resonance in converter-grid systems, IEEE Trans. Power Electron. 34 (4) (2018) 3025–3029.

[20] J. Sun, Small-signal methods for AC distributed power systems—a review, IEEE Trans. Power Electron. 24 (11) (2009) 2545–2554.

[21] M. Connor, A. Keane, S. K. Gurumurthy, D. Ryan und M. Ponce De Leon, „Extension to new approaches to high RES contexts," in EU H2020 Project RESERVE, Deliverable 3.2, 2017.

[22] S. K. Gurumurthy, A. Monti, A. Nouri, A. Souroudi, A. Keane, D. Ryan, M. Ponce De Leon, N. Grant, R. Murphy und J. Sandham, „Power electronics system-level stability criteria," in EU H2020 Project RESERVE, Deliverable 3.3, 2018.

[23] Z. Shuai, Y. Li, W. Wu, C. Tu, A. Luo, Z.J. Shen, Divided DQ small-signal model: a new perspective for the stability analysis of three-phase grid-tied inverters, IEEE Trans. Ind. Electron. 66 (8) (2018) 6493–6504.

[24] H. Gong, D. Yang, X. Wang, Impact analysis and mitigation of synchronization dynamics for DQ Impedance measurement, IEEE Trans. Power Electron. 34 (9) (2018) 8797–8807.

[25] M. Bienholz, G. Griepentrog, Wide-band impedance measurement for converter impedance determination in lv-grids, in: 2018 20th European Conference on Power Electronics and Applications (EPE'18 ECCE Europe), 2018.

[26] M. Jordan, F. Grumm, H. Langkowski, T. Do Thanh, D. Schulz, Online network impedance identification with wave-package and inter-harmonic signals, in: 2015 International School on Nonsinusoidal Currents and Compensation (ISNCC), 2015.

[27] S.K. Gurumurthy, R. Uhl, M. Pitz, F. Ponci, A. Monti, Non-invasive wideband-frequency grid impedance measurement device, in: 2019 IEEE 10th International Workshop on Applied Measurements for Power Systems (AMPS), 2019.

[28] A. Riccobono, M. Mirz, A. Monti, Noninvasive online parametric identification of three-phase AC power impedances to assess the stability of grid-tied power electronic inverters in LV networks, IEEE J. Emerg. Sel. Top. Power Electron. 6 (2) (2017) 629–647.

[29] A. Nouri, A. Souroudi, A. Keane, S.K. Gurumurthy, M. Mirz, A. Monti, Network impedance characterization for active distribution networks, in: EU H2020 Project RESERVE, Deliverable 3.4, 2018.

[30] T. Roinila, T. Messo, E. Santi, MIMO-identification techniques for rapid impedance-based stability assessment of three-phase systems in DQ domain, IEEE Trans. Power Electron. 33 (5) (2017) 4015–4022.

[31] A. Monti, S.K. Gurumurthy, R. Uhl, M. Pitz, Vorrichtung zur Bestimmung der Impedanz in Abhängigkeit der Frequenz eines zu messenden Versorgungsnetzes, (2019).

[32] S.K. Gurumurthy, M. Cupelli, A. Monti, A generalized framework for synthesizing virtual output impedance control of grid integrated power electronic converters, in: 2018 IEEE International Conference on Power Electronics, Drives and Energy Systems (PEDES), 2018.
[33] M. Cespedes, J. Sun, Mitigation of inverter-grid harmonic resonance by narrow-band damping, IEEE J. Emerg. Sel. Top. Power Electron. 2 (4) (2014) 1024–1031.
[34] R.N. Beres, X. Wang, F. Blaabjerg, M. Liserre, C.L. Bak, Optimal design of high-order passive-damped filters for grid-connected applications, IEEE Trans. Power Electron. 31 (3) (2015) 2083–2098.
[35] Y. He, H.S.-H. Chung, C.-T. Lai, X. Zhang, W. Wu, Active cancelation of equivalent grid impedance for improving stability and injected power quality of grid-connected inverter under variable grid condition, IEEE Trans. Power Electron. 33 (11) (2018) 9387–9398.
[36] D. Yang, X. Ruan, H. Wu, Impedance shaping of the grid-connected inverter with LCL filter to improve its adaptability to the weak grid condition, IEEE Trans. Power Electron. 29 (11) (2014) 5795–5805.
[37] S.K. Gurumurthy, A. Monti, A. Nouri, A. Souroudi, A. Keane, D. Ryan, M. Ponce De Leon, N. Grant, S. Bretzke, R. Murphy, J. Sandham, A. Löwen, Specification for an on-line system level monitoring system, in: EU H2020 Project RESERVE, Deliverable 3.5, 2018.
[38] S.K. Gurumurthy, M. Cupelli, A. Monti, R. Murphy, J. Sandham, Reserve Irish Demo–Virtual Output Impedance Control for Voltage Stability, CIRED Workshop, Ljubljana, 2018.
[39] R. Farac, Z. Mehdi, F. Williams, M. Ponce de Leon, D. Ryan, N. Grant, D. Lenniston, S. Gurumurthy, Report on requirements on scalable ICT to implement voltage control concepts, V2, in: EU H2020 Project RESERVE, Deliverable 3.7, 2019.
[40] H. Qin, J.W. Kimball, Solid-state transformer architecture using AC–AC dual-active-bridge converter, IEEE Trans. Ind. Electron. 60 (9) (2012) 3720–3730.
[41] M. Mogorovic, D. Dujic, Sensitivity analysis of medium-frequency transformer designs for solid-state transformers, IEEE Trans. Power Electron. 34 (9) (2018) 8356–8367.
[42] S.A.M. Saleh, C. Richard, X.F.S. Onge, K.M. McDonald, E. Ozkop, L. Chang, B. Alsayid, Solid-state transformers for distribution systems–part i: technology and construction, IEEE Trans. Ind. Appl. 55 (5) (2019) 4524–4535.
[43] R.V. Meshram, M. Bhagwat, S. Khade, S.R. Wagh, A.M. Stanković, N.M. Singh, Port-controlled phasor hamiltonian modeling and IDA-PBC control of solid-state transformer, IEEE Trans. Control Syst. Technol. 27 (1) (2017) 161–174.

Index

Note: Page numbers followed by *f* indicate figures, *t* indicate tables and *b* indicate boxes.

L

M

T

U

Kolibri 360 GmbH
Ferdinand-Jühlke-Straße 7
99095 Erfurt, Deutschland
produktsicherheit@kolibri360.de